Foundations of Logical Consequence

MIND ASSOCIATION OCCASIONAL SERIES

This series consists of carefully selected volumes of significant original papers on predefined themes, normally growing out of a conference supported by a Mind Association Major Conference Grant. The Association nominates an editor or editors for each collection, and may cooperate with other bodies in promoting conferences or other scholarly activities in connection with the preparation of particular volumes.

Director, Mind Association: M. Fricker
Publications Officer: Julian Dodd

Recently Published in the Series:

The Highest Good in Aristotle and Kant
Edited by Joachim Aufderheide and Ralf M. Bader

How We Fight
Ethics in War
Edited by Helen Frowe and Gerald Lang

The Morality of Defensive War
Edited by Cécile Fabre and Seth Lazar

Metaphysics and Science
Edited by Stephen Mumford and Matthew Tugby

Thick Concepts
Edited by Simon Kirchin

Wittgenstein's *Tractatus*
History and Interpretation
Edited by Peter Sullivan and Michael Potter

Philosophical Perspectives on Depiction
Edited by Catharine Abell and Katerina Bantinaki

Emergence in Mind
Edited by Cynthia Macdonald and Graham Macdonald

Empiricism, Perceptual Knowledge, Normativity, and Realism
Essays on Wilfrid Sellars
Edited by Willem A. deVries

Spheres of Reason
New Essays in the Philosophy of Normativity
Edited by Simon Robertson

Foundations of Logical Consequence

EDITED BY
Colin R. Caret
Ole T. Hjortland

Great Clarendon Street, Oxford, OX2 6DP,
United Kingdom

Oxford University Press is a department of the University of Oxford.
It furthers the University's objective of excellence in research, scholarship,
and education by publishing worldwide. Oxford is a registered trade mark of
Oxford University Press in the UK and in certain other countries

First Edition published in 2015

Impression: 1

Published in the United States of America by Oxford University Press
198 Madison Avenue, New York, NY 10016, United States of America

British Library Cataloguing in Publication Data
Data available

Library of Congress Control Number: 2014950809

ISBN 978–0–19–871569–6

Printed and bound by
CPI Group (UK) Ltd, Croydon, CR0 4YY

Preface

This volume is a result of talks given at events organized under the auspices of the AHRC-funded project *Foundations of Logical Consequence* (FLC). The project ran from 2009 to 2013 at the University of St Andrews, directed by Stephen Read (Principal Investigator), Graham Priest, and Stewart Shapiro (Co-Investigators). We were incredibly fortunate to work with them as Postdoctoral Research Fellows in the Arché Research Centre, and to participate in the management of the only large-scale project on the nature of logical consequence to date.

The project was a resounding success. It has produced high-level research on all of the topics that the project set out to cover. The present volume is only a small sample of the many excellent papers presented at FLC events. We are confident that these contributions will be of great value both to researchers and to graduate students in the field and in philosophy more broadly. In order to make the book more useful to those who are not familiar with all the debates in the philosophy of logic, we have written an introduction to current research trends, outlining the major arguments in the last couple of decades. Along the way we give indications of how the essays in the volume contribute to these debates.

We are immensely grateful to the authors of this volume. Their contributions stand as a testimony to the vibrant and productive workshops during the FLC project. They have all been extremely helpful and patient while the volume has been edited. We also want to thank the many visitors to the FLC project, and especially all the members of Arché who joined us in discussions in seminars and conferences. Special thanks also go to Toby Meadows, who helped usher the project through its final stages. Finally, we are grateful for wonderful help and advice from OUP's Assistant Commissioning Editor, Eleanor Collins.

Contents

List of Contributors

Jc Beall, University of Connecticut and University of Tasmania
Colin R. Caret, Yonsei University
Hartry Field, New York University
Michael Glanzberg, Northwestern University
Ole T. Hjortland, University of Bergen
Vann McGee, Massachusetts Institute of Technology
Graham Priest, City University of New York
Stephen Read, University of St Andrews
Greg Restall, University of Melbourne
David Ripley, University of Connecticut
Stewart Shapiro, Ohio State University
Heinrich Wansing, Ruhr-Universität Bochum
J. Robert G. Williams, University of Leeds
Elia Zardini, University of Lisbon

PART I

Introduction

1

Logical Consequence

Its Nature, Structure, and Application

Colin R. Caret and Ole T. Hjortland

1.1 Introduction

Recent work in philosophical logic has taken interesting and unexpected turns. It has seen not only a proliferation of logical systems, but new applications of a wide range of different formal theories to philosophical questions. As a result, philosophers have been forced to revisit the nature and foundation of core logical concepts, chief amongst which is the concept of logical consequence. This volume collects together some of the most important recent scholarship in the area by drawing on a wealth of contributions that were made over the lifetime of the AHRC-funded *Foundations of Logical Consequence* project. In the following introductory essay we set these contributions in context and identify how they advance important debates within the philosophy of logic.

Logical consequence is the relation that obtains between premises and conclusion(s) in a *valid argument*. Validity, most will agree, is a virtue of an argument, but what sort of virtue? Orthodoxy has it that an argument is valid if it must be the case that when the premises are true, the conclusion is true. Alternatively, that it is impossible for the premises to be true and the conclusion false simultaneously. In short, the argument is *necessarily truth preserving*.

These platitudes, however, leave us with a number of questions. How does the truth of the premises guarantee the truth of the conclusion in a valid argument? Is an argument valid in virtue of the meaning of logical constants? What constraints does logical consequence impose on rational belief? What are the formal properties of the consequence relation: compactness, reflexivity, transitivity, monotonicity? Can there be more than one conclusion in a valid argument? Are there phenomena, such as the semantic and set theoretic paradoxes, which ought to compel us to revise the standard theory of logical consequence? Is there, in some sense, more than one correct consequence relation, more than one way in which an argument can be valid? And so on.

The debate about the nature of logical consequence has traditionally divided along *model-theoretic* vs. *proof-theoretic* lines. Some of the essays in this volume contribute to that particular debate, while others represent views which do not neatly fall on either side of the traditional divide. By now, some philosophers even question the orthodox 'truth-preservation' view itself: that is, whether necessary truth preservation is really a necessary and sufficient condition for validity. A recent trend is the search for a 'third way' of understanding the nature of consequence. Some of the contributions to this volume address emerging *primitivist* and *deflationist* views about the concept of consequence. As early as Chapter 2, Hartry Field will argue that the orthodox definition—indeed, *any* definition of logical consequence—must be set aside in favour of an approach which emphasizes the role of validity in the cognitive economy.

1.2 Models and Consequence

The model-theoretic tradition is the heritage from Tarski (1936). It attempts to give a precise explication of the platitude that valid arguments are necessarily truth preserving. This is a reductive project: at the first stage, extensionally individuated constructs or 'models' are deployed to define the notion of *truth-in-a-model*. Valid arguments are then defined as those that preserve (from premises to conclusion) truth-in-a-model *in all models*:

> A is a logical consequence of a set of premises Γ ($\Gamma \models A$) if and only if, for every model M: whenever every $B \in \Gamma$ is true in M, A is also true in M.[1]

As a special case, we say that a sentence that is true in all models is *logically true* ($\models A$). Because the definition quantifies over all models, and because the range of available models is (in part) determined by the meanings of logical terms like $\neg$ and $\exists$, Tarski argues that his analysis satisfies the criteria of *necessity* and *formality* that are distinctive of the concept of logical consequence. In short, the truth of the premises *necessitates* the truth of the conclusion, and the truth preservation is a result of the *form* of the argument (as opposed to the content of non-logical expressions).[2]

[1] Equivalently, we can think of the logical consequence with a *no counter model* definition: $\Gamma \models A$ if and only if there is no (counter-)model M such that every $B \in \Gamma$ is true in M, but A is false in M.

[2] See for example Sher (1991) and MacFarlane (2000) for a discussion of formality.

Moreover, model-theoretic consequence is supposed to capture the modal nature of validity without reference to more philosophically controversial concepts such as *necessity*, *analyticity*, or *aprioricity*. It is this reductive feature, together with the extensibility of model-theoretic methods to all sorts of formal systems, that has made it the dominating theory of consequence for almost a century. In Etchemendy's words, Tarski's work has earned him the highest compliment: his definition of consequence is "no longer seen as the result of conceptual analysis [...] the need for analysis is forgotten, and the definition is treated as common knowledge" (Etchemendy 1990, 1).

There is a rich literature on the adequacy of Tarski's approach. While the contributions of the present volume do not engage with every aspect of this literature, several address the question of the demarcation and meaning of the logical terms. In Chapter 2, Field also offers a general criticism of the model-theoretic approach in the spirit of Moore's open question argument. Field argues that what we mean by 'valid' cannot, conceptually, be exhausted by the notion of truth(-preservation) in all models.

Etchemendy (1990) influentially argued that there is a systematic problem with the Tarskian definition of logical consequence. On his reading of Tarski, each valid argument is associated with a class of arguments that are materially truth preserving. Roughly speaking, the class is determined by the available reinterpretations of the non-logical vocabulary of the argument. In contemporary terms, we would say it is determined by the models.[3]

Etchemendy argues that if the class of models determines whether or not a sentence is a logical truth (and a conclusion a logical consequence of a set of premises), then the concept of logical consequence is hostage to non-logical facts. The existence of a model is not a purely logical matter, it depends on the existence of the members of its domain. The upshot is that *Tarski's Thesis*—viz. the claim that a sentence is logically true just in case it is true in all models—is itself contingently true. Yet this violates the commonplace assumption that logical truths are necessary truths. Etchemendy takes this to reveal a conceptual flaw in Tarski's account of logical truth and consequence.

This conceptual flaw, Etchemendy tells us, underlies an *extensional* problem with Tarskian consequence. The account overgenerates: It predicts that contingent sentences are logical truths. For example, in a finite universe containing exactly n

[3] Etchemendy contends that Tarski's original account deviated from the contemporary notion of a model. For Etchemendy it is critical that Tarski did not consider models where one varies the size of the domain. The historical accuracy of this interpretation of Tarski has been criticized, for example, in Gómez-Torrente (1996) and Ray (1996). See also Patterson (2012) for a more comprehensive historical account of Tarski's work.

things, the first-order sentence 'there exist no more than n things' will be logically true. Of course, this first-order sentence is not a logical truth according to standard first-order semantics. Etchemendy's point, rather, is that in the standard semantics the existence of any model which serves as a countermodel is guaranteed by a non-logical assumption, namely one about the size of the universe. This is a substantial metaphysical assumption, one which for example involves the set theory which serves as the background theory for the models.[4]

According to Etchemendy, as the expressive power of the language increases, the problem becomes more acute. In a higher-order logic we can express substantial set-theoretic claims, such as the continuum hypothesis.[5] The result is that on the Tarskian account, either the continuum hypothesis is a (second-order) logical truth, or its negation is a (second-order) logical truth. But, as Etchemendy suggests, many philosophers would consider it an extra-logical question whether a substantial mathematical sentence like the continuum hypothesis is true or false.[6]

Etchemendy's work produced a spate of replies. McGee (1992b) argues that although Etchemendy is right that the existence of models relies on non-logical facts, models exist as a matter of metaphysical or mathematical necessity.[7] Thus, contrary to Etchemendy's contention, Tarski's Thesis is not contingent. Even if it is not strictly logical, it holds necessarily, and that is good enough. Priest (1995) and Sher (1996) have objected to Etchemendy's construal of Tarski's account of consequence by suggesting that it allows for the universe of quantification to include non-existing (but possible) objects. The point is that if the domains can include possibilia as opposed to only actually existing objects, there is no reason to fear that finitism (about the actual world) will lead to a restriction on the models. Gómez-Torrente (1998) defends Tarski's Thesis by arguing that it does justice to important properties that we attribute to the ordinary concepts of logical truth and consequence. Hanson (1997) and Shapiro (1998) both attempt to amend Tarski's account of consequence with an explicit modal element. Sher (1996) proposes a 'formal-structural' understanding on which model theory is supposed to be grounded in the formal properties of the world.[8]

One of the crucial issues to emerge from this debate concerns the classification of 'logical' and 'non-logical' terms. Tarski's account of consequence—as he himself acknowledged—depends on fixing a class of logical constants. Only once

[4] See especially Etchemendy (1990, ch. 8).

[5] See, for example, Shapiro (1991) for details.

[6] See Etchemendy (1990, 122–4; 2008, 276–7). Soames (1998) and Badici (2003) give objections to Etchemendy's second-order case.

[7] See also McGee (1992a).

[8] See Hanson (1997), Sher (2001), and Hanson (2002) for a subsequent discussion.

a choice of logical constants is made can the logical form of an argument, and thus the associated class of arguments, be identified. Tarski's concept of logical consequence therefore invites us to answer the question of *logicality*: Is there a non-arbitrary property that determines the class of logical constants?

Tarski (1936) himself was initially critical of any non-arbitrary divide between logical and non-logical constants, but he later developed an influential proposal (see Tarski 1986). His suggestion is an *invariance* account for logical constants, in which expressions count as logical when they are invariant under a class of transformations. The view has received renewed interest lately, and has a number of proponents, (e.g., Sher 1991, Bonnay 2008, and Bonnay and van Benthem 2008).[9] Tempting as it may sound, this account is not free of problems. The invariance account is moderately successful for distinguishing the paradigmatic logical and non-logical terms, but there are borderline cases (identity, set membership, etc.) about which different precisifications of the invariance account disagree. Some even argue that the invariance approach allows clearly non-logical expressions to count as logical. See McCarthy (1987; 1989), Feferman (1998), Gómez-Torrente (2002), and Casanovas (2007) for some sample criticisms.

In the wake of Montague's influential work (1973; 1974a), model theory found a second home in the formal semantics of natural languages. One might hope that careful attention to the empirical properties of natural language would, in turn, provide insights into the adequacy of the model-theoretic definition of logical consequence. In Chapter 3, Michael Glanzberg raises a series of challenges to such optimism. He argues that the 'lexical entailments' which drive extensive portions of natural language competence are not of the same kind as logical entailments (consequences), and that the logical terms are not simply *given* but instead arise from a process of *abstracting* and *idealizing* away from natural language when we theorize about logic.

The very flexibility of the model-theoretic approach has also been cause for some degree of skepticism about its success. In particular, since the advent of Kripke models, also known as 'possible worlds semantics', most formal systems have been outfitted with something resembling a Tarskian model theory. In Chapter 4, Graham Priest addresses such a case of model-theoretic consequence that has been criticized for being 'merely algebraic' rather than philosophically illuminating. He offers a compelling interpretation of the Routley–Meyer ternary semantics for conditionals, on which conditionals express *functions* from antecedent propositions to consequent propositions.

[9] The invariance account of logical constants can be constrasted with the proof-theoretic tradition discussed below. See, for example, Hacking (1979) and Došen (1989).

1.3 Proof and Consequence

Despite the fruitfulness of Tarski's approach, model theory does not have a monopoly on logical consequence. Its most prominent rival is the proof-theoretic tradition that goes back to Gerhard Gentzen's (1935) idea that the meaning of a logical connective is 'defined' by its introduction rules (while the elimination rules are justified by respecting the stipulation made by the introduction rules). This meaning-theoretic claim is sometimes referred to as *logical inferentialism* or *inferential role semantics*. More recently the formal business end of the theory has been called *proof-theoretic semantics*, in contrast to the more familiar model-theoretic semantics. In a slogan, the meanings of logical connectives are *proof conditional* rather than *truth conditional.*

For the inferentialist, an argument is valid or invalid by virtue of the inferential meaning of the logical connectives occurring in the argument. Early versions of the view were defended by Popper (1946, 1947) and Carnap (1934), but the subsequent criticism in Prior (1960) was more influential.[10] Prior introduces the mock connective tonk, whose rules are supposed to be 'analytically valid':

$$\frac{A}{A \mathsf{tonk} B} \qquad \frac{A \mathsf{tonk} B}{B}$$

The problem for the inferentialist is that the rules for tonk lead straight to triviality. From any assumption A we may introduce AtonkB, and then use the elimination rule to conclude B. Starting with the seminal Belnap (1962), most contemporary work on proof-theoretic validity is therefore concerned with formulating appropriate constraints on the inference rules that can be meaning-determining. The constrains are intended to rule out tonk and other ill-behaved connectives, while including at least the laundry list of ordinary connectives.

The informal suggestion is that the inference rules for tonk do not follow Gentzen's recipe for standard connectives such as $\wedge$, $\rightarrow$, etc. The elimination rule for tonk is, in some sense yet to be specified, too strong for the introduction rule (or, alternatively, the introduction rule is too weak for the elimination rule). Dummett (1973; 1991) coins the term 'harmony' for the desired relationship between introduction rules and elimination rules. Following his suggestion, the foremost challenge to a proof-theoretic semantics is a precise formal account of harmony.

Dummett's (1991) own formulation of harmony is inspired by Prawitz's (1971, 1985, 2005) work on proof-theoretic validity. This proof-theoretic notion of logical consequence is in turn based on Prawitz's normalization theorem (Prawitz 1965). Prawitz defines proof-theoretic validity inductively on what he calls *arguments.*

[10] For a recent perspective on Popper's contribution, see Schroeder-Heister (1984).

An argument is typically taken to be a natural deduction tree consisting of a conclusion (labelling the end node), and a finite set of premises (labelling leaf nodes). An argument is *closed* if it has no undischarged (open) assumptions; otherwise it is open. An argument is *canonical* if its last rule application is an application of an introduction rule (e.g. $\vee$ introduction or conditional proof) and the arguments for its premises are valid (open or closed) arguments. An open argument is *valid* if the result of replacing its open assumptions with a valid closed argument is valid. A closed argument is *valid* if it is a canonical argument or it can be converted into a canonical argument with the same conclusion using *reductions*.

Thus, for example, the introduction and elimination rules for $\wedge$ are valid:

$$\dfrac{\begin{matrix}\Pi_0 & \Pi_1 \\ A & B\end{matrix}}{A \wedge B}\,(\wedge I) \qquad \dfrac{\begin{matrix}\Pi \\ A \wedge B\end{matrix}}{A}\,(\wedge E(i)) \qquad \dfrac{\begin{matrix}\Pi \\ A \wedge B\end{matrix}}{B}\,(\wedge E(ii))$$

It is straightforward to check that the introduction rule is valid. It is an open argument, so we let Π_0, Π_1 be closed arguments for the respective assumptions. Since $\wedge$ introduction is by definition a canonical argument, the result is a closed valid argument. For the elimination rules we need to apply a reduction. Since Π is a closed valid argument, it is either canonical or reducible to a canonical argument. We can therefore let Π be an argument that terminates with an application of an introduction rule:

$$\dfrac{\dfrac{\begin{matrix}\Pi_3 & \Pi_2 \\ A & B\end{matrix}}{A \wedge B}}{A}$$

We can then convert the above derivation into the simpler derivation $\begin{matrix}\Pi_3 \\ A\end{matrix}$, avoiding the detour through the conjunction rules completely. Since Π_3 is a closed valid argument for A, it is either canonical or reducible to a canonical argument. Similar argument can be given for other connectives by detour conversions known from the normalization theorem. For example:

$$\dfrac{\dfrac{\begin{matrix}[A]^u \\ \Pi_0 \\ B\end{matrix}}{A \to B}\,(u) \qquad \begin{matrix}\Pi_1 \\ A\end{matrix}}{B} \;\rightsquigarrow\; \begin{matrix}\Pi_1 \\ A \\ \Pi_0 \\ B\end{matrix}$$

$$\dfrac{\dfrac{\begin{matrix}\Pi \\ A_i\end{matrix}}{A_1 \vee A_2} \qquad \begin{matrix}[A_1]^u \\ \Pi_1 \\ C\end{matrix} \qquad \begin{matrix}[A_2]^u \\ \Pi_2 \\ C\end{matrix}}{C}\,(u) \;\rightsquigarrow\; \begin{matrix}\Pi \\ A_i \\ \Pi_i \\ C\end{matrix}$$

where $i \in \{1, 2\}$.

Prawitz's account of validity relies on two crucial assumptions. First, introduction rules are meaning-determining. Put differently, the introduction rules *implicitly define* the meaning of the connective in question. Applications of these rules are therefore considered immediately valid—they are canonical. Second, elimination rules must allow for the required reduction conversions. This latter part can be thought of as the harmony condition, and it easy to see that tonk does not allow for a reduction conversion. Finally, note that Prawitz's account, albeit proof-theoretic, does not rely on any particular proof system. Proof-theoretic validity is independent of derivation in any particular system.[11]

As we have seen, the notion of proof-theoretic validity hinges crucially on the presence of reduction procedures that can transform any argument into an argument which terminates with an introduction step. Interestingly, Prawitz and Dummett argue that the inference rules for classical negation do not possess the desired reduction property. They interpret this as a *revisionary argument* against classical logic (see especially Dummett 1991, 296–300 and Prawitz 1977, 34), an approach that has since been criticized by a number of philosophers. One example is Read (2000) who argues that there is a perfectly natural extension of Prawitz's reduction steps that also covers inference rules for full classical logic.[12]

In Chapter 5, Read continues the tradition of proof-theoretic validity by developing a more encompassing notion of harmony.[13] He takes on Prior's challenge by arguing that analytically valid arguments may yet fail to be truth preserving. Vann McGee's contribution in Chapter 6 also contributes to this debate. McGee suggests a way to bridge proof- and truth-conditional approaches on the 'realist' assumption that classical inferences are valid (for use in metatheory).[14]

1.4 One or Many?

Moving on from direct analyses of the nature of consequence, one thing that cannot be overlooked about contemporary *formal* logic is the proliferation of consequence relations. Some philosophers, such as Prawitz and Dummett above, have argued that there is one correct logic—in their case, intuitionistic logic. More frequently, philosophers have taken classical logic to be the one true logic (e.g. Quine 1986; Burgess 1992; Williamson 1994).

[11] For a detailed study of Prawitz on proof-theoretic validity, see Schroeder-Heister (2006).

[12] There are plenty of sensible replies from classical logicians. See also Weir (1986), Milne (1994), and Rumfitt (2000).

[13] See also Read (2010, 2008).

[14] For some important criticisms of the proof-theoretic approach to consequence, see Etchemendy (1990), Priest (1999), Williamson (2003), and the contribution of Field in Chapter 2.

Others strive for an inclusive position on which there are several *equally good* candidates for the extension of the consequence relation, a view now known as *logical pluralism*. One motive for this position is Carnap's (1934) conventionalism, encapsulated by the slogan 'In logic, there are no morals'. For Carnap the choice of logic is determined by our choice of language transformation rules, a choice that is in turn largely pragmatic. One might hold, for instance, that one language is better suited for mathematics, another for mereology, and yet another for vague expressions. In sum: different discourses, different logics. Carnapian pluralism is in other words openly *language dependent*. This framework, however, is not entirely unproblematic, since it appears that if the difference in logic only arises as a result of a difference in language, then disagreements about logic (and about logical consequence) are merely verbal disputes. Quine's (1986) *meaning-variance argument* makes heavy weather of the fact that non-classical logicians only seemingly disagree with the classical logician. In reality, Quine claims, they only 'change the subject'. Quine's argument, briefly stated, is that the logical rules associated with an expression (e.g., a negation) are constitutive of the meaning of that expression, and so revising the logical rules is tantamount to changing the expression itself.[15]

Quine's meaning-variance argument has not gone uncontested. It has been criticized, for example, in Restall (2002), Priest (2006a), and Paoli (2003; 2007). In Chapter 7, Stewart Shapiro describes a novel brand of pluralism (or *relativism*, as he prefers to say; see also Shapiro 2014) on which a given consequence relation has authority only relative to background assumptions about mathematical structure. From this perspective, for example, both classical and intuitionstic logic are correct systems for reasoning about classical and intuitionistic mathematics, respectively. Shapiro rejects the Quinean criticism since, for the relativist, there simply is no fact of the matter whether a logical term in one domain means the same thing as a logical term in a different domain.

Others have argued that logical pluralism can be maintained without Carnapian language dependence. The most developed account of pluralism in recent times is precisely such a langauge-independent pluralism. Beall and Restall (2000; 2001; 2006) develop a pluralist position according to which the concept of logical consequence is *schematic*, and therefore allows for more than one specific consequence relation. The heart of logical consequence, according to them, is that it is truth preservation across a range of *cases*. Cases are the schematic part of the

[15] See Haack (1974) for a discussion of the meaning-change argument. Unlike Quine, Haack coaches the argument in terms of the meaning of a logical expression. Most recent debates on the argument similarly reconstruct it in meaning-theoretic terms, even though that would have been unpalatable to Quine.

consequence relation, and can be instantiated by, for example, classical models, possible worlds, situations, or constructions. They call the schematic consequence relation the *Generalized Tarskian Thesis* (GTT):

> **GTT** an argument is valid$_x$ if, and only if, in every case$_x$ in which the premises are true the conclusion is true. (Beall and Restall 2006, 29)

The GTT view of consequence entails a form of pluralism as long as there is more than one permissible class of 'cases'. Beall and Restall themselves defend three such candidate classes: classical models, constructions, and situations, although they do not rule out others. Truth preservation across all models is the classical consequence relation, truth preservation across all constructions is the intuitionistic consequence relation, and truth preservation across all situations is the relevant consequence relation. That is to say, varying the cases gives rise to consequence relations with different extensions. The result is a logical pluralism that supports multiple consequence relations over a single language.

Whether GTT pluralism is ultimately tenable depends on the permissibility of the cases. Are there any conditions on when cases are permissible instantiations of the GTT schema? Beall and Restall offer no strict criterion, but they do suggest a guideline. Any proper consequence relation, they claim, must be necessary, formal, and normative. Hence, only instances of GTT that can be argued to satisfy at least these three conditions will be counted as proper consequence relations. That nevertheless leaves a worry about whether or not GTT is too inclusive (e.g. Goddu 2002). Without clear restrictions on which structures may count as a case, the result might be that GTT trivially gives us pluralism. For example, two distinct classes of possible world models may yield two different consequence relations (e.g. **S4** and **K**), but we do not want to say that they support logical pluralism.

The GTT pluralism has met with a number of other objections in recent literature. Both Priest (2006a, ch. 12) and Read (2006) have come to the defence of logical monism. Priest, for example, has made the case that GTT is too restrictive to capture the many different consequence relations that are advocated in philosophical debates. It does not, for example, allow for relations which preserve more values than truth (e.g. Priest's own Logic of Paradox), nor does it allow substructural consequence relations.[16] Others have argued that, contrary to Beall and Restall's contention, their pluralism does not succeed in being language independent (e.g. Hjortland 2013). Interpretations of logical constants will typically be

[16] It is easy to see that Beall and Restall's formulation of GTT builds in a number of structural properties, all of which are rejected in some substructural logic or other.

sensitive to the same class of cases that determine the logical consequence relation. As a result, moving from one class of cases to another will affect not only the extension of the consequence relation, but also the meaning of the involved logical constants.[17]

Alternative variants of logical pluralism has also been offered in response to Beall and Restall. Field (2009a) argues that truth preservation is not the best approach to logical consequence, a position that is echoed in his contribution to this volume (Chapter 2). Instead Field defends a form of logical pluralism that is premised directly on the normativity of logic. Logical consequence is tied to agents' rational management of (full and partial) beliefs, and a logical pluralism follows, he claims, from the existence of a number of equally good epistemic norms.[18] Russell (2008) explores a logical pluralism which arises from varying the relata of the consequence relation: for example, whether they are sentences, propositions, or utterances. Whether or not a consequence relation is 'correct' will depend on what we take it to be a relation of. Shapiro (2006) and Cook (2002; 2010) have argued that logics model reasoning practices, but that indeterminacy about the target practice can lead to multiple, incompatible models of the same phenomenon. Finally, Restall (2014) moves away from the model-theoretic approach to pluralism, and investigates a form of pluralism in a proof-theoretic setting.

1.5 The Structure of the Consequence Relation

The debate about the extension of logical consequence also involves a number of difficult questions about the structure of a consequence relation. For instance, an argument (valid or invalid) is usually taken to be a pair consisting of a set of premises and a conclusion. But debates in the philosophy of logic frequently involve various forms of generalized consequence relations. A fairly well-known example is multiple-conclusion consequence. An argument can be extended to a pair of sets, that is, a relation between a set of premises and a set of conclusions. Such multiple-conclusion arguments also give rise to a truth-preservational consequence relation, one for which the set of conclusions is interpreted disjunctively:

> The set of conclusions Δ is a logical consequence of Γ ($\Gamma \vDash \Delta$) if and only if, for every model M: whenever every $B \in \Gamma$ is true in M, some $A \in \Delta$ is also true in M.

[17] See also Wyatt (2004), Bueno and Shalkowski (2009), Griffiths (2013), and Keefe (2014) for other discussions of Beall and Restall's pluralism. Russell (2014) is an excellent overview of logical pluralism.

[18] See section 1.7 below for more about the normativity of logical consequence.

Multiple-conclusion arguments have been standard in proof-theory for a long time.[19] Gentzen's (1935) work on sequent calculus shows that a highly attractive axiomatization of classical logic can be given using multiple-conclusion arguments. The calculus has the added advantage that intuitionistic logic turns out to be exactly the same calculus, but with the conclusions restricted to either a singleton set (single conclusion) or the empty set.[20] Furthermore, it turns out that for some logical systems, the difference between single- and multiple-conclusion arguments teases out important differences. Supervaluational logic, for instance, has the same valid arguments as classical logic for single-conclusion arguments, but not for multiple-conclusion arguments.[21]

Multiple-conclusion consequence relations have, however, generated significant philosophical controversy. Intuitionists have argued that they are objectionable from a constructivist perspective (e.g. Dummett 1991; Tennant 1997). But the main point of contention is whether there is such a thing as multiple-conclusion arguments at all. After all, human agents do not appear to infer from premises to a set of conclusions. Rumfitt (2008) and Steinberger (2011) have both developed wholesale arguments against using multiple-conclusion arguments, on the ground that they have no natural counterpart in reasoning, or have no sensible interpretation. Shoesmith and Smiley (1978) is perhaps the most influential study of multiple-conclusion arguments, and contains a passionate defence of their logical significance. Restall (2005) provides a novel interpretation of multiple-conclusion consequence in terms of assertion and denial.

A further issue about the structure of logical consequence is the *nature* of the relata. Regardless of whether the conclusions are multiple or not, philosophers want to know what the premises and conclusion are. Syntactically, they are merely formulae, but as interpreted entities they could be propositions, characters, statements, sentence types, judgements, or even utterances. The choice of relata for logical consequence will have a significant impact on what we ought to take its extension to be. A number of entities are candidates for being truth-bearers, and therefore relata of consequence, but the resulting consequence relations will differ in terms of what they deem valid. Russell (2008) and Zardini (2012) are two recent discussions of the question, connecting the choice of relata to context-

[19] Unsurprisingly, there are also single-premise (multiple-conclusion) arguments, but these are less prevalent. The best-known calculus of this sort is dual-intuitionistic logic (see Urbas 1996).

[20] For details about such systems and a number of interesting proof-theoretic results, see Troelstra and Schwichtenberg (2000).

[21] Importantly, although supervaluational logic is classical with respect to arguments, it is non-classical with respect to meta-inferences: that is, inferences from arguments to arguments. Both conditional proof and classical *reductio ad absurdum* are examples of this. See Priest (2008) and Williamson (1994).

sensitivity and Kaplanian semantics (cf. Kaplan 1989). Russell (2008) gives a simple example to illustrate how, for example, contexts will affect the truth-preservation relation between utterances. Two different sentences uttered in the same context might express the same proposition: for example, 'I am in Paris' as uttered by Napoleon, and 'You are in Paris' as uttered by Wellington (to Napoleon) both express the proposition that *Napoleon is in Paris*. The proposition is a logical consequence of itself, of course, but it appears counterintuitive that 'You are in Paris' follows from 'I am in Paris'. Russell suggests that the example shows how an apparently invalid argument can have a conclusion expressing a proposition that is a consequence of the propositions expressed by the premises. Conversely, consider the argument 'Achilles is running now, therefore Achilles is running now'. Here we have seemingly uncontroversial consequence (reflexivity), but one that comes into doubt if we allow shifts of context of utterance from premise to conclusion. Finally, Russell observes that 'I am here now' is a logical truth in Kaplan's semantics, but if so it is a logical truth that cannot have inherited its validity from the proposition it expresses. The sentences expresses a contingent proposition when uttered by a speaker in a context, and is logically true only in virtue of the indexical nature of its expressions.

(These examples suggest that if we take natural language semantics seriously, the orthodox account of logical consequence comes under renewed scrutiny. For more on this theme, see Glanzberg's contribution in Chapter 3.)

Just as the relata of the consequence relation are open to discussion, so is what is *preserved* in the consequence relation. Again, the orthodoxy has it that logical consequence preserves truth from premises to conclusion, but that is not unproblematic. First, we have already seen that model-theoretic consequence preserves truth-in-a-model, a theoretic construct that is not necessarily a good model of truth simpliciter. Second, with the popularity of many-valued logic it is commonplace to talk about preservation of *designated values* rather than just the truth value *true*.[22] This reflects the fact that many-valued logics might have consequence relations that are defined as preservation of any of a set of values. Let the set of truth values be $\mathcal{V}$ and let the designated values be $\mathcal{D} \subseteq \mathcal{V}$:

> The set of conclusions Δ is a logical consequence of Γ ($\Gamma \models \Delta$) if and only if, for every model $\mathcal{M}$: whenever every $B \in \Gamma$ has a value from $\mathcal{D}$ in $\mathcal{M}$, some $A \in \Delta$ also has a value from $\mathcal{D}$ in $\mathcal{M}$.

An example of this is Graham Priest's *Logic of Paradox* where logical consequence is preservation of either the value *true* or the value *true-and-false*. In other

[22] See Priest (2008, ch. 7) for more about many-valued logics.

words, logical consequence is preservation of (at least) truth, not preservation of merely truth. The result is a paraconsistent non-classical logic, but one that shares the above schematic definition of logical consequence with classical logic. The difference consists only in what the logics treat as designated values preserved in consequence. The generalization of logical consequence also leads to a host of other considerations about preservation. Some consequence relations are defined by *anti-preservation* of non-designated values instead of preservation, and these may not have equivalent formulations in terms of preservation.[23] Finally, it is not clear that consequence is best understood in terms of preservation of truth values, or at least not as preservation of truth values alone. Perhaps, as some have argued, consequence requires preservation of *warrant*, or a relevance relation between the content of the premises and the content of the conclusion, such as variable sharing (cf. Read 1988, Restall 2004).

Part of the discussion of the structural properties of logical consequence is also about a range of purely formal properties. The traditional analysis of logical consequence has inherited some formal properties from Tarski's work: a *Tarskian* or *standard* consequence relation is reflexive, transitive, and monotonic.[24] That is, the following properties all hold:

$A \models A$;
if $A \models B$ and $B \models C$, then $A \models C$;
if $\Gamma \models A$, then $\Gamma, B \models A$.

However, contemporary logic, and especially proof theory, has frequently challenged these formal properties. In a sequent calculus, for example, reflexivity and monotonicity correspond to the identity axiom (Id) and the structural rule of weakening (K), respectively. Similarly, transitivity corresponds roughly to the cut rule:

$$\frac{}{A \Rightarrow A}\ (\mathrm{Id}) \qquad \frac{A \Rightarrow B \quad B \Rightarrow C}{A \Rightarrow C}\ (\mathrm{Cut}) \qquad \frac{\Gamma \Rightarrow A}{\Gamma, B \Rightarrow A}\ (\mathrm{K})$$

In addition, there is the question of whether the relata of a consequence relation are sets, multisets, or even lists (ordered sets). Although the former is often assumed by standard consequence relations, proof theories frequently come equipped with multisets or lists instead. Indeed, Gentzen's original sequent calculi for classical and intuitionistic logic had lists, and only yielded standard relations in the presence of additional structural rules, such as contraction (W) and exchange (C):[25]

[23] See Cook (2005) for a neat example. [24] See especially Tarski (2002).

[25] A catalogue of non-standard consequence relations, both semantic and proof-theoretic, is discussed in Avron (1991).

$$\frac{\Gamma, A, A \Rightarrow B}{\Gamma, A \Rightarrow B}\ (W) \qquad \frac{\Gamma, A, B, \Sigma \Rightarrow C}{\Gamma, B, A, \Sigma \Rightarrow C}\ (C)$$

Although these structural properties strike some philosophers and logicians as obvious, systems without them abound and they have found a range of applications in recent times. Logics which omit structural rules like weakening and contraction are, for example, commonplace. Such logics—called *substructural logics*—play a significant role in a number of debates. In philosophical logic, substructural systems are already familiar in the form of relevant logics, paraconsistent logics, and non-monotonic logics, while others such as linear logic and affine logics are more recent additions to the philosophical *organon*.[26]

But why would one drop structural rules from a logic? One reason, discussed for example in Paoli (2002), is that the presence of structural rules restricts the expressive power of a logic. A common example will serve as an illustration:

$$\frac{\Gamma, A_i \Rightarrow C}{\Gamma, A_0 \wedge A_1 \Rightarrow C}\ (L\wedge) \qquad \frac{\Gamma \Rightarrow A \quad \Gamma \Rightarrow B}{\Gamma \Rightarrow A \wedge B}\ (R\wedge)$$

$$\frac{\Gamma, A, B \Rightarrow C}{\Gamma, A \wedge B \Rightarrow C}\ (L\otimes) \qquad \frac{\Gamma \Rightarrow A \quad \Sigma \Rightarrow B}{\Gamma, \Sigma \Rightarrow A \wedge B}\ (R\otimes)$$

The additive conjunction $\wedge$ and the multiplicative conjunction $\otimes$ ('o-times') are equivalent in the presence of weakening and contraction. In linear logic, however, the distinct logics of the two conjunctions can be used for various applications. In Barker (2010), for example, the linear logic distinction of the conjunctions (and the dual distinction for disjunction) is used to give an account of free choice permission in natural language. Linear logic has also been used for set-theoretic purposes, in particular to give consistent theories of unrestricted comprehension (cf. Shiharata 1994). Finally, linear logic, because of its lack of contraction, is used to model various resource-sensitive phenomena.

In philosophical debates, substructural theories of semantic paradoxes have become fairly common in recent years. (See section 1.6 for more about semantic paradoxes.) It is well known that dropping structural contraction leaves room for an unrestricted truth predicate, as explored for example in Petersen (2000), Brady (2006), Shapiro (2011a), Zardini (2011). Similarly, non-transitive theories have received attention in the literature on both vagueness and truth. Zardini (2008) and Cobreros et al. (2012) use tolerant logics as the backdrop for a theory of vague predicates that can avoid the sorites paradox. Ripley (2013), Ripley (2012),

[26] For an introduction to substructural logic, see Restall (2000) or Paoli (2002). Paoli in particular gives a series of interpretations and motivations for dropping structural rules of various sorts. Relevant logics are treated in, for instance, Anderson and Belnap (1975), Routley et al. (1982), Read (1988), and Mares (2004). For more on non-monotonic logics, see Makinson (2005).

and Cobreros et al. (2013) have also developed a paradox-free theory of truth based on a non-transitive consequence relation. Finally, substructural logics also appear in formal epistemology (e.g. Sequoiah-Grayson 2009).

In Chapter 8, Elia Zardini provides further reasons for adopting non-transitive consequence relations for different philosophical applications. He gives an exhaustive analysis of what non-transitive consequence relations mean for the orthodox understanding of consequence, both in terms of what the relata are and in terms of the epistemological significance of consequence.[27]

1.6 Paradox and Revision

Logic is often thought of as immutable and apriori. In Frege's less prosaic phrase, the laws of logic are "boundary stones set in an eternal foundation, which our thought can overflow, but never displace". But is logic really immune to revision? And what would it even mean to revise logic? Should we, as Putnam (1969) argues, consider logic just another empirical discipline, subject to the same scientific calibrations as physics or chemistry, or is justification of logical laws a rationalist project (Dummett 1978, Wagner 1987)?

Quine, a seminal figure in this debate, is often taken to have changed his mind about the status of logic. In *Two Dogmas* (Quine 1960) he insists that logic is no more immune to revision than any other part of our sciences. Yet, in his later *Philosophy of Logic* (Quine 1986) he appears to take the meaning-variance argument to show that one cannot have rational grounds for a 'change of logic'. (See section 1.4 for more on meaning-variance.) Priest (2006a, ch. 10) offers a defence of the earlier Quinean position, arguing both concretely that classical logic ought to be revised and that it is not just possible to rationally revise logic, but historically evident that this has already occurred several times. According to Priest, part of the confusion in this debate rests on conflating revision of the *theory of logic* with revision of the *practice of logic*.[28]

By 'revision of logic' philosophers typically mean setting aside the classical logic of Frege and Russell in favour of some weaker system. This dialectic presumes that classical logic is prima facie plausible, but allows that it may be ultimately insufficient when all details of the role of logic are considered. Some revisionary arguments are premised on general views about the nature of logic, such as the intuitionist programme of Dummett and Prawitz discussed in section 1.3.[29]

[27] Other proponents of non-transitive systems include Weir (1998) and Tennant (1994).
[28] For another take on revision of logic, see Resnik (2004).
[29] See also Wright (1986, 2000) for a more general argument about revisionism.

A more frequent source of revisionary arguments, however, is the existence of particular domains of discourse to which classical logic appears unsuited. Revision of classical logic for quantum theory is one traditional, albeit somewhat dated, example. More fashionable is the growing number of theories of semantic and set-theoretic paradoxes that rely on non-classical reasoning: for example, theories about vague expressions or paradox-free theories of unrestricted truth. Many of these theories have in common that they do not advocate revision of logic as a universal rejection of classical logic, but only as a piecemeal reaction to particular—and perhaps isolated—philosophical problems.

Paradoxical reasoning typically involves two different kinds of principle: those governing a specific concept of philosophical interest such as truth or set membership, and logical principles that sanction reasoning involving, for example, quantifiers or negation. The idea behind revisionary solutions to paradox is that we can hang onto the intuitive principles of truth or membership by making certain logical inferences impermissible within the relevant discourse.[30] Formally speaking, a problematic operator or predicate might have desirable axioms or inference rules that are inconsistent with classical logic, but that can nonetheless consistently extend non-classical logics. The best-known example is formal theories of an unrestricted or naïve truth predicate T, as captured for example by the unrestricted Tarski biconditional:[31]

$$T(\ulcorner A \urcorner) \Leftrightarrow A$$

where $\ulcorner\urcorner$ is a form of quotation device, for example, the Gödel code of A. We know that that no classical theory sufficient to express its own syntax can consistently include the unrestricted Tarski biconditionals.

One variation on this theme is directly pertinent to *validity* itself. Suppose, as in section 1.1, that the orthodox 'truth-preservation' platitude really is constitutive of the concept of logical consequence. Then such a concept would seem to be inexpressible on pain of violating Gödel's Second Incompleteness theorem. As Field (2006; 2008) argues, if we had the resources to express, unrestrictedly, that all of our valid reasoning is truth preserving, then we could truly assert that every sentence of mathematics is true and all of its inferences are truth preserving and

[30] This should not be confused with *inconsistent meaning theory*. Rather than revise either the intuitive truth principles or intuitive principles of logic, the inconsistent meaning theorist holds that all such principles *are genuinely valid*. What paradoxes show, on this view, is that natural language suffers from an incoherent clash of meaning-constitutive principles. This view has its roots in Chihara (1979). Other advocates include Barker (1998); Eklund (2002); Azzouni (2003, 2007); Scharp (2007) and Patterson (2006, 2009).

[31] But not the only one, of course. Revisionary responses to Russell's paradox are also a major source of motivation for non-classical consequence relations.

thereby establish that the entire theory is true, violating Gödel's theorem. Field bites the bullet and accepts that this poses an absolute barrier to defining validity in terms of truth preservation. (See his Chapter 2 for more.)

In Chapter 9, Beall offers an alternative perspective on this problem. He suggests that we can stratify the concept of validity: for any order validity$_n$ there is a stronger notion validity$_{n+1}$ *which can assert* that all valid$_n$ arguments are truth-preserving. Beall argues that this approach is compatible with a deflationary stance toward truth. His argument draws on Lionel Shapiro's (2011a) novel *deflationary* account of logical consequence.

The contentious question for revisionary solutions to paradox is precisely which classical axioms or inference to reject: *paracomplete* ('gap') theories, for example, reject the law of excluded middle (LEM); while *paraconsistent* ('glut') theories reject the law of explosion (EFQ).[32] As noted above, there are also substructural theories that maintain all of the classical operational rules for the connectives, but reject one or more of the structural rules.[33]

There is no widespread agreement about how to adjudicate between the consequence relations of different revisionary theories. Nonetheless, there are some challenges that most such theories face. A typical kind of objection to revisionary theories is an objection from *revenge paradoxes*. These arise from the desire to increase the expressive power of our non-classical truth theory. A non-classical theory treats some sentences as *true* and others as *false*, but paradoxical sentences may not fall neatly into either category, so for certain explanatory purposes we might feel compelled to classify them in some other way. Gap theorists, for example, would like to say that paradoxical sentences are *indeterminate*, while glut theorists would like to say that paradoxical sentences are distinct from ordinary *consistent* sentences. Yet, enhancing a revisionary theory of truth with the resources to express such concepts is a non-trivial task. It is well known that predicates of this sort are susceptible to paradoxes which threaten the breakdown of the salient non-classical theories.[34] Indeed, it is one of the most resilient problems for theories of truth.

One way that revisionary theorists have tried to sidestep worries about revenge is to demarcate the paradoxes not in terms of a further semantic category (indeterminacy, consistency, etc.) but in terms of the norms or attitudes uniquely fitting

[32] For the former, see Kripke (1975), Martin and Woodruff (1975), Field (2002, 2007, 2008), and Horsten (2009, 2011). For the latter, see Priest (2004, 2006a, 2006b, 2010) and Beall (2009).

[33] There are, of course, also a variety of classical theories of *restricted* truth. For broadly Tarskian approaches see, for example, Parsons (1974), Burge (1979), Barwise and Etchemendy (1987), Simmons (1993), Glanzberg (2004a, 2005). See also Halbach (2011) for axiomatic approaches.

[34] For discussion see for example (Beall 2007) and Shapiro (2011b).

paradoxical sentences. Because revisionary theories typically appeal to a subclassical account of negation, critics have been quick to point out that asserting $\neg A$ does not rationally require rejecting A within the context of such a theory. Advocates have turned the tables and argued that we might get a grip on classifying paradoxical sentences by recognizing that this affords a specially *non-classical stance* toward permissible belief states.

In Chapter 10, David Ripley explores the prospects for making this rigorous by adding a *rejection* or *denial* operator to the language.[35] He considers whether this introduces a revenge paradox—the *denier*—and he argues that revisionary theorists actually can and should keep negation and denial closely linked together. Greg Restall argues in Chapter 11 that a familiar revenge objection to gap theories of truth can be turned into a symmetrical objection against glut theories. This leaves the debate at a stalemate without further tie-breaking considerations.

In Chapter 12, Heinrich Wansing discusses the knowability paradox. He argues that in its usual incarnation, due to F. Fitch and A. Church, this paradox provides the impetus for rethinking the meaning of the so-called anti-realist thesis that 'truth is knowable'. The quantificational pattern in the natural language explication of this thesis is reassessed and shown to be free of paradox in a standard interpretation in branching-time semantics.

1.7 Logic and Rationality

It is often assumed that there is a connection between logical consequence and rational management of beliefs. One might think, for example, that consistency is a minimal normative constraint on belief states.[36] Similarly, if one thinks that truth is the aim of belief, and, furthermore, that logical consequence necessarily preserves truth, then we ought to believe anything that is an obvious logical consequence of our beliefs.

Nevertheless, formulating the connections between rational belief and logical consequence turns out to be difficult. For example, we might locate the connection at the level of *degrees of belief* rather than full beliefs, and argue that there are logical constraints on the distribution of an agent's credences. Indeed, if *probabilism* for degrees of belief is true (i.e. the norm that any belief state ought to satisfy the axioms of the probability calculus), then it might be argued that some minimal logical constraints on beliefs are already in place.[37] Another question is

[35] See, for example, Parsons (1984) and Tappenden (1999).

[36] The AMG theory of belief revision is one tradition in which logic and belief management are intimately related (Alchourrón et al. 1985).

[37] See especially Christensen (2004).

what the normative force of these norms are: which deontic operator is present in normative connections between validity (or inconsistency) and believing? Is it one of permissibility or one of obligation? Moreover, the scope of attitudes constrained by logic is also in dispute. Do the connections only govern belief, or do they equally constrain *non-believing* and *dis-believing*?[38] Indeed, does logic equally constrain other propositional attitudes such as acceptance, rejection, and suspension of belief?[39]

Worse, the very existence of such a connection between logic and rational belief has been denied. Both Harman (1986) and Maudlin (2004), for example, insist that logic is merely a theory of truth preservation, whereas the question of which beliefs we *ought* or *ought not* to have is an independent issue. Harman gives a series of arguments against deductive closure and consistency as requirements on rational belief. For example, he argues that the Preface Paradox shows that inconsistent belief states can be rational. Consequently, whatever the constraint on rational belief states is, it has to be weaker than consistency. Harman also rejects the idea that logical consequence is a guide to rational inference. Just because B is a logical consequence of A, and an agent believes A, does not mean that the agent ought to believe B. After all, the agent might have independent reasons to disbelieve B (say, B is a contradiction), and so rational belief management should involve disbelieving A as well.[40]

Kolodny (2007) argues that there is no consistency requirement on belief states, since these are independent of the *aims* of belief. In Chapter 13, Williams takes on such critics by arguing that recent work on probabilism for degrees of belief provides an argument for logical constraints. Williams suggests that Joyce's (1998, 2009) accuracy-domination argument can be interpreted as an explanation of the value of consistency norms for belief.[41]

Field (2009b) goes further when he argues that, rather than logical consequence informing us about how to manage our beliefs, the bridge principles between rational belief and logical consequence are our best shot at explicating the nature of consequence itself.[42] This is a *primitivist* view about logical consequence, developed as a reaction to problems with giving a formal theory of truth that *declares* valid arguments truth preserving (see especially Field 2006). Field argues that any reasonable theory of truth (including his own) cannot consistently be extended with theorems that say that the axioms of the theory are true and

[38] See, for example, Broome (1999) and MacFarlane (unpublished).

[39] Some authors even propose that the very relata of consequence are propositional attitudes rather than propositions. See section 1.5 on the structure of consequence.

[40] See Broome (1999) for a related discussion with alternative formulations of the connection between consequence and rational belief.

[41] See also Williams (2012).

[42] For discussion, see Milne (2009) and Harman (2009).

the rules truth preserving. Field takes this to indicate that any formal theory of valid argument as necessary truth preservation will fail. The bridge principles between doxastic norms and logical consequence are further explored in Field's contribution in Chapter 2.

References

Alchourron, Carlos E., Gärdenfors, Peter, and Makinson, David. 1985. "On the logic of theory change: partial meet contraction and revision functions." *Journal of Symbolic Logic* 50:510–30.

Anderson, Alan Ross and Belnap, Nuel D. 1975. *Entailment: The Logic of Relevance and Necessity*, volume I. Princeton, NJ: Princeton University Press.

Avron, Arnon. 1991. "Simple consequence relations." *Information and Computation* 92: 105–39.

Azzouni, Jody. 2003. "The strengthened liar, the expressive strength of natural languages, and regimentation." *The Philosophical Forum* 34:329–50.

Azzouni, Jody. 2007. "The inconsistency of natural languages: how we live with it." *Inquiry* 50:590–605.

Badici, Emil. 2003. "A defense of the standard account of logical truth." In L. Behounek (ed.), *The Logica Yearbook 2003*, 45–58. Prague: Filosofia.

Barker, Chris. 2010. "Free choice permission as resource-sensitive reasoning." *Semantics & Pragmatics* 3:1–38.

Barker, John. 1998. "The inconsistency theory of truth". Ph.D. thesis, Princeton University.

Barwise, John and Etchemendy, John. 1987. *The Liar: An Essay on Truth and Circularity*. Oxford: Oxford University Press.

Beall, Jc. 2007. *Revenge of the Liar: New Essays on the Paradox*. Oxford: Oxford University Press.

Beall, Jc. 2009. *Spandrels of Truth*. Oxford: Oxford University Press.

Beall, Jc and Restall, Greg. 2000. "Logical pluralism." *Australasian Journal of Philosophy* 78:475–93.

Beall, Jc and Restall, Greg. 2001. "Defending logical pluralism." In John Woods and Bryson Brown (eds.), *Logical Consequence: Rival Approaches, Proceedings of the 1999 Conference of the Society of Exact Philosophy*. Stanmore: Hermes.

Beall, Jc and Restall, Greg. 2006. *Logical Pluralism*. Oxford: Oxford University Press.

Belnap, Nuel. 1962. "Tonk, plonk and plink." *Analysis* 22:30–4.

Bonnay, Denis. 2008. "Logicality and invariance." *Bulletin of Symbolic Logic* 14:29–68.

Bonnay, Denis and van Benthem, Johan. 2008. "Modal logic and invariance." *Journal of Applied Non-Classical Logics* 18:153–73.

Brady, Ross. 2006. *Universal Logic*. Stanford, CA: CSLI Publications.

Broome, John. 1999. "Normative requirements." *Ratio* 12:398–419.

Bueno, Otavio and Shalkowski, Scott A. 2009. "Modalism and logical pluralism." *Mind* 118:295–321.

Burge, Tyler. 1979. "Semantical paradox." *Journal of Philosophy* 76: 169–98.

Burgess, John. 1992. "Proofs about proofs: a defense of classical logic." In *Proof, Logic, and Formalization*, 8–23. London: Routledge.

Carnap, Rudolf. 1934. *Logische Syntax der Sprache*. Vienna: J. Springer. Reprinted as *The Logical Syntax of Language*, trans. Amethe Smeaton, Paterson, NJ: Littlefield, Adams and Co., 1937.

Casanovas, Enrique. 2007. "Logical operations and invariance." *Journal of Philosophical Logic* 36:33–60.

Chihara, Charles. 1979. "The semantic paradoxes: a diagnostic investigation." *Philosophical Review* 88:590–618.

Christensen, David. 2004. *Putting Logic in Its Place: Formal Constraints on Rational Belief*. Oxford: Clarendon Press.

Cobreros, Pablo, Egré, Paul, Ripley, David, and van Rooij, Robert. 2012. "Tolerant, classical, strict." *Journal of Philosophical Logic* 41:347–85.

Cobreros, Pablo, Egré, Paul, Ripley, David, and van Rooij, Robert. 2013. "Reaching transparent truth." *Mind* 122:841–66.

Cook, Roy. 2002. "Vagueness and mathematical precision." *Mind* 111:225–47.

Cook, Roy. 2005. "What's wrong with tonk(?)" *Journal of Philosophical Logic* 34:217–26.

Cook, Roy. 2010. "Let a thousand flowers bloom: a tour of logical pluralism." *Philosophy Compass* 5:492–504.

Došen, Kosta. 1989. "Logical constants as punctuation marks." *Notre Dame Journal of Formal Logic* 30:362–81.

Dummett, Michael. 1973. *Frege: Philosophy of Language*. London: Duckworth.

Dummett, Michael. 1978. "Is logic empirical?" In *Truth and Other Enigmas*, 269–89. London: Duckworth.

Dummett, Michael. 1991. *The Logical Basis of Metaphysics*. Cambridge, MA: Harvard University Press.

Eklund, Matti. 2002. "Inconsistent languages." *Philosophy and Phenomenological Research* 64:251–75.

Etchemendy, John. 1990. *The Concept of Logical Consequence*. Cambridge, MA: Harvard University Press. Reprinted in the David Hume Series, Stanford, CA: CSLI Publications, 1999.

Etchemendy, John. 2008. "Reflections on consequence." In D. Patterson (ed.), *New Essays on Tarski and Philosophy*, 263–99. Oxford: Oxford University Press.

Feferman, Solomon. 1998. *In the Light of Logic*. Oxford: Oxford University Press.

Field, Hartry. 2002. "Saving the truth schema from paradox." *Journal of Philosophical Logic* 31:1–27.

Field, Hartry. 2006. "Truth and the unprovability of consistency." *Mind* 115:567.

Field, Hartry. 2007. "Solving the paradoxes, escaping revenge." In Jc Beall (ed.), *Revenge of the Liar: New Essays on the Paradox*. Oxford: Oxford University Press.

Field, Hartry. 2008. *Saving Truth from Paradox*. Oxford: Oxford University Press.

Field, Hartry. 2009a. "Pluralism in logic." *Review of Symbolic Logic* 2:342–59.

Field, Hartry. 2009b. "What is the normative role of logic?" *Aristotelian Society Supplementary Volume* 83:251–68.

Gentzen, Gerhard. 1935. "Untersuchungen über das logische Schliessen." *Mathematische Zeitschrift* 39:175–210, 405–31. English translation "Investigations concerning logical deduction" in Szabo (1969, pp. 68–131).

Glanzberg, Michael. 2004. "A contextual-hierarchical approach to truth and the Liar paradox." *Journal of Philosophical Logic* 32: 27–88.

Glanzberg, Michael. 2005. "Presumptions, truth values, and expressing propositions." In Gerhard Preyer and Georg Peter (eds.), *Contextualism in Philosophy*, 349–96. Oxford: Clarendon Press.

Goddu, G.C. 2002. "What exactly is logical pluralism?" *Australasian Journal of Philosophy* 80:218–30.

Gómez-Torrente, Mario. 1996. "Tarski on logical consequence." *Notre Dame Journal of Formal Logic* 37:125–51.

Gómez-Torrente, Mario. 1998. "Logical truth and Tarskian logical truth." *Synthese* 117: 375–408.

Gómez-Torrente, Mario. 2002. "The problem of logical constants." *The Bulletin of Symbolic Logic* 8:1–37.

Griffiths, Owen. 2013. "Problems for logical pluralism." *History and Philosophy of Logic* 34:170–82.

Haack, Susan. 1974. *Deviant Logic*. Cambridge: Cambridge University Press.

Hacking, Ian. 1979. "What is logic?" *Journal of Philosophy* 76:285–319.

Halbach, Volker. 2011. *Axiomatic Theories of Truth*. Cambridge: Cambridge University Press.

Hanson, William H. 1997. "The concept of logical consequence." *Philosophical Review* 106:365–409.

Hanson, William H. 2002. "The formal-structural view of logical consequence: a reply to Gila Sher." *Philosophical Review* 243–58.

Harman, Gilbert. 1986. *Change in View*. Cambridge, MA: MIT Press.

Harman, Gilbert. 2009. "Field on the normative role of logic." *Proceedings of the Aristotelian Society* 109:333–5.

Hjortland, Ole Thomassen. 2013. "Logical pluralism, meaning-variance, and verbal disputes." *Australasian Journal of Philosophy* 91:355–73.

Horsten, Leon. 2009. "Levity." *Mind* 118:555–81.

Horsten, Leon. 2011. *The Tarskian Turn: Deflationism and Axiomatic Truth*. Cambridge, MA: MIT Press.

Joyce, James M. 1998. "A nonpragmatic vindication of probabilism." *Philosophy of Science* 65:575–603.

Joyce, James M. 2009. "Accuracy and coherence: prospects for an alethic epistemology of partial belief." In F. Huber and C. Schmidt-Petri (eds.), *Degrees of Belief*. 263–98. Dordrecht: Springer.

Kaplan, David. 1989. "Demonstratives." In Joseph Almog and Howard Wettstein (eds.), *Themes from Kaplan*. Oxford: Oxford University Press.

Keefe, Rosanna. 2014. "What logical pluralism cannot be." *Synthese* 191:1375–90.

Kolodny, Niko. 2007. "How does coherence matter?" *Proceedings of the Aristotelian Society* 107:229–63.

Kripke, Saul. 1975. "Outline of a theory of truth." *Journal of Philosophy* 72:690–716.

MacFarlane, John. 2000. "What does it mean to say that logic is formal?" Ph. D. dissertation, University of Pittsburgh.

MacFarlane, John. "In what sense (if any) is logic normative for thought?" Unpublished paper delivered at the American Philosophical Association Central Division conference, 2004.

Makinson, David. 2005. *Bridges from Classical to Nonmonotonic Logic*. London: King's College Publications.

Mares, Edwin. 2004. *Relevant Logic: A Philosophical Interpretation*. Cambridge: Cambridge University Press.

Martin, Robert L. and Woodruff, Peter W. 1975. "On representing 'true-in-L' in L." *Philosophia* 5:213–17.

Maudlin, Tim. 2004. *Truth and Paradox*. Oxford: Oxford University Press.

McCarthy, Timothy. 1987. "Modality, invariance, and logical truth." *Journal of Philosophical Logic* 16:423–43.

McCarthy, Timothy. 1989. "Logical form and radical interpretation." *Notre Dame Journal of Formal Logic* 30:401–19.

McGee, Vann. 1992a. "Review of Etchemendy: the concept of logical consequence." *Journal of Symbolic Logic* 57:329–32.

McGee, Vann. 1992b. "Two problems with Tarski's theory of consequence." *Proceedings of the Aristotelian Society* 92:273–92.

Milne, Peter. 1994. "Classical harmony: rules of inference and the meaning of the logical constants." *Synthese* 100:49–94.

Milne, Peter. 2009. "What is the normative role of logic?" *Aristotelian Society Supplementary Volume* 83:269–98.

Montague, Richard. 1973. "The proper treatment of quantification in ordinary English." In J. Hintikka, J. Moravcsik, and P. Suppes (eds.), *Approaches to Natural Language*, 221–42. Dordrecht: Reidel. Reprinted in Montague (1974).

Montague, Richard. 1974a. "English as a formal language." In Richard H. Thomason (ed.), *Formal Philosophy: Selected Papers of Richard Montague*. New Haven, CT: Yale University Press.

Montague, Richard. 1974b. *Formal Philosophy: Selected Papers of Richard Montague*. New Haven, CT: Yale University Press. Edited by R. H. Thomason.

Paoli, Francesco. 2002. *Substructural Logics: A Primer*. Dordrecht: Kluwer.

Paoli, Francesco. 2003. "Quine and Slater on paraconsistency and deviance." *Journal of Philosophical Logic* 32:531–48.

Paoli, Francesco. 2007. "Implicational paradoxes and the meaning of logical constants." *Australasian Journal of Philosophy* 85:553–79.

Parsons, Charles. 1974. "The Liar paradox." *Journal of Philosophical Logic* 3: 381–412.

Parsons, Terence. 1984. "Assertion, denial, and the liar paradox." *Journal of Philosophical Logic* 13:137–52.

Patterson, Douglas. 2006. "Tarski, the Liar, and inconsistent languages." *The Monist* 89: 150–77.

Patterson, Douglas. 2009. "Inconsistency theories of semantic paradox." *Philosophy and Phenomenological Research* 74:387–422.

Patterson, Douglas (ed.). 2012. *Alfred Tarski: Philosophy of Language and Logic*. History of Analytic Philosophy. London: Palgrave Macmillan.

Petersen, Uwe. 2000. "Logic without contraction as based on inclusion and unrestricted abstraction." *Studia Logica* 64:365–403. ISSN 0039-3215.

Popper, K.R. 1946. "Logic without assumptions." *Proceedings of the Aristotelian Society* 47:251–92.

Popper, K.R. 1947. "New foundations for logic." *Mind* 56:193–235.

Prawitz, Dag. 1965. *Natural Deduction: A Proof-theoretical Study*. Stockholm: Almqvist and Wiksell.

Prawitz, Dag. 1971. "Ideas and results in proof theory." In J. E. Fenstad (ed.), *Proceedings of the 2nd Scandinavian Logic Symposium*. 237–309. Amsterdam: North-Holland.

Prawitz, Dag. 1977. "Meaning and proofs: on the conflict between classical and intuitionistic logic." *Theoria* 43:2–40.

Prawitz, Dag. 1985. "Remarks on some approaches to the concept of logical consequence." *Synthese* 62:153–71.

Prawitz, Dag. 2005. "Logical consequence from a constructivist point of view." In Stewart Shapiro (ed.), *The Oxford Handbook of Philosophy of Mathematics and Logic*. 671–95. Oxford: Oxford University Press.

Priest, Graham. 1995. "Etchemendy and logical consequence." *Acta Philosophica Fennica* 25:283–92.

Priest, Graham. 1999. "Validity." In Achille C. Varzi (ed.), *European Review of Philosophy: The Nature of Logic*. 183–205. Stanford, CA: CSLI Publications.

Priest, Graham. 2006a. *Doubt Truth to be a Liar*. Oxford: Oxford University Press.

Priest, Graham. 2006b. *In Contradiction*. Oxford: Oxford University Press, 2nd edition.

Priest, Graham. 2008. *An Introduction to Non-Classical Logic: From Ifs to Is*. Cambridge: Cambridge University Press, 2nd edition.

Priest, Graham. 2010. "Hopes fade for saving truth." *Philosophy* 85:109–40.

Priest, Graham, Beall, Jc, and Armour-Garb, Bradley. 2004. *The Law of Non-Contradiction*. Oxford: Oxford University Press.

Prior, Arthur N. 1960. "The runabout inference-ticket." *Analysis* 21:38–9.

Putnam, Hilary. 1969. "Is logic empirical?" *Boston Studies in the Philosophy of Science* 216.

Quine, W.V. 1960. *Word and Object*. Cambridge, MA: MIT Press.

Quine, W.V. 1986. *Philosophy of Logic*. Cambridge, MA: Harvard University Press, 2nd edition.

Ray, Greg. 1996. "Logical consequence: a defence of Tarski." *Journal of Philosophical Logic* 25:617–77.

Read, Stephen. 1988. *Relevant Logic*. Oxford: Blackwell.

Read, Stephen. 2000. "Harmony and autonomy in classical logic." *Journal of Philosophical Logic* 29:123–54.

Read, Stephen. 2006. "Monism: the one true logic." In D. DeVidi and T. Kenyon (eds.), *A Logical Approach to Philosophy: Essays in Honour of Graham Solomon*, 193–209. Dordrecht: Springer.

Read, Stephen. 2008. "Harmony and modality." In L. Keiff, C. Dégremont, and H. Rückert (eds.), *On Dialogues, Logics and other Strange Things*. London: King's College Publications.

Read, Stephen. 2010. "General-elimination harmony and the meaning of the logical constants." *Journal of Philosophical Logic* 39:557–76.

Resnik, Michael D. 2004. "Revising logic." In Graham Priest, Jc Beall, and Bradley Armour-Garb (eds.), *The Law of Non-Contradiction: New Philosophical Essays*. Oxford: Oxford University Press.

Restall, Greg. 2000. *An Introduction to Substructural Logics*. London: Routledge.

Restall, Greg. 2002. "Carnap's tolerance, meaning, and logical pluralism." *Journal of Philosophy* 99:426–43.

Restall, Greg. 2004. "Logical pluralism and the preservation of warrant." *Logic, Epistemology, and the Unity of Science* 163–73.

Restall, Greg. 2005. "Multiple conclusions." In P. Hajek, L. Valdez-Villanueva, and D. Westerståhl (eds.), *Proceedings of the Twelfth International Congress on Logic, Methodology and Philosophy of Science*, 189–205. London: King's College Publications.

Restall, Greg. 2014. "Pluralism and proofs." *Erkenntnis* 79:279–91.

Ripley, David. 2012. "Conservatively extending classical logic with transparent truth." *Review of Symbolic Logic* 5:354–78.

Ripley, David. 2013. "Paradoxes and failures of cut." *Australasian Journal of Philosophy* 91:139–64.

Routley, Richard, Plumwood, Val, Meyer, Robert K., and Brady, Ross. 1982. *Relevant Logics and Their Rivals*, volume I. Atascadero, CA: Ridgeview.

Rumfitt, Ian. 2000. " 'Yes' and 'No'. " *Mind* 109:781–820.

Rumfitt, Ian. 2008. "Knowledge by Deduction." *Grazer Philosophische Studien* 77:61–84.

Russell, Gillian. 2008. "One true logic?" *Journal of Philosophical Logic* 37:593–611.

Russell, Gillian. 2014. "Logical pluralism." In Edward N. Zalta (ed.), *The Stanford Encyclopedia of Philosophy*. Stanford, CA: CSLI Publications, spring 2014 edition.

Scharp, Kevin. 2007. "Replacing truth." *Inquiry* 50:606–21.

Schroeder-Heister, Peter. 1984. "Popper's theory of deductive inference and the concept of a logical constant." *History and Philosophy of Logic* 5:79–110.

Schroeder-Heister, Peter. 2006. "Validity concepts in proof-theoretic semantics." *Synthese* 148:525–71.

Sequoiah-Grayson, Sebastian. 2009. "A positive information logic for inferential information." *Synthese* 167:409–31.

Shapiro, Lionel. 2011a. "Deflating logical consequence." *Philosophical Quarterly* 61:320–42.

Shapiro, Lionel. 2011b. "Expressibility and the liar's revenge." *Australasian Journal of Philosophy* 89:297–314.

Shapiro, Stewart. 1991. *Foundations without Foundationalism: A Case for Second-order Logic*. Oxford: Oxford University Press.

Shapiro, Stewart. 1998. "Logical consequence: models and modality." In M. Schirn (ed.), *Philosophy of Mathematics Today: Proceedings of an International Conference in Munich*. 131–256. Oxford: Oxford University Press.

Shapiro, Stewart. 2006. *Vagueness in Context*. Oxford: Oxford University Press.

Shapiro, Stewart. 2014. *Varieties of Logic*. Oxford: Oxford University Press.

Sher, Gila. 1991. *The Bounds of Logic*. Cambridge, MA: MIT Press.

Sher, Gila. 1996. "Did Tarski commit 'Tarski's Fallacy'?" *Journal of Symbolic Logic* 61: 653–86.

Sher, Gila. 2001. "The formal-structural view of logical consequence." *Philosophical Review* 241–61.

Shiharata, Masaru. 1994. "Linear set theory". Ph.D. thesis, Stanford University.

Shoesmith, D.J. and Smiley, T.J. 1978. *Multiple-Conclusion Logic*. Cambridge: Cambridge University Press.

Simmons, Keith. 1993. *Universality and the Liar*. Cambridge: Cambridge University Press.

Soames, Scott. 1998. *Understanding Truth*. Oxford: Oxford University Press.

Steinberger, Florian. 2011. "Why conclusions should remain single." *Journal of Philosophical Logic* 40:333–55.

Szabo, M.E. (ed.). 1969. *The Collected Papers of Gerhard Gentzen*. Amsterdam and London: North-Holland.

Tappenden, Jamie. 1999. "Negation, denial and language change in philosophical logic." In D. Gabbay and H. Wansing (eds.), *What is Negation?* 261–98. Dordrecht: Kluwer.

Tarski, Alfred. 1936. "Über den Begriff der Logischen Folgerung." *Actes du Congrés International de Philosophie Scientifique* 7:1–11. Reprinted as "On the concept of logical consequence" in *Logic, Semantics, Metamathematics*, trans. J.H. Woodger, Oxford: Oxford University Press, 1956.

Tarski, Alfred. 1986. "What are logical notions?" *History and Philosophy of Logic* 7:143–54. Ed. by John Corcoran.

Tarski, Alfred. 2002. "On the concept of following logically." *History and Philosophy of Logic* 23:155–96. Translated by Magda Stroinska and David Hitchcock.

Tennant, Neil. 1994. "The transmission of truth and the transitivity of deduction." In D.M. Gabbay (ed.), *What is a Logical System*? 161–78 Oxford: Oxford University Press.

Tennant, Neil. 1997. *The Taming of the True*. Oxford: Oxford University Press.

Troelstra, Anne S. and Schwichtenberg, Helmut. 2000. *Basic Proof Theory*. Cambridge: Cambridge University Press.

Urbas, Igor. 1996. "Dual-intuitionistic logic." *Notre Dame Journal of Formal Logic* 37: 440–51.

Wagner, Steven J. 1987. "The rationalist conception of logic." *Notre Dame Journal of Formal Logic* 28:3–35.

Weir, Alan. 1986. "Classical harmony." *Notre Dame Journal of Formal Logic* 27:459–82.

Weir, Alan. 1998. "Naive set theory is innocent!" *Mind* 107:763–98.

Williams, J. Robert G. 2012. "Generalized probabilism: Dutch books and accuracy domination." *Journal of Philosophical Logic* 41:811–40.

Williamson, Timothy. 1994. *Vagueness*. London: Routledge.

Williamson, Timothy. 2003. "Blind reasoning." *Aristotelian Society Supplementary Volume* 77:249–93.

Wright, Crispin. 1986. *Realism, Meaning and Truth*. Oxford: Blackwell.

Wright, Crispin. 2000. "On being in a quandry: relativism, vagueness, logical revisionism." *Mind* 110:45–98.

Wyatt, Nicole. 2004. "What are Beall and Restall pluralists about?" *Australasian Journal of Philosophy* 82:409–20.

Zardini, Elia. 2008. "A model of tolerance." *Studia Logica* 90:337–68.

Zardini, Elia. 2011. "Truth without contra(di)ction." *Review of Symbolic Logic* 4:498–535.

Zardini, Elia. 2012. "Truth preservation in context and in its place." In C. Dutilh-Novaes and O.T. Hjortland (eds.), *Insolubles and Consequences*. 249–71. London: College Publications.

PART II

Consequence: Models and Proofs

2

What is Logical Validity?

Hartry Field

2.1 Introduction

Whatever other merits proof-theoretic and model-theoretic accounts of validity may have, they are not remotely plausible as accounts of the meaning of 'valid'. And not just because they involve technical notions like 'model' and 'proof' that needn't be possessed by a speaker who understands the concept of valid inference. The more important reason is that competent speakers may agree on the model-theoretic and proof-theoretic facts, and yet disagree about what's valid.

Consider for instance the usual model-theoretic account of validity for sentential logic: an argument is sententially valid iff any total function from the sentences of the language to the values T, F that obeys the usual compositional rules and assigns T to all the premises of the argument also assigns T to the conclusion. Let's dub this *classical sentential validity*.[1] There's no doubt that this is a useful notion, but it couldn't possibly be what we *mean* by 'valid' (or even by 'sententially valid', i.e. 'valid by virtue of sentential form'). The reason is that even those who reject classical sentential logic will agree that the sentential inferences that the classical logician accepts are valid in *this* sense. For instance, someone who thinks that "disjunctive syllogism" (the inference from $A \vee B$ and $\neg A$ to B) is not a valid form of inference will, if she accepts a bare minimum of mathematics,[2] agree that the

[1] Note that this is very different from saying that validity consists in necessarily preserving truth; that account will be considered in section 2.2. The model-theoretic account differs from the necessary truth-preservation account in being purely mathematical: it invokes functions that assign the object T or F to each sentence (and that obey the compositional rules of classical semantics), without commitment to any claims about how T and F relate to truth and falsity. For instance, it involves no commitment to the claim that each sentence is either true or false and not both, or that the classical compositional rules *as applied to truth and falsity* are correct.

[2] And there's no difficulty in supposing that the non-classical logician does so, or even that she accepts classical mathematics across the board: she may take mathematical objects to obey special non-logical assumptions that make classical reasoning "effectively valid" within mathematics.

inference is *classically* valid, and will say that that just shows that classical validity outruns genuine *validity*. Those who accept disjunctive syllogism don't just believe it *classically* valid, which is beyond serious contention; they believe it *valid*.

This point is in no way peculiar to classical logic. Suppose an advocate of a sentential logic without disjunctive syllogism offers a model theory for her logic—for example, one on which an argument is sententially valid iff any assignment of one of the values T, U, F to the sentences of the language that obeys certain rules and gives the premises a value other than F also gives the conclusion a value other than F ("*LP-validity*"). This may make only her preferred sentential inferences come out "valid", but it would be subject to a similar objection if offered as an account of the meaning of 'valid': classical logicians who accept more sentential inferences, and other non-classical logicians who accept fewer, will agree with her as to what inferences meet this definition, but will disagree about which ones are valid. Whatever logic L one advocates, one should recognize a distinction between the concept 'valid-in-L' and the concept 'valid'.[3]

The same point holds (perhaps even more obviously) for provability in a given deductive system: even after we're clear that a claim does or doesn't follow from a given deductive system for sentential logic, we can disagree about whether it's valid.

I don't want to make a big deal about definition or meaning: the point I'm making can be made in another way. It's that advocates of different logics presumably disagree about something—and something more than just how to use the term 'valid', if their disagreement is more than verbal. It would be nice to know what it is they disagree about. And they don't disagree about what's classically valid (as defined either model-theoretically or proof-theoretically); nor about what's intuitionistically valid, or LP-valid, or whatever. So *what do they disagree about?* That is the main topic of this chapter, and will be discussed in sections 2.2–2.5.

Obviously model-theoretic and proof-theoretic accounts of validity are important. So another philosophical issue is to explain what their importance is, given that it is not to explain the concept of validity. Of course, one obvious point can be made immediately: the model theories and proof theories for classical logic, LP, etc. are effective tools for ascertaining what is and isn't classically valid, LP-valid,

[3] I'm tempted to call my argument here a version of Moore's Open Question Argument: a competent speaker may say "Sure, that inference is *classically* valid, but is it *valid*?" (or, "Sure, that inference is *LP-invalid*, but is it *invalid*?"). But this can be done only on a sympathetic interpretation of Moore, in which he isn't attempting a general argument against there being an acceptable naturalistic definition but rather is simply giving a way to elicit the implausibility of particular naturalistic definitions. In section 2.2 I will consider a proposal for a naturalistic definition of 'valid' which (though I oppose) I do not take to be subject to the kind of "open question" argument I'm employing here.

etc.; so to someone convinced that one of these notions extensionally coincides with genuine validity, the proof-theory and model-theory provide effective tools for finding out about validity. But there's much more than this obvious point to be said about the importance of model-theoretic and proof-theoretic accounts; that will be the topic of sections 2.6 and 2.7.

2.2 Necessarily Preserving Truth

One way to try to explain the concept of validity is to define it in other (more familiar or more basic) terms. As we've seen, any attempt to use model theory or proof theory for this purpose would be hopeless; but there is a prominent alternative way of trying to define it. In its simplest form, validity is explained by saying that an inference (or argument)[4] is valid iff it preserves truth by logical necessity.

It should be admitted at the start that there are non-classical logics (e.g. some relevance logics, dynamic logics, linear logic) whose point seems to be to require more of validity than logically necessary preservation of truth. Advocates of these logics may want their inferences to necessarily preserve truth, but they want them to do other things as well: for example, to preserve conversational relevance, or what's settled in a conversation, or resource use, and so forth. There are other logics (e.g. intuitionist logic) whose advocates may or may not have such additional goals. Some who advocate intuitionistic logic (e.g. Dummett) think that reasoning classically leads to error; which perhaps we can construe as, possibly fails to preserve truth. But others use intuitionistic logic simply in order to get proofs that are more informative than classical, because constructive; insofar as *those* intuitionists reserve 'valid' for intuitionistic validity, they too are imposing additional goals of quite a different sort than truth preservation.

While it is correct that there are logicians for whom truth preservation is far from the sole goal, this isn't of great importance for my purposes. That's because my interest is with *what people who disagree in logic are disagreeing about*; and if proponents of one logic want that logic to meet additional goals that proponents of another logic aren't trying to meet, and reject inferences that the other logic accepts only because of the difference of goals, then the apparent disagreement in logic seems merely verbal.

[4] The term 'inference' can mislead: as Harman has pointed out many times, inferring is naturally taken as a dynamic process in which one comes to new beliefs, and inference in that sense is not directly assessable in terms of validity. But there is no obvious non-cumbersome term for what it is that's valid that would be better—'argument' has the same problem as 'inference', and other problems besides. (A cumbersome term for what's valid is "pair $\langle \Gamma, B \rangle$ where B is a formula and Γ a set of formulas".)

I take it that logically necessary truth preservation is a good first stab at what advocates of classical logic take logic to be concerned with. My interest is with those who share the goals of the classical logician, but who are in non-verbal disagreement as to which inferences are valid. This probably doesn't include any advocates of dynamic logics or linear logic, but it includes some advocates of intuitionist logic and quantum logic, and most advocates of various logics designed to cope with vagueness and/or the semantic paradoxes. So these will be my focus. The claim at issue in this section is that *genuine* logical disagreement is disagreement about which inferences preserve truth by logical necessity.

Having set aside linear logic and the like, a natural reaction to the definition of validity as preservation of truth by logical necessity is that it isn't very informative: logical necessity looks awfully close to validity, indeed, logically necessary truth is just the special case of validity for 0-premise arguments. One can make the account slightly more informative by explaining logical necessity in terms of some more general notion of necessity together with some notion of logical form, yielding that an argument is valid iff (necessarily?) *every argument that shares its logical form* necessarily preserves truth.[5] Even so, it could well be worried that the use of the notion of necessity is helping ourselves to something that ought to be explained.

This worry becomes especially acute when we look at the way that logical necessity needs to be understood for the definition of validity in terms of it to get off the ground. Consider logics according to which excluded middle is not valid. Virtually no such logic accepts of any instance of excluded middle that it is not true: that would seem tantamount to accepting a sentence of form $\neg(B \vee \neg B)$, which in almost any logic requires accepting $\neg B$, which in turn in almost any logic requires accepting $B \vee \neg B$ and hence is incompatible with the rejection of this instance of excluded middle. To say that $B \vee \neg B$ is not *necessarily* true would seem to raise a similar problem: it would seem to imply that it is *possibly* not true, which would seem to imply that there's a possible state of affairs in which $\neg(B \vee \neg B)$; but then, by the same argument, that would be a possible state of affairs in which $\neg B$ and hence $B \vee \neg B$, and we are again in contradiction. Given this, how is one who

[5] The idea of one inference "sharing the logical form of" another requires clarification. It is easy enough to explain 'shares the *sentential* form of', 'shares the *quantificational* form of', and so on, but explaining 'shares the logical form of' is more problematic, since it depends on a rather open-ended notion of what aspects of form are logical. But perhaps, in the spirit of the remarks in Tarski (1936) on there being no privileged notion of logical constant, we should say that we don't really need a general notion of validity, but only a notion of validity relative to a given choice of which terms count as logical constants (so that validity then subdivides into sentential validity, first-order quantificational validity, first-order quantificational-plus-identity validity, and so on). And in explaining, for example, sentential validity in the manner contemplated, we need no more than the idea of sharing sentential form.

regards some instances of excluded middle as invalid to maintain the equation of validity with logically necessary truth? The only obvious way is to resist the move from 'it isn't logically necessary that p' to 'there's a possible state of affairs in which $\neg p$'. I think we must do that; but if we do, I think we remove any sense that we were dealing with a sense of necessity that we have a grasp of independent of the notion of logical truth.[6]

But let's put aside any worry that the use of necessity in explaining validity is helping ourselves to something that ought to be explained. I want to object to the proposed definition of validity in a different way: that it simply gives the wrong results about what's valid. That is: it gives results that are at variance with our ordinary notion of validity. Obviously it's possible to simply insist that by 'valid' one will simply mean 'preserves truth by logical necessity'. But as we'll see, this definition would have surprising and unappealing consequences, which I think should dissuade us from using 'valid' in this way.

Let $A_1, \ldots, A_1 \Rightarrow B$ mean that the argument from $A_1, \ldots, A_n$ to B is valid.[7] The special case $\Rightarrow B$ (that the argument from no premises to B is valid) means in effect that B is a valid *sentence*: that is, is in some sense logically necessary. The proposed definition of valid argument tries to explain

(I) $A_1, \ldots, A_n \Rightarrow B$

as

(II_T) $\Rightarrow True(\langle A_1 \rangle) \wedge \ldots \wedge True(\langle A_n \rangle) \rightarrow True(\langle B \rangle)$.

This is an attempt to explain validity of inferences in terms of the validity (logical necessity) of single sentences. I think that any attempt to do this is bound to fail.

The plausibility of thinking that (I) is equivalent to (II_T) depends, I think, on two purported equivalences: first, between (I) and

[6] An alternative move would be to say that one who rejects excluded middle doesn't believe that excluded middle isn't valid, but either (i) merely fails to believe that it is, or (ii) *rejects* that it is, without believing that it isn't. But (i) seems clearly wrong: it fails to distinguish the advocate of a logic that rejects excluded middle with someone agnostic between that logic and classical. (ii) is more defensible, though I don't think it's right. (While I think an opponent of excluded middle does reject some instances of excluded middle while not believing their negation, I think that he or she regards excluded middle as not valid.) But in any case, the distinction between rejection and belief in the negation is not typically recognized by advocates of the necessary truth-preservation account validity, and brings in ideas that suggest the quite different account of validity to be advanced in section 2.3.

[7] I take the A_i s and B in '$A_1, \ldots, A_n \Rightarrow B$' to be variables ranging over sentences, and the '$\Rightarrow$' to be a predicate. (So in a formula such as $\Rightarrow A \rightarrow B$, what comes after the '$\Rightarrow$' should be understood as a complex singular term with two free variables, in which the '$\rightarrow$' is a function symbol. A claim such as $A \Rightarrow True(\langle A \rangle)$ should be understood as saying that (for each sentence A) the inference from A to the result of predicating 'True' of the structural name of A is valid.) In other contexts I'll occasionally use italicized capital letters as abbreviations of formulas or as schematic letters for formulas; I don't think any confusion is likely to result.

(II) $\Rightarrow A_1 \wedge \ldots \wedge A_n \rightarrow B$;

second, between (II) and (II_T).[8]

An initial point to make about this is that while (I) is indeed equivalent to (II) in classical and intuitionist logic, there are many non-classical logics in which it is not. (These include even supervaluational logic, which is sometimes regarded as classical.) In *most* standard logics, (II) requires (I). But there are many logics in which conditional proof fails, so that (I) does not require (II). (Logics where $\wedge$-Elimination fails have the same result.) In such logics, we wouldn't expect (I) to require (II_T), so validity would not require logically necessary truth preservation.

Perhaps this will seem a quibble, since *many* of those who reject conditional proof want to introduce a notion of "supertruth" or "super-determinate truth", and will regard (I) as equivalent to

(II_{ST}) $\Rightarrow \text{Supertrue}(\langle A_1 \rangle) \wedge \ldots \wedge \text{Supertrue}(\langle A_n \rangle) \rightarrow \text{Supertrue}(\langle B \rangle)$.

In that case, they are still reducing the validity of an inference to the validity of a conditional, just a different conditional, and we would have a definitional account of validity very much in the spirit of the first. I will be arguing, though, that the introduction of supertruth doesn't help: (I) not only isn't equivalent to (II_T), it isn't equivalent to (II_{ST}) either, whatever the notion of supertruth. Validity isn't the preservation of either truth or "supertruth" by logical necessity.

To evaluate the proposed reduction of validity to preservation of truth or supertruth by logical necessity, we need first to see how well validity so defined coincides in extension with validity as normally understood. Here there's good news and bad news. The good news is that (at least insofar as vagueness can be ignored, as I will do) there is *very* close agreement; the bad news is that where there is disagreement, the definition in terms of logically necessary preservation of truth (or supertruth) gives results that seem highly counterintuitive.

The good news is implicit in what I've already said, but let me spell it out. Presumably for at least a wide range of sentences $A_1, \ldots, A_n$ and B, claim (II) above is equivalent to (II_T), and claim (I) is equivalent to the (I_T) of note 8. (I myself think these equivalences holds for *all* sentences, but I don't want to presuppose controversial views. Let's say that (II) is equivalent to (II_T) (and (I) to (I_T)) *at least* for all "ordinary" sentences $A_1, \ldots, A_n$, B, leaving unspecified where exceptions

[8] Or alternatively, first between (I) and

(I_T) $\text{True}(\langle A_1 \rangle), \ldots, \text{True}(\langle A_n \rangle) \Rightarrow \text{True}(\langle B \rangle)$;

second, between (I_T) and (II_T). The latter purported equivalence is, of course, a special case of the purported equivalence between (I) and (II), so the discussion in the text still applies.

might lie if there are any.) And presumably when $A_1, \ldots, A_n$ and B are "ordinary", (I) is equivalent to (II) (and (I_T) to (II_T)).[9] In that case, we have

(GoodNews) The equivalence of (I) to (II_T) holds at least for "ordinary" sentences: for those, validity does coincide with preservation of truth by logical necessity.

(Presumably those with a concept of supertruth think that for sufficiently "ordinary" sentences it coincides with truth; if so, then the good news also tells us that (I) coincides with the logically necessary preservation of *supertruth*.)

Despite this good news for the attempt to define validity in terms of logically necessary truth preservation, the bad news is that the equivalence of (I) to either (II_T) or (II_{ST}) can't plausibly be maintained for *all* sentences. The reason is that in certain contexts, most clearly the semantic paradoxes but possibly for vagueness too, this account of validity requires a wholly implausible divorce between which inferences are declared valid and which ones are deemed acceptable to use in reasoning (even static reasoning, for instance in determining reflective equilibrium in one's beliefs). In some instances, the account of validity would require having to reject the validity of logical reasoning that one finds completely acceptable and important. In other instances, it would even require declaring reasoning that one thinks leads to error to be nonetheless valid!

I can't give a complete discussion here, because the details will depend on how one deals either with vagueness or with the "non-ordinary" sentences that arise in the semantic paradoxes. I'll focus on the paradoxes, where I'll sketch what I take to be the two most popular solutions and show that in the context of each of them, the proposed definition of validity leads to very bizarre consequences. (Of course the paradoxes themselves force some surprising consequences, but the bizarre consequences of the proposal for validity go way beyond that.)

► **Illustration 1:** It is easy to construct a "Curry sentence" K that is equivalent (given uncontroversial assumptions) to "If True(⟨K⟩) then 0=1". This leads to an apparent paradox. The most familiar reasoning to the paradox first argues from the assumption that True(⟨K⟩) to the conclusion that 0=1, then uses conditional proof to infer that *if* True(⟨K⟩) *then 0=1*, then argues from that to the conclusion that True(⟨K⟩); from which we then repeat the original reasoning to '0=1', but this time

[9] This actually may not be so in the presence of vagueness: in many logics of vagueness (e.g. Lukasiewicz logics), (I) can hold when (II) doesn't. Admittedly, in Lukasiewicz logics there is a notion of super-determinate truth for which (I) does correspond to something close to (II), viz. (II_{ST}). But as I've argued elsewhere (e.g. Field 2008, ch. 5), the presence of such a notion of super-determinate truth in these logics is a crippling defect: it spoils them as logics of vagueness. In an adequate non-classical logic of vagueness, (I) won't be equivalent to anything close to (II).

with True(⟨K⟩) as a previously established result rather than as an assumption. Many theories of truth (this includes most supervaluational theories and revision theories as well as most non-classical theories) take the sole problem with this reasoning to be its use of conditional proof. In particular, they agree that the reasoning from the assumption of 'True(⟨K⟩)' to '0=1' is perfectly acceptable (given the equivalence of K to "If True(⟨K⟩) then 0=1"), and that the reasoning from "If True(⟨K⟩) then 0=1" to K and from that to "True(⟨K⟩)" is acceptable as well. I myself think that the best solutions to the semantic paradoxes take this position on the Curry paradox.

But what happens if we accept such a solution, but define 'valid' in a way that requires truth preservation? In that case, though we can legitimately reason from K to '0=1' (via the intermediate 'True(⟨K⟩)', we can't declare the inference "valid". For to say that it is "valid" in this sense is to say that True(⟨K⟩) → True(⟨0=1⟩), which yields True(⟨K⟩) → 0=1, which is just K; and so calling the inference "valid" in the sense defined would lead to absurdity. That's very odd: this theorist accepts the reasoning from K to 0=1 as completely legitimate, and indeed *it's only because he reasons in that way that he sees that he can't accept K*; and yet on the proposed definition of 'valid' he is precluded from calling that reasoning "valid". ◀

▶ **Illustration 2:** Another popular resolution of the semantic paradoxes (the truth-value gap resolution) has it that conditional proof is fine, but it isn't always correct to reason from A to True(⟨A⟩). Many people who hold this (those who advocate "Kleene-style gaps") do think you can reason from True(⟨A⟩) to True(⟨True(⟨A⟩)⟩); and so, by conditional proof, they think you should accept the conditional True(⟨A⟩) → True(⟨True(⟨A⟩)⟩). Faced with a Curry sentence, or even a simpler Liar sentence L, their claim is that L isn't true, and that the sentence ⟨L⟩ *isn't true* (which is equivalent to L) isn't true either. There is an obvious oddity in such resolutions of the paradoxes: in claiming that one should believe L but not believe it true, the resolution has it that truth isn't the proper object of belief.[10] But odd or not, this sort of resolution of the paradoxes is quite popular.

But the advocate of such a theory *who goes on to define "valid" in terms of necessary preservation of truth* is in a far odder situation. First, this theorist accepts the reasoning to the conclusion ¬True(⟨L⟩)—he regards ¬True(⟨L⟩) as essentially a theorem. But since he regards the conclusion as not true, then he regards the

[10] Others, even some who accept the resolution of the truth paradoxes in Illustration 1, introduce a notion of supertruth on which it isn't always correct to reason from A to Supertrue(⟨A⟩) but is correct to assert Supertrue(⟨A⟩) → Supertrue(⟨Supertrue⟨A⟩⟩). This resolution leads to a Liar-like sentence L*, and asserts that L* isn't supertrue and that '⟨L*⟩ isn't supertrue' isn't supertrue either. This may be slightly less odd, since it says only that the technical notion of supertruth isn't the proper object of belief.

(0-premise) reasoning to it as "invalid", on the definition in question: he accepts a conclusion on the basis of reasoning, while declaring that reasoning "invalid". This is making an already counterintuitive theory sound even worse, by a perverse definition of validity.

But wait, there's more! Since the view accepts both $\neg\text{True}(\langle L\rangle)$ and $\neg\text{True}(\langle\neg\text{True}(\langle L\rangle)\rangle)$, and doesn't accept contradictions, it obviously doesn't accept the reasoning from $\neg\text{True}(\langle L\rangle)$ to $\text{True}(\langle\neg\text{True}(\langle L\rangle)\rangle)$ as good. But on the proposed definition of "valid", the view does accept it as valid! For on the proposed definition, that simply means that $\text{True}(\langle\neg\text{True}(\langle L\rangle)\rangle) \rightarrow \text{True}(\langle\text{True}(\langle\neg\text{True}(\langle L\rangle)\rangle)\rangle)$, and as remarked at the start, the view does accept all claims of form $\text{True}(\langle A\rangle) \rightarrow \text{True}(\langle\text{True}(\langle A\rangle)\rangle)$. On the definition of validity, not only can good logical reasoning come out invalid, but fallacious reasoning can come out valid. ◀

These problems for defining validity in terms of necessary preservation of truth are equally problems for defining validity in terms of necessary preservation of supertruth: we need only consider paradoxes of supertruth constructed in analogy to the paradoxes of truth (e.g. a modified Curry sentence that asserts that if it's supertrue then $0 = 1$, and a modified Liar that asserts that it isn't supertrue). Then given any resolution of such paradoxes, we reason as before to show the divorce between the supertruth definition of validity and acceptable reasoning.

I've said that as long as we put vagueness aside (see n. 9), it's only for fairly "non-ordinary" inferences that the definition of validity as preservation of truth (or supertruth) by logical necessity is counterintuitive: for most inferences that don't crucially involve vague terms, the definition gives extremely natural results. But that is because it is only for such non-ordinary inferences that the approach leads to different results than the approach I'm about to recommend.

In the next section I'll recommend a different approach to validity, whose central idea is that validity attributions *regulate our beliefs*. Considerations about whether an inference preserves truth are certainly highly *relevant* to the regulation of belief. Indeed, on my own approach to the paradoxes and some others, the following all hold:[11]

(A) Logically necessary truth-preservation *suffices for* validity in the regulative sense; for example, if $\Rightarrow \text{True}(\langle A\rangle) \rightarrow \text{True}(\langle B\rangle)$ then one's degree of belief in B should be at least that of A.

[11] (A) follows from the equivalence between $\text{True}(\langle C\rangle)$ and C for arbitrary C, modus ponens, and the principle (VP) to be given in section 2.3. (B) holds since it's only for inferences with at least one premise that conditional proof is relevant. (C) and (D) follow from what I've called "restricted truth preservation", (e.g. in Field 2008, 148): on a theory like mine, valid arguments with unproblematically true premises preserve truth.

(B) The validity of sentences coincides with logically necessary truth: it is only for inferences with at least one premise that the implication from validity to truth preservation fails.

(C) If an argument is valid, there can be no *clear* case of its failing to preserve truth.

(D) If an argument is valid, then we should believe that it is truth preserving to at least as high a degree as we believe the conjunction of its premises.

This collection of claims seems to me to get at what's right in truth-preservation definitions of validity, without the counterintuitive consequences.

2.3 Validity and the Regulation of Belief

The necessary truth-preservation approach to explaining the concept of validity tried to define that concept in other (more familiar or more basic) terms. I'll briefly mention another approach that takes this form, in section 2.4; but first I'll expound what I think a better approach, which is to leave 'valid' undefined but to give an account of its "conceptual role". That's how we explain negation, conjunction, etc.; why not 'valid' too?

The basic idea for the conceptual role is

(VB)$_a$ To regard an inference or argument as valid is (in large part anyway) to accept a constraint on belief: one that prohibits fully believing its premises without fully believing its conclusion.[12] (At least for now, let's add that the prohibition should be "due to logical form": for any other argument of that form, the constraint should also prohibit fully believing the premises without fully believing the conclusion.[13] This addition may no longer be needed once we move to the expanded version in section 2.3.4.)

The underlying idea here is that *a disagreement about validity (insofar as it isn't merely verbal) is a disagreement about what constraints to impose on one's belief system.*

It would be natural to rewrite this principle as saying that to regard an inference as valid is to hold that one *shouldn't* fully believe its premises without fully believing its conclusion. And it's then natural to go from the rewritten principle about what *we regard as* valid to the following principle about what *is* valid:

[12] Note that the constraint on belief in (VB)$_a$ is a static constraint, not a dynamic one: it doesn't dictate that if a person discovers new consequences of his beliefs he should believe those consequences (rather than abandoning some of his current beliefs).

[13] The remarks on logical form in note 5 then apply here too.

$(VB)_n$ If an argument is valid, then we shouldn't fully believe the premises without fully believing the conclusion.

(The subscripts on $(VB)_a$ and $(VB)_n$ stand for 'attitudinal' and 'normative'.) I'll play up the differences between $(VB)_a$ and $(VB)_n$ in section 2.5, and explain why I want to take a formulation in the style of $(VB)_a$ as basic. But in the rest of the chapter the difference will play little role; so until then I'll work mainly with the simpler formulation $(VB)_n$. And since the distinction won't matter until then, I'll usually just leave the subscript off.

In either form, (VB) needs both qualification and expansion. One way in which it should be qualified is an analog of the qualification already made for necessary truth-preservation accounts: (VB) isn't intended to apply to logics (such as linear logic) whose validity relation is designed to reflect matters such as resource use that go beyond what the classical logician is concerned with; restricting what counts as "valid" merely because of such extra demands on validity isn't in any non-verbal sense a disagreement with classical logic. This is a point on which the legitimacy of belief approach to validity and the necessary truth-preservation approach are united; they diverge only on whether the core concern of classical logic (and many non-classical logics too, though not linear logic) is to be characterized in terms of legitimacy of belief or necessary truth preservation.

The other qualifications of (VB) mostly concern (i) the computational complexity of logic and (ii) the possibility of logical disagreement among informed agents. The need for such qualifications is especially clear when evaluating other people. To illustrate (i), suppose that I have after laborious effort proved a certain unobvious mathematical claim, by a proof formalizable in standard set theory, but that you don't know of the proof; then (especially if the claim is one that seems prima facie implausible),[14] there seems a clear sense in which I think you should *not* believe the claim without proof, even though you believe the standard set-theoretic axioms; that is, you should violate my prohibition. To illustrate (ii), suppose that you and I have long accepted restrictions on excluded middle to handle the semantic paradoxes, and that you have developed rather compelling arguments that this is the best way to go; but suppose that I have recently found new considerations, not known to you, for rejecting these arguments and for insisting on one of the treatments of the paradoxes within classical logic. On this scenario, I take excluded middle to be valid; but there seems a clear sense in which I think you *shouldn't* accept arguments which turn on applying excluded

[14] For example, the existence of a continuous function taking the unit interval onto the unit square, or the possibility of decomposing a sphere into finitely many pieces and rearranging them to get two spheres, each of the same size as the first.

middle in a way that isn't licensed *by your theory*, which again means violating my prohibition.

To handle such examples, one way to go would be to suppose that to regard an inference as valid is to accept the above constraint on belief only as applied to those who recognize it as valid. But that is awkward in various ways. A better way to go (as John MacFarlane convinced me a few years back)[15] is to say that we recognize multiple constraints on belief, which operate on different levels and may be impossible to satisfy simultaneously. When we are convinced that a certain proof from premises is valid, we think that in some "non-subjective" sense another person *should* either fully believe the conclusion or fail to fully believe all the premises—even if we know that he doesn't recognize its validity (either because he's unaware of the proof or because he mistakenly rejects some of its principles). That doesn't rule out our employing other senses of 'should' (other kinds of constraint) that take account of his logical ignorance and that point in the other direction.

A somewhat similar issue arises from the fact that we may *think* an inference valid, but not be completely sure that it is. (Again, this could be either because we recognize our fallibility in determining whether complicated arguments are, say, classically valid, or because we aren't totally certain that *classically* valid arguments are really *valid*.) In that case, though we think the argument valid, there's a sense in which we should take account of the possibility that it isn't in deciding how firmly to believe a conclusion given that we fully believe the premises. But the solution is also similar: to the extent we think it valid, we think that there's a non-subjective sense in which we should either not fully believe the premises or else fully believe the conclusion; at the same time, we recognize that there are other senses of what we "should" believe that take more account of our logical uncertainty.

To summarize, we should qualify (VB) by saying that it concerns the core notion of validity, in which "extra" goals such as resource use are discounted; and also by adding that the notion of constraint (or 'should') in it is not univocal, and that (VB) is correct only on what I've called the non-subjective reading of the term.[16]

More interesting than the qualifications for (VB) is the need to expand it, and in three ways: (a) to cover not only full belief but also partial belief; (b) to cover not only belief (or acceptance) but also disbelief (or rejection); (c) to cover conditional belief. It turns out that these needed expansions interact in interesting ways. I'll consider them in order.

[15] In an email exchange about MacFarlane (unpublished) that I discuss in Field (2009).

[16] A still less subjective sense of 'should' is the sense in which we shouldn't believe anything false. In this "hyper-objective" sense, (VB) is too weak to be of interest.

2.3.1 *Constraints on partial belief*

(VB) is stated in terms of full belief, but often we only have partial belief. How we generalize (VB) to that case depends on how we view partial belief. I'm going to suppose here that a useful (though at best approximate) model of this involves degrees of belief, taken to be real numbers in the interval [0,1]; so, representing an agent's actual degrees of belief (credences) by Cr,

(1) $0 \leqslant Cr(A) \leqslant 1$

(I don't assume that an agent has a degree of belief in every sentence of his language—that would impose insuperable computational requirements. We should understand (1) as applying when the agent has the credence in question.) I do *not* suppose that degrees of belief obey all the standard probabilistic laws, for any actual person's system of beliefs is probabilistically incoherent. I don't even suppose that a person's degrees of belief *should* obey all the standard probabilistic laws. Obviously that would fail on senses of 'should' that take into account the agent's computational limitations and faulty logical theory, but even for what I've called the non-subjective sense it is contentious: for instance, since it prohibits believing any theorem of classical sentential logic to degree less than 1, it is almost certainly objectionable if those theorems aren't all really valid. (As we'll soon see, it is also objectionable on some views in which all the classical theorems are valid, e.g. supervaluationism.)

What then do I suppose, besides (1)? The primary addition is

(VP) Our degrees of belief should (non-subjectively) be such that

(2) If $A_1, \ldots, A_n \Rightarrow B$ then $Cr(B) \geqslant \Sigma_i Cr(A_i) - n + 1$.

To make this less opaque, let's introduce the abbreviation Dis(A) for 1 − Cr(A); we can read Dis as "degree of disbelief". Then an equivalent and more immediately compelling way of writing (2) is

(2_{equiv}) If $A_1, \ldots, A_n \Rightarrow B$ then $Dis(B) \leqslant \Sigma_i Dis(A_i)$.

That (2_{equiv}) and hence (2) is a compelling principle has been widely recognized, at least in the context of classical logic: see for instance Adams (1975) or Edgington (1995).

But (2_{equiv}) and hence (2) also seem quite compelling in the context of non-classical logics, or at least many of them.[17] They explain many features of the constraints on degrees of belief typically associated with those logics.

[17] We would seem to need to generalize it somehow to deal with typical substructural logics. However, many logics that are often presented as substructural can be understood as obtained from an

To illustrate this, let's look first at logics that accept the classical principle of explosion

(EXP) $A \wedge \neg A \Rightarrow B$,

that contradictions entail everything. Or equivalently given the usual rules for conjunction,

(EXP*) $A, \neg A \Rightarrow B$.[18]

Since we can presumably find sentences B that it's rational to believe to degree 0, (2) applied to (EXP) tells us that $Cr(A \wedge \neg A)$ should always be 0, in the probability theory[19] for these logics as in classical probability theory; and (2) as applied to (EXP*) tells us that $Cr(A) + Cr(\neg A)$ shouldn't ever be greater than 1. These constraints on degrees of belief are just what we'd expect for a logic with these forms of explosion.

There are also logics that accept the first form of explosion but not the second. (This is possible because they don't contain $\wedge$-Introduction.) The most common one is *subvaluationism* (the dual of the better-known supervaluationism, about which more shortly): see Hyde (1997). A subvaluationist might, for instance, allow one to simultaneously believe that a person is bald and that he is not bald, since on one standard he is and on another he isn't; while prohibiting belief that he is both since there is no standard on which he's both. On this view, it would make sense to allow the person's degrees of belief in A and in $\neg A$ to add to more than 1, while still requiring his degree of belief in $A \wedge \neg A$ to be 0: just what one gets from (2), in a logic with (EXP) but not (EXP*).

Let's also look at logics that accept excluded middle:

(LEM) $\Rightarrow A \vee \neg A$.

(2) tells us that in a logic with (LEM), $Cr(A \vee \neg A)$ should always be 1. Interestingly, we don't have a full duality between excluded middle and explosion, in the current context: there is no obvious (LEM)-like requirement that in conjunction with (2) leads to the requirement that $Cr(A) + Cr(\neg A)$ shouldn't ever be less than 1. For this we would need a principle (LEM*) that bears the same relation to (LEM) that

ordinary (non-substructural) logic by redefining validity in terms of a non-classical conditional in the underlying language. For instance, $A_1, \ldots, A_n \Rightarrow_{substruc} B$ might be understood as $\Rightarrow_{ord} A_1 \rightarrow (A_2 \rightarrow \ldots (A_n \rightarrow B))$; or for other logics, as $\Rightarrow_{ord} A_1 \circ (A_2 \circ \ldots (A_{n-1} \circ A_n)) \rightarrow B$, where $C \circ D$ abbreviates $\neg(C \rightarrow \neg D)$. I think that these logics are best represented in terms of $\Rightarrow_{ord}$, and that principle (2) in terms of $\Rightarrow_{ord}$ is still compelling.

[18] (EXP*) is equivalent to disjunctive syllogism, given a fairly minimal though not wholly uncontentious background theory. (For instance, the argument from (EXP*) to disjunctive syllogism requires reasoning by cases.)

[19] Taking 'probability theory' to mean: theory of acceptable combinations of degrees of belief.

(EXP*) bears to (EXP), but the notation of implication statements isn't general enough to formulate such a principle.

Indeed, one view that accepts excluded middle is supervaluationism,[20] and the natural way to model appropriate degrees of belief for supervaluationism is to allow degrees of belief in A and in $\neg A$ to add to less than 1 (e.g. when A is 'Joe is bald' and one considers Joe to be bald on some reasonable standards but not on others). More fully, one models degrees of belief in supervaluationist logic by Dempster–Shafer probability functions, which allow $Cr(A) + Cr(\neg A)$ to be anywhere in the interval [0,1] (while insisting that $Cr(A \vee \neg A)$ should be 1 and $Cr(A \wedge \neg A)$ should be 0).[21] Obviously this requires that we *not* accept the classical rule that we should have

$$\text{(3?)} \quad Cr(A \vee B) = Cr(A) + Cr(B) - Cr(A \wedge B).$$

(We do require that the left-hand side should be no less than the right-hand side; indeed we require a more general principle, for disjunctions of arbitrary length, which you can find in the references in the first sentence of the previous note.) So it is unsurprising that we can't get that $Cr(A) + Cr(\neg A) \geqslant 1$ from (LEM) all by itself.

[20] For supervaluationism, see Fine (1975). But note that while early in the article Fine builds into the supervaluationist view that truth is to be identified with what he calls supertruth, at the end of the article he suggests an alternative on which a new notion of determinate truth is introduced, and supertruth is equated with that, with "$\langle A \rangle$ is true" taken to be equivalent to A. I think the alternative an improvement. But either way, I take the idea to be that supertruth is the proper goal of belief, so that it may be allowable (and perhaps even compulsory) to fully believe a given disjunction while fully *dis*believing both disjuncts.

For instance, it would be natural for a supervaluationist who doesn't identify truth with supertruth to hold that neither the claim that a Liar sentence is true, nor the claim that it isn't true, is supertrue; in which case it's permissible (and indeed in the case of non-contingent Liars, mandatory) to believe that such a sentence is either true or not true while disbelieving that it's true and disbelieving that it's not true. According to the supervaluationist, this wouldn't be ignorance in any normal sense: ignorance is a fault, and in this case what would be faulty (even in the non-subjective sense) would be to believe one of the disjuncts. (If I think I'm ignorant as to whether the Liar sentence is true, I'm not a supervaluationist; rather, I think that either a theory on which it isn't true (e.g. a classical gap theory) is correct, or one on which it is true (e.g. a glut theory) is correct, but am undecided between them. The supervaluationist view is supposed to be an alternative to that.)

[21] See Shafer (1976); and for a discussion of this in the context of a supervaluationist view of vagueness (in the sense of supervaluationism explained in the previous footnote), chapter 10 of Field (2001). (I no longer accept the supervaluationist view of vagueness or the associated Dempster–Shafer constraints on degrees of belief: see Field (2003).)

Some (e.g. Schiffer 2003; MacFarlane 2010) have thought that taking $Cr(A)$ and $Cr(\neg A)$ to be specific numbers that in the case of crucial vagueness and the paradoxes add to less than 1 doesn't do justice to our feeling "pulled in both directions" by crucially vague and paradoxical sentences. I think this objection misfires: a view that represents our attitude by a pair of numbers $Cr(A)$ and $Cr(\neg A)$ that add to less than 1 can be equivalently represented by an assignment of the interval $[Cr(A), 1 - Cr(\neg A)]$ to A and $[Cr(\neg A), 1 - Cr(A)]$ to $\neg A$; and this latter representation does clear justice to our being "pulled both ways".

If we do keep the principle (3?) in addition to (2), then if our logic includes (EXP) we must have

$$Cr(A \vee \neg A) = Cr(A) + Cr(\neg A),$$

and if our logic has (LEM) we must have

$$Cr(A \wedge \neg A) = Cr(A) + Cr(\neg A) - 1.$$

The first is part of the natural probability theory for "strong Kleene logic", the latter part of the natural probability theory for Priest's "logic of paradox" LP. Both these logics also take $\neg\neg A$ to be equivalent to A, which given (2) yields the additional constraint

$$Cr(\neg\neg A) = Cr(A).$$

Note that in the contexts of any of these logics (supervaluationist and subvaluationist as well as Kleene or Priest), we can keep the definition of "degree of disbelief" Dis as 1 minus the degree of belief Cr.[22] What we must do, though, is to reject the equation that we have in classical probability, between degree of disbelief in A and degree of belief in $\neg A$. In Kleene and supervaluational logic, $Dis(A)$ (that is, $1 - Cr(A)$) can be greater than $Cr(\neg A)$; in Priest and subvaluation logic it can be less.

The point of all this is to illustrate that (VP) (which incorporates Principle (2)) is a powerful principle applicable in contexts where we don't have full classical logic, and leads to natural constraints on degrees of belief appropriate to those logics. And (VP) is a generalization of (VB): (VB) is simply the very special case where all the $Cr(A_i)$ are 1.

Incidentally, I've formulated (VP) in normative terms; the attitudinal variant is

> **(VP)$_a$** To regard the argument from $A_1, \ldots, A_n$ to B as valid is to accept a constraint on degrees of belief: one that prohibits having degrees of belief where $Cr(B)$ is less than $\Sigma_i Cr(A_i) - n + 1$; i.e. where $Dis(B) > \Sigma_i Dis(A_i)$.

2.3.2 Constraints on belief and disbelief together

Let us now forget about partial belief for a while, and consider just full belief and full disbelief. Even for full belief and disbelief, (VB) is too limited. Indeed,

[22] If we do so, then the situation we have in Kleene logic and supervaluationist logic, that the degrees of belief in A and $\neg A$ can add to less than 1, can be equivalently put as the situation where the degrees of *dis*belief add to *more* than 1. Analogously, in Priest logic and subvaluation logic, the degrees of disbelief can add to *less* than 1.

it doesn't directly deal with disbelief at all. We can derive a very limited principle involving disbelief from it, by invoking the assumption that it is impossible (or at least improper) to believe and disbelieve something at the same time, but this doesn't take us far. Can we, without bringing in partial belief, build in disbelief in a more significant way? Yes we can, and doing so gives information that the probabilistic generalization above doesn't.

The idea here is (as far as I know) due to Greg Restall (2005). He proposes an interpretation of Gentzen's sequent calculus in terms of belief (or acceptance) and disbelief (or rejection). The idea is that the sequent $A_1, \ldots, A_n \Rightarrow B_1, \ldots, B_m$ directs you *not to fully believe all the* A_i *while fully disbelieving all the* B_j. (Restall doesn't explicitly say 'fully', but I take it that that's what he means: otherwise the classically valid sequent $A_1, \ldots, A_n \Rightarrow A_1 \wedge \ldots \wedge A_n$ would be unacceptable in light of the paradox of the preface.) The idea doesn't depend on the underlying logic being classical. And it has the nice feature that disbelief is built into the system completely on par with belief.

To illustrate, one of the principles in the classical sequent calculus is

$$\text{(RC)} \quad A_1 \vee A_2 \Rightarrow A_1, A_2.$$

("Reasoning by cases".) On Restall's interpretation, this is a prohibition against fully believing a disjunction while fully disbelieving each disjunct; something which (VB) doesn't provide. Of course, one might reject that prohibition—supervaluationists do (on the interpretation offered in note 20, on which our goal should be to believe supertruths and to disbelieve things that aren't supertrue). But if one does so, one is rejecting the sequent (RC).

An apparently serious problem with Restall's proposal is that when applied to sequents with a single sentence in the consequent, it yields less information than (VB). That is, $A_1, \ldots, A_n \Rightarrow B$ would direct us *not to fully reject* B while fully accepting $A_1, \ldots, A_n$; whereas what (VB) directs, and what I assume we want, is *to fully accept* B whenever one fully accepts $A_1, \ldots, A_n$, at least if the question of B arises. In other words, if $A_1, \ldots, A_n \Rightarrow B$ (and one knows this), then to fully accept all of $A_1, \ldots, A_n$ while *refusing to accept* B seems irrational; but the Restall account fails to deliver this. By contrast, the approach in terms of partial belief, offered in section 2.3.1, handles this: it tells us that when the $Cr(A_i)$ are 1, $Cr(B)$ should be too.

Should we then abandon Restall's approach for the partial belief approach? When I first read Restall's paper, that's what I thought. But I now see that the proper response is instead to combine them—or rather, to find a common generalization of them.

2.3.3 *The synthesis*

What we need is to generalize the formula (2) to multiple-conclusion sequents. The best way to formulate the generalized constraint, so as to display the duality between belief and disbelief, is to either use degrees of belief Cr for sentences in the antecedent of a sequent and degrees of disbelief (Dis = 1−Cr) for sentences in the consequent, or the other way around. In this formulation, the constraint is: our degrees of belief should satisfy

(2^+) If $A_1, \ldots, A_n \Rightarrow B_1, \ldots, B_m$ then $\Sigma_i \mathrm{Cr}(A_i) + \Sigma_j \mathrm{Dis}(B_j) \leqslant n + m - 1$;
or equivalently,

If $A_1, \ldots, A_n \Rightarrow B_1, \ldots, B_m$ then $\Sigma_i \mathrm{Dis}(A_i) + \Sigma_j \mathrm{Cr}(B_j) \geqslant 1$.

(That's the normative form, call it $(\mathrm{VP}^+)_n$. The attitudinal form is:

$(\mathrm{VP}^+)_a$ To regard the sequent $A_1, \ldots, A_n \Rightarrow B_1, \ldots, B_m$ as valid is to accept the consequent of (2^+) as a constraint on degrees of belief.)

Note the following:

(i) When there's only one consequent formula, the right-hand side of (2^+) reduces to

$$\mathrm{Cr}(A_1) + \mathrm{Cr}(A_n) + 1 - \mathrm{Cr}(B) \leqslant n,$$

so (2^+) yields precisely the old (2).

(ii) When we fully reject each B_j, i.e. when each $\mathrm{Dis}(B_j)$ is 1, the right-hand side of (2^+) yields

$$\mathrm{Cr}(A_1) + \ldots + \mathrm{Cr}(A_n) \leqslant n - 1;$$

that is,

$$\mathrm{Dis}(A_1) + \ldots + \mathrm{Dis}(A_n) \geqslant 1.$$

(iii) As a special case of (ii), when we fully reject each B_j and fully accept $n - 1$ of the antecedent formulas, you must *fully* reject the other one. This is a stronger version of Restall's constraint, which was that you shouldn't fully reject each B_j while fully accepting each A_i. So (2^+) generalizes Restall's constraint as well as generalizing (2).

These seem to be intuitively the right results, and I leave it to the reader to convince him- or herself that the results in every other case are just what we should expect.

Because this formulation yields the one in section 2.3.1 as a special case, it is immediate that it resolves the problem of excessive weakness that I raised in section 2.3.2 for using the unprobabilized sequent calculus as the desired constraint on belief and disbelief.

And because it also yields Restall's constraint as a special case, it also has the advantage that that account has over the simpler probabilistic account: it yields the constraint on our attitudes that the acceptance of reasoning by cases provides.[23]

And indeed it does considerably better than the unprobabilized Restall on this. Recall that in a sequent formulation, reasoning by cases is represented by the sequent

(RC) $A_1 \vee A_2 \Rightarrow A_1, A_2$.

On Restall's constraint, this means that we shouldn't fully accept a disjunction while fully rejecting each disjunct—information that the old constraint (2) of section 2.3.1 didn't provide. But the generalized constraint tells us still more: it tells us that

$Cr(A_1 \vee A_2)$ should be less than or equal to $Cr(A_1) + Cr(A_2)$.

Recall that that's a constraint accepted in the probability theory for Kleene logic (it follows from (3?) plus (2) plus (EXP)). But the constraint is rejected in the probability theory for supervaluationist logic (i.e. the Dempster–Shafer theory): the latter even allows $Cr(A_1 \vee A_2)$ to be 1 when $Cr(A_1)$ and $Cr(A_2)$ are each 0. (Thus if A_1 is the claim that a Liar sentence is true and A_2 is the negation of A_1, a supervaluationist demands that the disjunction be fully accepted, but will allow that both disjuncts be fully rejected—indeed, will probably demand this, in the case of a non-contingent Liar.) The reason this can happen in supervaluationist logic is that that logic rejects (RC). The generalized constraint thus shows how this dramatic difference between two probability theories that allow $Cr(A) + Cr(\neg A)$ to be less than 1 arises out of a difference in their underlying logic.

2.3.4 *Conditional belief*

There is an obvious strengthening of (VP) and (VP$^+$) that I've been suppressing: we should make them principles governing not only belief, but also conditional belief. For instance, we might strengthen (VP$^+$) by strengthening (2$^+$) to

[23] At least, this is so for reasoning by cases in the form that takes you from $\Gamma, A \Rightarrow X$ and $\Delta, B \Rightarrow Y$ to $\Gamma, \Delta, A \vee B \Rightarrow X, Y$. To get the form that takes you from $\Gamma, A \Rightarrow X$ and $\Gamma, B \Rightarrow X$ to $\Gamma, A \vee B \Rightarrow X$ one also needs structural contraction. But the generalization to be considered in section 2.3.4 delivers that.

If $A_1, \ldots, A_n \Rightarrow B_1, \ldots, B_m$ then for all C, $\Sigma_{i \leqslant n}\mathrm{Dis}(A_i \mid C) + \Sigma_{j \leqslant m}\mathrm{Cr}(B_j \mid C) \geqslant 1$.

Part of my reason for not having made this strengthening earlier is to show how much can be done without it. Another part of the reason is that conditional degree of belief is more complicated in non-classical logics than in classical. Indeed, since in typical non-classical logics a distinction is made between rejection and acceptance of the negation, the simple form of conditional credence $\mathrm{Cr}(A \mid C)$ is inadequate: $\mathrm{Cr}(A \mid C)$ is the credence in A conditional on full acceptance of C, but we'll need a more general notion $\mathrm{Cr}(A \mid C/D)$ of the credence in A conditional on full acceptance of C and full rejection of D.[24] (In classical logic this will just be $\mathrm{Cr}(A \mid C \wedge \neg D)$, but if rejection isn't just acceptance of the negation they will differ.) So the above strengthening of (2^+) should be adjusted:

(2^+_{cond}) If $A_1, \ldots, A_n \Rightarrow B_1, \ldots, B_m$ then for all C and D, $\Sigma_{i \leqslant n}\mathrm{Dis}(A_i \mid C/D) + \Sigma_{j \leqslant m}\mathrm{Cr}(B_j \mid C/D) \geqslant 1$.

While as I've illustrated, much can be done to illuminate the connection between validity and degrees of belief using only (2^+), one needs (2^+_{cond}) to go much further. For instance, one thing I haven't done is to use generalized probabilistic constraints to derive (3?) from the logical principles of those logics where (3?) is appropriate (e.g. Kleene logic and LP, as well as classical). With (2^+_{cond}) we can do this. This follows from two more general consequences of (2^+_{cond}): first, that for logics with $\wedge$-Introduction (and a few other uncontroversial laws),

(I) $\mathrm{Cr}(A \mid C/D) + \mathrm{Cr}(B \mid C/D) \leqslant \mathrm{Cr}(A \wedge B \mid C/D) + \mathrm{Cr}(A \vee B \mid C/D)$;

second, that for logics with $\vee$-Elimination (and a few other uncontroversial laws),

(II) $\mathrm{Cr}(A \mid C/D) + \mathrm{Cr}(B \mid C/D) \geqslant \mathrm{Cr}(A \wedge B \mid C/D) + \mathrm{Cr}(A \vee B \mid C/D)$.

[24] The usual central law of conditional probability, generalized to the extra place, is

$$\mathrm{Cr}(A \mid B \wedge C/D) \cdot \mathrm{Cr}(B \mid C/D) = \mathrm{Cr}(A \wedge B \mid C/D).$$

We impose also the dual law

$$\mathrm{Dis}(A \mid C/B \vee D) \cdot \mathrm{Dis}(B \mid C/D) = \mathrm{Dis}(A \vee B \mid C/D).$$

Some policy is needed for how to understand $\mathrm{Cr}(A \mid C/D)$ and $\mathrm{Dis}(A \mid C/D)$ when the logic dictates that C can't be coherently accepted or D can't be coherently rejected; I tentatively think the best course is to let both $\mathrm{Cr}(A \mid C/D)$ and $\mathrm{Dis}(A \mid C/D)$ be 1 in these cases. Spelling it out, the idea would be to call A *absurd* if for all X and Y, $\mathrm{Dis}(A \mid X/Y) = 1$, and *empty* if for all X and Y, $\mathrm{Cr}(A \mid X/Y) = 1$; then the requirement that $\mathrm{Dis}(A \mid C/D) = 1 - \mathrm{Cr}(A \mid C/D)$ holds only for *acceptable pairs* $\langle C, D \rangle$, pairs where C isn't absurd and D isn't empty. (Some derived laws will then be subject to the same restriction.) But this is only one of several possible policies for dealing with these "don't care" cases.

I leave the proofs to a footnote.[25] I'd conjecture that $(2^+{}_{cond})$ is enough to get from any logic (of the kinds under consideration here) to *all* reasonable laws about credences appropriate to that logic.

I've been talking about how to derive constraints on degrees of belief from the logic, but it would be natural to argue that, conversely, we can obtain the logic from the probability theory by turning a principle in the general ballpark of (2^+) into a biconditional. But the converse of $(2^+{}_{cond})$ is far more plausible than that of (2^+); moreover, we need it to be $(2^+{}_{cond})$ rather than (2^+) to carry out the idea technically. For to fully derive the logic, one of the things one needs to derive is the Gentzen structural rules. Of these, Reflexivity, Thinning, and Permutation are immediately obvious from the biconditional form of either (2^+) or $(2^+{}_{cond})$. The same holds for Cut in the form

$$\frac{\Sigma_1 \Rightarrow \Delta_1, C \qquad \Sigma_2, C \Rightarrow \Delta_2}{\Sigma_1, \Sigma_2 \Rightarrow \Delta_1, \Delta_2}$$

But structural contraction requires the generalization over C and D in $(2^+{}_{cond})$: one must vary the C in the biconditional form of $(2^+{}_{cond})$ to get contraction on the left, and the D to get contraction on the right.[26]

[25] $\wedge$-I gives:

$$Cr(A \mid C\wedge(A\vee B)/D)+Cr(B \mid C\wedge(A\vee B)/D) \leqslant Cr(A\wedge B \mid C\wedge(A\vee B)/D)+1.$$

Multiplying each term by $Cr(A \vee B \mid C/D)$, and using the central law of conditional probability (note 24), we get

$$Cr(A \wedge (A \vee B) \mid C/D) + Cr(B \wedge (A \vee B) \mid C/D) \leqslant Cr((A \wedge B) \wedge (A \vee B) \mid C/D) + Cr(A \vee B \mid C/D).$$

Using obvious equivalences this yields (I).

$\vee$-E (i.e. RC) gives:

$$Dis(A \vee B \mid C/D \wedge (A \wedge B)) + 1 \geqslant Dis(A \mid C/D \wedge (A \wedge B)) + Dis(B \mid C/D \wedge (A \wedge B)).$$

Multiplying each term by $Dis(A \wedge B \mid C/D)$, and using the dualized central law of conditional probability (note 24), we get

$$Dis((A \vee B) \vee (A \wedge B) \mid C/D) + Dis(A \wedge B \mid C/D) \geqslant Dis(A \vee (A \wedge B) \mid C/D) + Dis(B \vee (A \wedge B) \mid C/D).$$

Using obvious equivalences this yields:

$$Dis(A \mid C/D) + Dis(B \mid C/D) \leqslant Dis(A \vee B \mid C/D) + Dis(A \wedge B \mid C/D),$$

which yields (II).

[26] We also need that $Cr(A \wedge A \mid C/D) = Cr(A \mid C/D) = Cr(A \vee A \mid C/D)$ and $Cr(A \wedge B \mid C/D) \leqslant Cr(A \mid C/D) \leqslant Cr(A \vee B \mid C/D)$; the first of these and the laws in note 24 yield (i) $Cr(A \mid C \wedge A/D) = 1$ and (ii) $Cr(A \mid C/D \vee A) = 0$.

Contraction on the left (leaving out side formulas for simplicity): To establish the meta-rule $A, A \Rightarrow B / A \Rightarrow B$, we must get from

$$\forall C \forall D[Cr(B \mid C/D) \geqslant 2Cr(A \mid C/D) - 1]$$

to

$$\forall C \forall D[Cr(B \mid C/D) \geqslant Cr(A \mid C/D)]$$

Once one has the structural rules, there is no problem getting the other logical meta-rules appropriate to given laws of credence: as remarked above, reasoning by cases then falls out whenever the laws on credences include

$$Cr(A_1 \vee A_2 \mid C/D) \leqslant Cr(A_1 \mid C/D) + Cr(A_2 \mid C/D)$$

(see note 23), and it's easy to see that conditional proof falls out whenever they include that, $Cr(A \vee \neg A \mid C/D) = 1$, and $Cr(A \rightarrow B \mid C/D) \geqslant \max\{Cr(\neg A \mid C/D), Cr(B \mid C/D)\}$.

Details aside, I think this discussion shows that it is illuminating to view validity as providing constraints on our (conditional) degrees of belief. And debates about validity are in effect debates about which constraints on (conditional) degrees of belief to adopt. The next two sections deal with some issues about how to understand this.

2.4 Normative Definitions or Primitivism

The view I've been advocating has it that instead of trying to define validity in other terms, as the necessary truth-preservation account does, we should take it as a primitive, and explain its conceptual role in terms of how it constrains our (conditional) beliefs.

We should contrast this with another proposal: that we define validity, but in normative terms, in a way that reflects the connection between validity and belief. That alternative proposal is that we define $A_1, \ldots, A_n \Rightarrow B$ as "One shouldn't (in the non-subjective sense) fully believe $A_1, \ldots, A_n$ without fully believing B";

But there's no problem doing so: rewrite the bound C in the assumption as C^*, and instantiate it by $C \wedge A$, to get $\forall C \forall D[Cr(B \mid C \wedge A/D) \geqslant 2Cr(A \mid C \wedge A/D) - 1]$, which by (i) yields $\forall C \forall D[Cr(B \mid C \wedge A/D) = 1]$. Multiplying both sides by a common factor we get $\forall C \forall D[Cr(B \mid C \wedge A/D) \cdot Cr(A \mid C/D) = Cr(A \mid C/D)]$, which by laws in note 24 yields $\forall C \forall D[Cr(A \wedge B \mid C/D) = Cr(A \mid C/D)]$, and the left-hand side is $\leqslant Cr(B \mid C/D)$; so $\forall C \forall D[Cr(A \mid C/D) \leqslant Cr(B \mid C/D)]$, as desired.

Contraction on the right (again leaving out side formulas): To establish the meta-rule $A \Rightarrow B, B / A \Rightarrow B$, we must get from

$$\forall C \forall D[2Cr(B \mid C/D) \geqslant Cr(A \mid C/D)]$$

to

$$\forall C \forall D[Cr(B \mid C/D) \geqslant Cr(A \mid C/D)].$$

Rewrite the bound D in the assumption as D^*, and instantiate it by $D \vee B$, to get $\forall C \forall D[2Cr(B \mid C/D \vee B) \geqslant Cr(A \mid C/D \vee B)]$, which by (ii) yields $\forall C \forall D[Cr(A \mid C/D \vee B) = 0]$, i.e. $\forall C \forall D[Dis(A \mid C/D \vee B) = 1]$. Multiplying both sides by a common factor we get $\forall C \forall D[Dis(A \mid C/D \vee B) \cdot Dis(B \mid C/D) = Dis(B \mid C/D)]$, which by laws of note 24 yield $\forall C \forall D[Dis(A \vee B \mid C/D) = Dis(B \mid C/D)]$, and the left-hand side is $\leqslant Dis(A \mid C/D)$; so $\forall C \forall D[Dis(A \mid C/D) \geqslant Dis(B \mid C/D)]$, as desired.

or a variant of this based on (VP) or (VP^+) or (VP_{cond}) or ($VP^+{}_{cond}$) instead of (VB). I won't attempt a serious discussion of this, but I think it would sully the purity of logic to define validity in normative terms whose exact content is less than clear.

But is the approach in section 2.3 significantly different? I think so. Compare the notion of chance. I take it that no one would want to say that a claim such as

The chance of this atom decaying in the next minute is 0.02

is equivalent in meaning to

You ought to believe to degree 0.02 that the atom will decay in the next minute.

The first claim, unlike the second, is not literally about what we should believe. On the other hand, there seems to be no prospect of a reductive account of chance. And it seems clear that an important part of our understanding of the first claim lies in its conceptual ties to the second: what would we make of someone who claimed to accept the first claim but thought that it was rational to disbelieve strongly that the atom will decay in the next minute? While claims about chance aren't literally claims about what we should believe, it's hard to deny that debates about the chance of an atom decaying in the next minute are intimately related to debates about what degree of belief to have in such a decay. I think the situation with validity is much like that with chance.

In the chance case as in the validity case, the "ought" in question is in an important sense "non-subjective": it does not take into consideration the agent's evidence. If the chance of decay is actually 0.02, but a person has reason to think it is far higher, then there's a subjective sense of 'ought' in which the person ought to believe in decay to that higher degree, not to degree 0.02. But as in the case of validity, the existence of the idea of "ought based on one's evidence" doesn't drive out the existence of a non-subjective ought, in which one ought to believe in decay to degree 0.02.

It seems, then, that a large part of our understanding of the notion of chance is our acceptance of something like the following principle:

$(C)_n$ If the chance of A is p, then (in the non-subjective sense of 'ought') our degree of belief in A ought to be p.

A more accurate formulation would be:

$(C_{cond})_n$ If the chance of A happening in situation B is p, then (in the non-subjective sense of 'ought') our conditional degree of belief in A given B ought to be p.

That something like this is so seems relatively uncontroversial.

$(C)_n$ and its variant $(C_{cond})_n$ seem closely analogous in form to the $(V)_n$ principles that I have suggested for validity. It seems to me that for validity as for chance, a primitivist (i.e. non-definitional) conceptual role account, rather than either a non-normative reduction or a normative reduction, is the way to go, and that it is no more problematic for validity than it is for chance.

2.5 Realism vs. Projectivism

There is, I admit, a worry about the primitivist line: not just in the case of validity, but in the case of chance as well. The worry is that the $(C)_n$ principles and the $(V)_n$ principles employ non-subjective 'ought's, and one might reasonably wonder what sense is to be made of them. If someone were to ask what's wrong with having degrees of belief that violate one of these principles, is there no answer to be made beyond saying in one's most serious voice "*It is forbidden!*"?

Suppose I know that the chance of an atom decaying in the next minute, given its nature and circumstances, is 0.02. Naturally enough, I believe to degree 0.02 that it will so decay; but Jones believes to degree 0.9 that the atom will decay in the next minute. (He might have the corresponding view about the chances, but needn't: for example, he might think it isn't a chance process.)[27] Even if his degree of belief in the decay is reasonable given his evidence, there seems to be a sense in which it's wrong given the actual chances—and that's so even if the atom *does* decay in the next minute.

But if asked *what's* wrong with having a degree of belief that doesn't accord with the chances, what can I say? To say "There's a high chance of what one believes being false" is obviously unhelpful: why is the *chance* of its being false, as opposed to its actually being false, relevant? It is also no help to say

> "What's wrong is that if he had a policy of betting in terms of that degree of belief in similar situations, *the chance is very high that* he'd lose money."

There are two reasons for this, both familiar. The first is relevance: the person with degree of belief 0.9 in this instance needn't have a long-term policy of believing to degree 0.9 in all similar instances, so why would a problem with the long-term strategy imply a problem with the individual instance? But the more basic reason is that the imagined answer merely pushes the question back: what's wrong with Jones thinking he'll win with this long-term betting strategy, given merely that the

[27] Or more fancifully, there might be an odd disconnect between his belief about the chance and his degree of belief about the decay.

actual *chance* of winning is low? It seems obvious that there is something wrong with it, even if it was reasonable given his evidence and in addition he happens to be very lucky and wins; but that's just another instance of the idea that our degrees of belief ought to correspond to the chances, which is what it was attempting to explain.

So this attempt to explain the sense in which there's something wrong with degrees of belief that don't accord with the chances is hopeless. If there's no possibility of doing better, we will have to either give up some natural 'ought' judgements or settle for a primitive non-subjective ought on which nothing more can be said than that you are forbidden to believe to any degree other than 0.02. But *who* forbids it, and why should we care?

There have been various closely related proposals for a more informative answer to the "What's wrong?" question in the case of chance (See, for instance, Jeffrey 1965, Blackburn 1980, and Skyrms 1984). Roughly put, the common theme is that we regard chances as projections from our epistemic state. They aren't just credences, of course, but rather "de-subjectivized credences"[28]. The de-subjectivization means that the 'ought's based on them don't merely involve what an agent's degrees of belief ought to be *given her evidence, beliefs, etc.*, but are quite a bit more objective than that; still, the chances are ultimately based on credences, in a way that makes the corresponding 'ought's unmysterious.

I think that the general idea has considerable appeal, and what I want to suggest is that an analogous idea has a similar appeal in the case of validity. Indeed, it seems to me that whatever difficulties the idea has for the case of validity it equally has for the case of chance. I will not attempt a precise formulation, in either the case of chance or validity: indeed I'm deliberately presenting the general idea in such a way as to allow it to be filled out in more than one way. The goal of the section is simply to pursue the analogy between the two cases. (And I should say that, while I'm sympathetic to the projectivist view in both cases, nothing outside this section of the chapter requires it.)

In the case of chance, the projectivist idea involves temporarily shifting the subject from chance itself to attributions of chance. Instead of directly asking

(i) What is it for there to be chance p that under conditions B, A will happen?

we start out by asking

(ii) What is it for someone *to take* there to be chance p that under conditions B, A will happen?

[28] 'Credence' as often understood is already a somewhat de-subjectified notion, in that constraints of coherence are imposed, but the idea is to de-subjectivize it further in a way that makes it less dependent on evidence.

We then argue that to answer (ii), *we needn't use the notion of chance to characterize the content of the person's mental state.* (This is the first stage of the "projectivism".) For instance, the answer to (ii) might be that the person has a kind of resilient conditional degree of belief in A given B (and, perhaps, that he recommends this conditional degree of belief to others): for any C of a certain kind, $Cr(A \mid B \wedge C) = p$.[29] The restriction that C be "of a certain kind" is hard to spell out precisely, but a very rough stab might be that C contains no information about times later than the conditions referred to in B.[30] This is only very rough: one important reason for this is that evidence about past experiments might lead one to alter our physical theory in a way that led us to change our conditional probabilities, and this should count as *changing our views about what the chances are* rather than *showing that our conditional probabilities weren't resilient enough to count as views about chance.* I doubt that the idea of resilience can be made perfectly precise, but also doubt that we need total precision for the notion to be useful in an account of chance-attribution.

In short, the idea is that the concept of chance is illuminated by the following principle:

> $(C_{cond})_a$ To regard the chance of A happening, under condition B, as p is to accept a constraint on belief: it is to demand the conditional degree of belief p in A, given B and C, for any admissible C.

This of course is broadly analogous to our $(V)_a$ principles for validity: those characterize what it is for someone to take an argument to be valid without using the notion of validity to characterize the content of her mental state.

$(C_{cond})_a$ explains what would be wrong with someone *taking* the chance of A in situation B to be 0.02 and yet (despite having no other relevant information) knowingly having credence 0.9 of A given B: such a person would be seriously unclear on the concept of chance. Similarly, the $(V)_a$ principles explain what would be wrong with someone (e.g. Lewis Carroll's Achilles) taking modus ponens to be valid in all instances, and yet (in full awareness of what he was doing) fully accepting the premises of a modus ponens while refusing to accept its conclusion.

Does this account of chance-*judgements* and validity-*judgements* shed any light on what chance and validity *are*? The key second stage of projectivism is to argue that the answer is yes. The idea isn't to reject the normative principles $(C_{cond})_n$

[29] Alternatively, it might be that the conditional degree of belief that best fits his overall system of degrees of belief is like this, even though his actual conditional degree of belief is different.

[30] The issue is similar to the issue of clarifying "admissible evidence" in Lewis's "Principal Principle": see Lewis (1980).

and $(V)_n$; rather, these principles properly express the projectivist's attitudes. But the projectivist will see the acceptance of $(C_{cond})_n$ as founded on the more basic truth $(C_{cond})_a$, and take this to remove worries about the 'ought's appealed to in $(C_{cond})_n$; and similarly for $(V)_n$.

I won't try to spell out this second stage of the projectivist account; there are different ways in which the details could go. But in the chance case, it should be clear that if I myself have a resilient conditional degree of belief p in A given B, and know that someone else has evidence that supports a very different conditional degree of belief *q*, I can distinguish between what her conditional degree of belief should be given her evidence (viz., *q*) and what it "objectively" should be (viz., p, ignoring the possibility that I myself am misled). Similarly in the validity case: if I myself believe in the law of explosion but I know that a friend doesn't, I can distinguish between what she should believe about a matter *given her logical views* and what she "objectively" should believe about it. Judging what she "objectively" should believe about validity is straightforward on the view: it's the same as judging what's valid. Judging what she should believe *given her logical views* involves some sort of projection, something like an "off-line decision" of what to believe about the matter in question on the pretense that one's views are in key respects like hers. The distinction between what X should believe given his or her current evidence or beliefs and what he or she "objectively" should believe is equally unproblematic when X is my past self, or my counterpart in another possible world. And it isn't a whole lot more problematic even when X is my actual current self. If I have a resilient conditional degree of belief p in A given B, and am sure that it's appropriate given my evidence, I can still make sense of the possibility that, despite my evidence, the chance is different: in doing so I am projecting to a different epistemic state than the one I now occupy, perhaps one that might be obtained were I to learn new truths about frequencies or underlying mechanisms, or unlearn some falsehoods about these things. Similarly, such a view for validity can allow for a divergence between what my degrees of belief ought to be *given my current opinions about validity* and what they "*objectively*" ought to be: I can project to a dimly imagined epistemic state where I learn things about logical space that would lead me to revise some of my current epistemic policies.

Similarly, I can make sense of both unknown chances and unknown validities: I'm projecting to judgements I'd make were I to acquire additional knowledge of certain sorts.

In the last few paragraphs I've been putting 'objectively' in quotes, because there's a question as to whether this rather vague term is fully appropriate. And my view is that the "level of objectivity" we get is greater in some cases than in others. If I differ with someone on the chance of rain tomorrow, it's very likely that this difference is *largely* due to differences in our information (information expressible without the term 'chance'); to the extent that this is so, there is certainly no threat to objectivity. There is also no threat to what I'd call objectivity if his belief differs from mine not because of a difference in evidence but because he takes seriously the predictions in some ancient text, and his taking such predictions seriously is impervious to evidence. In such a case, there are strong advantages for my conditional degrees of belief over his, and that seems to me to give all the "objectivity" we should ask for. (Of course he is likely to disagree with me about the advantages. In some cases that further disagreement might be based on differences in background information, but even if it isn't, I don't think any interesting notion of objectivity will count this case non-objective.)

Objectivity is a flexible notion. In the case of chance, it does not seem unreasonable to stipulate that any attribution of chance that departs drastically from actual stable frequencies is "objectively wrong", whatever the attributor's degrees of belief: one can say that the actual stable frequencies give a "metaphysical basis" for ruling out projections based on some credences. Still, the impossibility of an actual reduction of chances to frequencies means that there is some limitation on the "objectivity" achievable in this way. It is hard to find much non-objectivity about chance in examples such as radioactive decay, where the frequencies are so exceptionally stable; but most of us also speak of chance in other cases, such as coin tosses or weather patterns. And here, it is far from obvious that all disagreements about the chance of rain are fully explainable on the basis of either difference of evidence or gross irrationality (as it was in the ancient texts example). If not, I don't think it unreasonable to declare some such disagreements "non-objective". The "projectivism" I'm suggesting for chance has it that a judgement of chance carries with it a judgement of what a person ought to believe under appropriate circumstances, in a non-subjective sense: that is, independent of what evidence the person has; but it may be that *to whatever extent person-independent facts such as frequencies don't settle the chances*, there is nothing to legitimize the judgement over various alternatives.

The role that frequencies play for chance can be played by truth preservation for validity. Just as an attribution of chance that departs drastically from actual frequencies seems objectively wrong, an attribution of validity to an argument that definitely fails to preserve truth seems objectively wrong. (Recall (C) at the end of section 2.2.) *Prima facie*, this suffices for making most disputes about validity objective; though there are issues about sameness of meaning across logical theories that might undermine that claim. But even putting those issues aside, the impossibility of an actual reduction of validity to logically necessary truth preservation is a sign that there may be *some* degree of non-objectivity in the choice of logic.

It may be, for instance, that a view that locates the failure of the Curry argument in modus ponens and a view that locates it in conditional proof can't be distinguished in terms of how closely validity corresponds to truth preservation. I don't say that this is the actual situation, but suppose it is. In that case, the difference between the views is irreducibly a matter of normative policy. The proponent of unrestricted modus ponens will say that we ought to conform our degrees of belief to it, in the sense I've described, and the proponent of unrestricted conditional proof will say that we ought to conform our degrees of belief to a different standard. And each will take their 'oughts' to be non-subjective, in the sense that they aren't merely claims about what we ought to do given our logical theory. That wouldn't be a serious lack of objectivity if there were strong advantages to one view over the other; but it may not be entirely obvious that there are. Perhaps each party in the dispute recognizes that despite his personal preference for one route over the other, there is no really compelling advantage on either side. (Again, I don't say that this is the actual situation, but suppose it is.) Given Curry's paradox, we can't declare *both* valid, without restricting central principles of truth; and let us assume for the sake of argument that restricting the truth principles in the required way would have far worse disadvantages than restricting either modus ponens or conditional proof. Moreover, since a logic with neither principle would be far too weak for serious use, only Buridan's ass would use the symmetry of the situation to argue that *neither* is valid. And so, in the circumstances I'm imagining, where there really is no strong advantage for one or the other, each party to the dispute uses whichever of the principles he or she is more comfortable with, while recognizing that the choice isn't objective in any serious sense.

In cases like this where we recognize that we need to be permissive about alternatives, our validity judgements will presumably be somewhat nuanced. If my logic of choice includes modus ponens, I will typically evaluate arguments in accordance with it; but in evaluating the arguments of someone who has a worked out system which restricts modus ponens to get conditional proof, it will sometimes be appropriate to project to that person's point of view. Sometimes, not always: for example, if I find another person's arguments for an implausible conclusion prima facie compelling, I'll want to figure out where they go wrong on my own standards. In some situations it's best to relativize explicitly to one standard over the other (though at least if the non-objectivity is sufficiently widespread it is not possible always to do this: there wouldn't be enough of a common logic to be able to draw conclusions about the relativized concepts).

I think that projectivism gets all of this right. But this is not the place to explore it further.[31]

2.6 Soundness and Completeness

I argued at the very start of this paper that real validity is neither model-theoretic nor proof-theoretic, and have been offering an account of what it is instead. This has an impact on how to understand soundness and completeness proofs.

What are usually called soundness and completeness results for logics are results that relate the proof theory to the model theory. Suppose we have a system S of derivations for, say, classical sentential logic, which enables us to define a derivability relation $\Gamma \vdash_S B$ (or a multiple conclusion analog $\Gamma \vdash_S \Delta$). Suppose we also have a model theory M for that logic, say the usual one in terms of 2-valued truth tables, which enables us to define a model-theoretic validity relation $\Gamma \vDash_M B$ (or $\Gamma \vDash_M \Delta$). Then the usual soundness and completeness theorems relate these. In what I consider the best-case scenario,[32] these go as follows (focusing on the single consequent case, but the generalization to multiple consequent is obvious):

Formal Soundness Theorem: For any set Γ of sentences and any sentence B, if $\Gamma \vdash_S B$ then $\Gamma \vDash_M B$.

Formal Completeness Theorem: For any set Γ of sentences and any sentence B, if $\Gamma \vDash_M B$ then $\Gamma \vdash_S B$.

But calling these theorems "soundness theorems" and "completeness theorems" seems slightly misleading, for they merely connect two different notions about each of which it could be asked whether they are sound or complete. What

[31] The "projectivism" I'm advocating for deductive validity seems more obviously attractive in the case of inductive goodness. (1) Whereas those who ignore the paradoxes (and vagueness) might think that deductive validity is just logically necessary preservation of truth, no one could think that in the inductive case. And for many reasons, the thought that one could find a notion of reliability that does an analogous job doesn't withstand scrutiny. (The main reason, I think, is that there simply is no useful notion of reliability for methods that "self-correct" in the way that induction does: see section 4 of Chapter 13 of Field (2001).) (2) Whereas in the deductive case the idea that there is some degree of non-objectivity about which logic to adopt may seem surprising, it is far less so in the inductive case: any attempt to formulate an inductive method in detail will show that there are a large number of parameters affecting such matters as how quickly one adjusts one's degrees of beliefs on the basis of observed frequencies; the idea that there's exactly one "right" value to such a parameter seems absurd.

[32] For some logics it is common to employ proof procedures for which formal soundness must be either stated in a more complicated way than below, or else restricted to the case where Γ is empty (e.g., proof procedures containing a generalization rule for the universal quantifier or a necessitation rule for a modal operator.) Use of such a proof procedure often goes with defining entailment in terms of logical truth, something which isn't possible in every logic. It seems always to be possible, though, to convert a proof procedure that is formally sound only in the more complicated or restricted sense to one of intuitively equivalent power which is formally sound in the sense given below.

we really want to know is whether the model theory is sound and complete with respect to the validity relation, and whether the proof theory is sound and complete with respect to the validity relation; and neither of these questions is directly answered by the formal soundness and completeness theorems. Thus (using $\Gamma \Rightarrow B$ to mean that the argument from Γ to B is logically valid, as understood in the "primitivist" way suggested earlier in the chapter)[33] we can formulate genuine soundness and completeness as follows:

(P-Sound) [Genuine Soundness of the proof theory]: For any set Γ of sentences and any sentence B, if $\Gamma \vdash_S B$ then $\Gamma \Rightarrow B$.

(P-Comp) [Genuine Completeness of the proof theory]: For any set Γ of sentences and any sentence B, if $\Gamma \Rightarrow B$ then $\Gamma \vdash_S B$.

(M-Sound) [Genuine Soundness of the model theory]: For any set Γ of sentences and any sentence B, if $\Gamma \vDash_M B$ then $\Gamma \Rightarrow B$.

(M-Comp) [Genuine Completeness of the model theory]: For any set Γ of sentences and any sentence B, if $\Gamma \Rightarrow B$ then $\Gamma \vDash_M B$.

These four soundness and completeness claims involve a notion $\Rightarrow$ of validity that we're taking to be undefined, so there's no question of formally proving these "genuine" soundness and completeness claims. But to what extent can we convincingly argue for them nonetheless? The question, and the answer to follow, is I think a generalization of a question asked and answered in Kreisel (1967). (Kreisel's discussion is sometimes understood as solely concerning "set-sized interpretations" of quantification theory versus "interpretations with domains too big to be a set"; but I take him to have been implicitly concerned with the point here.)

Kreisel, as I'm understanding him, showed how formal completeness theorems bear on this question. Suppose we are antecedently convinced of (P-Sound) and (M-Comp), i.e.

Whenever $\Gamma \vdash_S B$, $\Gamma \Rightarrow B$

and

Whenever $\Gamma \Rightarrow B$, $\Gamma \vDash_M B$.

A formal completeness theorem then tells us that whenever $\Gamma \vDash_M B$, $\Gamma \vdash_S B$; so it follows that the three notions $\Gamma \vdash_S B$, $\Gamma \Rightarrow B$, and $\Gamma \vDash_M B$ all coincide in extension.

[33] 'Logically valid' is actually a contextually relative notion. If it's sentential logic that is in question, then we should take 'logically valid' here to mean 'valid on the basis of sentential structure'; if quantificational logic, 'valid on the basis of quantificational structure'; etc.

Of course, this argument for their extensional equivalence (which is called the "squeezing argument") turns not just on the formal completeness theorem, but on the assumptions (P-Sound) and (M-Comp).

This Kreiselian account makes no use of the formal soundness proof. We could imagine a parallel situation, where we are antecedently convinced of (M-Sound) and (P-Comp), i.e.

Whenever $\Gamma \vDash_M B$, $\Gamma \Rightarrow B$

and

Whenever $\Gamma \Rightarrow B$, $\Gamma \vdash_S B$.

Then a parallel argument using formal soundness would give you the same conclusion, that the three notions $\Gamma \vdash_S B$, $\Gamma \Rightarrow B$, and $\Gamma \vDash_M B$ all coincide in extension. But this situation has a strong air of unreality, because it's hard to imagine a situation where we are antecedently convinced of the completeness of a typical proof procedure.

Thus a Kreiselian analysis provides some explanation of the significance we attach to formal completeness proofs, but not of the significance we attach to formal soundness proofs.

Even without the completeness assumption (P-Comp) for the proof procedure, the formal soundness result plus the soundness assumption (M-Sound) for the model theory yields the soundness (P-Sound) of the proof procedure. But two considerations tend to undermine the significance of this.

First and most obviously, it won't help persuade a typical person who thinks the logic too powerful (e.g. an intuitionist is not going to be convinced that classical logic is sound by a formal soundness theorem): for such a person won't regard the model theory as genuinely sound.

Second, even an advocate of the logic in question is unlikely to find the model theoretic soundness *obvious*, except in very simple cases. Indeed, Kreisel's whole point was that it isn't obvious in the case of classical quantification theory. It isn't obvious there because it says that an argument that preserves truth *in all models* is valid; but because models have sets as their domains, they must misrepresent reality, which is too big to be a set. So preserving truth in all models doesn't obviously guarantee preserving truth in the real world, let alone in all logically possibilities. On Kreisel's analysis, we need to argue for (M-Sound), which we do by means of (P-Sound) and formal completeness. Given this, it would seem blatantly circular to use (M-Sound) to argue for (P-Sound).[34]

[34] I'll qualify this conclusion later in this section (at the end of its next-to-last paragraph).

This is not to deny that trying to give a formal soundness proof for a proof procedure can expose technical errors in its formulation (such as forgetting to impose restrictions on the substitution of terms for variables, or forgetting to impose existence assumptions at certain points in a proof procedure for a free logic). By the same token, a successful formal soundness proof offers reassurance that one hasn't made that sort of technical error. But this is a very minimal role for soundness proofs.

Is there any prospect of proving either (P-Sound) or (M-Sound) without relying on the other?

Obviously not, if the proof is supposed to persuade adherents of other logics; but suppose the question is just whether we can provide a proof *to ourselves*, so that the proof can use the full logic for which S is a proof procedure and M a model theory.

Even on this liberal interpretation, it's obvious that no such proof is possible if we leave the genuine validity relation $\Rightarrow$ as a primitive relation and don't make any assumptions about it: without putting some soundness assumption about it in, there's no getting one out.

But suppose (contrary to my argument earlier) that we were to understand validity as necessary truth preservation? The question then is whether there's any hope of proving

> (P-Sound*) For any set Γ of sentences and any sentence B, if $\Gamma \vdash_S B$ and all members of Γ are true then B is true.

The fact is that there is no hope whatever of proving this when S contains all the logical principles we employ, *even if we're not restricted to a proof that would convince advocates of other logics, but are allowed to use the full logic codified in S in our proof.* One might think it could be done, on the basis of some sort of inductive argument, but it can't (as long as the theory in which one is doing the arguing is absolutely consistent, i.e. non-trivial). This follows from Gödel's Second Incompleteness Theorem, as I've argued elsewhere.[35]

I will not repeat that argument here, except to note that it established that *we can't even prove the special case of (P-Sound*) in which Γ is required to be empty.* This is quite significant, since despite the limitations on the connection of validity of inferences to truth preservation, there is little doubt that it is part of the intuitive notion of validity that valid *sentences* are all *true*. (See point (B) at the end of

[35] I discuss it on the assumption of classical logic in sections 2.3, 11.5, and 12.4 of Field (2008), and for various alternative logics in other places in the book. (See my response to Vann McGee in Field (2010) for some clarification.)

section 2.2.) Thus the inability to prove, *even using our logic*, that (P-Sound) and (M-Sound) hold *even in the restricted case where Γ is empty*, means that there is a really serious limitation in our ability to prove the soundness of our own logic.

In summary: To establish that genuine validity lines up with the model theoretical and proof theoretical notions, we have to rely on one of the soundness assumptions (P-Sound) and (M-Sound). Not only is there no hope of proving that assumption to advocates of other logics (which is hardly surprising), but *there is no hope of proving it even to ourselves, using our own logic freely.* (Assuming, again, that what S and M are proof procedures and model theories for is our full logic, not just a fragment.) For Kreiselian reasons, (P-Sound) is at least in some ways more intuitively compelling than (M-Sound), at least as regards the quantificational component of the logic. But perhaps each is compelling in a different way; and since (M-Comp) transmits whatever independent credibility it has to (P-Comp) via the formal soundness proof, this would give a bit of a philosophical role for formal soundness proofs.

As an aside, the situation is interestingly different for completeness assumptions. As I've mentioned, for standard proof procedures (P-Comp) has no antecedent claim to credence independent of the formal completeness theorem; on the other hand, there is often a reasonably compelling (I don't say airtight) argument for the completeness claim (M-Comp). I sketch it in a footnote.[36]

[36] Recall that *logical* validity is something like validity on the basis of logical form. That means that it can be valid only if all possible arguments of that form are valid (sentential form if it's sentential logic in question, quantificational form if quantificational logic, and so on). For (M-Comp) to fail, there would then have to be an argument Γ/B such that all possible arguments Γ*/B* of the same form are valid and yet Γ/B is not model-theoretically valid. But model-theoretic validity is usually specified in such a way that

(a) For Γ/B not to be model-theoretically valid is for there to be a model-theoretic interpretation in which all members of Γ are "designated" and B is not "designated".

And model-theoretic interpretations are usually defined in such a way that we can obtain from any such model-theoretic interpretation a substitution function from the given language into it or an expansion of it, which "preserves logical form" and takes sentences that come out designated into well-behaved determinate truths and sentences that don't come out designated into well-behaved sentences that aren't determinate truths. (The hedge 'well-behaved' here is intended to keep you from substituting in "paradoxical" sentences.) If so, then

(b) If Γ/B (in language L) is not model-theoretically valid then there is an argument Γ*/B* (in L or an expansion of it) of the same form, such that all members of Γ* are well behaved and determinately true and B* is well behaved but not determinately true.

But if all members of Γ* are well behaved and determinately true and B* is well-behaved but not determinately true, then presumably the argument from Γ* to B* isn't valid; and since it has the form of the argument Γ/B, that argument isn't valid either. So 'not model-theoretically valid' implies 'not valid'; contraposing, we get (M-Comp). To repeat, this argument is not airtight, but I think it has considerable force.

2.7 More on Model-Theory and Proof Theory

The question of soundness and completeness proofs is only part of a more general question: if as I've claimed the notion of validity is neither model-theoretic nor proof-theoretic, then why are model-theoretic and proof-theoretic analogs of the notion of validity important?

I think they are important for several reasons, and I will not try to offer an exhaustive account. One, implicit in the previous section, is that they provide a useful means for investigating the notion of direct interest, real validity. For instance, to the extent that a "squeezing argument" is available to show that real validity coincides extensionally with a certain proof-theoretic relation and a certain model-theoretic relation, then certain mathematical features of those relations (e.g. decidability or effective generability) extend to the former. Those features might be hard to establish for validity directly since it is not defined in mathematical terms.

A second and still more obvious reason is that proof theory and model theory provide useful tools for finding out what's valid and what isn't. This is especially obvious in situations where the question of which logic is correct isn't at issue; then the question of whether a complicated inference is valid reduces to the question of whether it's valid *in the agreed-on logic*. A proof theory for a logic seems especially useful in showing what *is* valid in the logic, a model theory especially useful for showing what *isn't*. (Even in the absence of complete agreement about logic, proofs in a natural proof theory can be highly compelling: they persuade us to constrain our degrees of belief in accordance with them. Similarly, counter-models in a natural model theory can persuade us not to constrain our degrees of belief in a way that would automatically rule them out. This is a main reason why *natural* proof and model theories are better than mere algebraic tools that yield the same verdicts on validity.)

My third reason for the importance of model theory and proof theory, and the one I most want to emphasize, is that they provide a useful means of communication between adherents of different logics. And without such means of communication, there is no chance of intelligent debate between adherents of different logics.

The reason they aid communication is that many adherents of non-classical logics think that classical logic is "in effect valid" throughout much of mathematics—enough mathematics to do proof theory and usually enough to do model theory as well. For instance, an adherent of quantum logic is likely to think that mathematical objects such as proofs and models can't undergo superpositions, and that for objects that can't, the distributive laws are correct: so the distributive laws should be accepted as (not strictly logical) laws *within mathematics*. And quantum reasoning *from those laws* is in effect the same as classical reasoning. The same holds for many other non-classical positions: for instance, if excluded middle is problematic only for sentences containing 'true' and related terms, then since such terms aren't part of proof theory or model theory,[37] excluded middle is in effect valid within proof theory and model theory.[38] Given this, proof theory and model theory provide a ground that is neutral between many adherents of different logics.

This marks an important contrast between model theory and proof theory on the one hand and the alleged definition of validity in terms of necessary truth preservation. I've agreed that that alleged definition gives the right results in most instances, and at the moment I'm not worried about the cases where it fails; so for present purposes I could even concede that it is a correct definition. But even if correct—and even if it better captures the meaning of 'valid' than either proof-theoretic accounts or model-theoretic accounts do—it would be useless for certain purposes. For instance, it would be useless to employ it *in trying to explain one's logic to an adherent of a different logic*. The reason is that the adherents of the different logics disagree about which arguments preserve truth by logical necessity, so if I tell you that the valid inferences are (plus or minus a bit) those that preserve truth by logical necessity, that will convey very little about my logic. Whereas if I provide you with a model theory or proof theory for my logic, you will almost certainly reason from this information in the same way that I do, so there is real communication.[39]

This points up a fundamental difference between the concept of truth and the model-theoretic property of "designatedness" or "truth in a model". A typical model theory M for a non-classical logic will be one in which certain classical arguments (say, excluded middle) don't preserve the property of being "designated in M-models", where that property is specified in purely mathematical terms. There won't be serious disagreement as to which ones do and which ones don't: that's a purely mathematical question. The question of which ones preserve truth has an entirely different character, for two reasons.

[37] Obviously 'true in model M' occurs in model theory, but it is to be distinguished from 'true': for example, 'true in M' is set theoretically definable, but 'true (in set theory)' isn't. See the discussion that starts in the paragraph after next.

[38] And if excluded middle is problematic only there or when vagueness is relevant, then *to the extent that* there is no vagueness in the mathematics employed in proof theory and model theory, excluded middle is effectively valid within those disciplines.

[39] In principle, one party could also tell the other party his or her constraints on degrees of belief; that too is unlikely to be understood in the wrong way. But giving a model theory or proof theory is easier.

First, even the advocate of the logic whose model theory is M usually won't identify being true with being designated in some particular M-model. For even one who doubts the general applicability of excluded middle is likely to hold that it applies to precise properties such as being designated in a given M-model. But such a person will presumably think that it doesn't apply to the property of truth: if it's wrong to accept a given instance of excluded middle 'A or not-A', then presumably it's equally wrong to accept 'either ⟨A⟩ is true or ⟨A⟩ is not true'. Since the advocate of the non-classical logic thinks that designatedness obeys excluded middle and truth doesn't, she can't identify truth with designatedness. (Similarly for laws other than excluded middle: a logic that allows assertions of 'both A and not-A' should presumably equally allow corresponding assertions '⟨A⟩ is both true and not true', which prevents truth from being identified with any property definable in a model theory for that logic if the model theory is stated in classical logic.)[40]

This first reason is *slightly* controversial, in that it depends on a view that isn't universally accepted about how we should use the term 'true' if we advocate a non-classical logic. The second reason is quite independent of this: it is that even if we could somehow identify being true with being designated in an appropriate model *of the correct logic*, still the presence of the word 'correct' would preclude using this to explain one's logic to someone else. (If I say to you, "The arguments I take to be correct are those that preserve designatedness in models of the correct logic", that won't give you a clue as to what logic I accept.) Model theory, and proof theory too, are useful in explaining logic since they provide specifications of which arguments are valid that can be applied in the same way by adherents of different views as to what's correct.

There are some limits here: a person with a logic so weird that he reasoned in a very different way than the rest of us do *from premises within mathematics* couldn't really understand our proof-theoretic or model-theoretic explanations of our logic, for he would reason from what we said in very different ways than we do. But cases of that sort are mostly a philosopher's fiction: for the kind of disputes about logic that people actually have, proof theory and model theory provide an extremely useful tool of clarification.[41]

[40] A related point is that even a classical logician won't identify truth with designatedness in models of quantificational logic: this is because of the fact (used to a different purpose above in discussing Kreisel) that there is no model big enough to contain everything, so no model corresponding to the actual world, so that there is no model such that being designated in (or true in) that model corresponds to being true.

[41] Thanks to Paul Boghossian, Daniel Boyd, Josh Dever, Sinan Dogramaci, Kit Fine, Paul Horwich, Jeff Russell, and an anonymous reviewer for some criticisms that have led to improvements.

References

Adams, Ernest. 1975. *The Logic of Conditionals*. Dordrecht: Reidel.

Blackburn, Simon. 1980. "Opinions and chances." In D. H. Mellor (ed.), *Prospects for Pragmatism*. 175–960. Cambridge: Cambridge University Press.

Edgington, Dorothy. 1995. "On conditionals." *Mind* 104:235–329.

Field, Hartry. 2001. *Truth and the Absence of Fact*. Oxford: Oxford University Press.

Field, Hartry. 2003. "No fact of the matter." *Australasian Journal of Philosophy* 81:457–80.

Field, Hartry. 2008. *Saving Truth from Paradox*. Oxford: Oxford University Press.

Field, Hartry. 2009. "What is the normative role of logic?" *Aristotelian Society Supplementary Volume* 83:251–68.

Field, Hartry. 2010. "Reply to Vann McGee." *Philosophical Studies* 147:457–60.

Fine, Kit. 1975. "Vagueness, truth and logic." *Synthese* 30:265–300.

Hyde, Dominic. 1997. "From heaps and gaps to heaps of gluts." *Mind* 106:641–60.

Jeffrey, Richard. 1965. *The Logic of Decision*. New York: McGraw-Hill.

Kreisel, Georg. 1967. "Informal rigour and completeness proofs." In I. Lakatos (ed.), *Problems in the Philosophy of Mathematics*. 138–57. Amsterdam and London: North-Holland Publishing Co.

Lewis, David K. 1980. "A subjectivist's guide to objective chance." In R. Jeffrey (ed.), *Studies in Inductive Logic and Probability*, volume II. 83–132. Oakland, CA: University of California Press.

MacFarlane, John. 2010. "Fuzzy epistemicism." In R. Dietz and S. Moruzzi (eds.), *Cuts and Clouds: Vagueness, its Nature, and its Logic*. 438–63. Oxford: Oxford University Press.

MacFarlane, John. unpublished. "In what sense (if any) is logic normative for thought?" Paper delivered at the American Philosophical Association Central Division conference, 2004.

Restall, Greg. 2005. "Multiple conclusions." In P. Hajek, L. Valdez-Villanueva, and D. Westerståhl (eds.), *Proceedings of the Twelfth International Congress on Logic, Methodology and Philosophy of Science*, 189–205. London: King's College.

Schiffer, Stephen. 2003. *The Things We Mean*. Oxford: Oxford University Press.

Shafer, Glen. 1976. *A Mathematical Theory of Evidence*. Princeton, NJ: Princeton University Press.

Skyrms, Brian. 1984. *Pragmatics and Empiricism*. New Haven, CT: Yale University Press.

Tarski, Alfred. 1936. "Über den Begriff der Logischen Folgerung." *Actes du Congrés International de Philosophie Scientifique* 7:1–11. Reprinted as "On the concept of logical consequence" in *Logic, Semantics, Metamathematics*, trans. J.H. Woodger, Oxford: Oxford University Press, 1956.

3

Logical Consequence and Natural Language

Michael Glanzberg

3.1 Introduction

One of the great successes of the past fifty or so years of the study of language has been the application of formal methods. This has yielded a flood of results in many areas, of both linguistics and philosophy, and has spawned fruitful research programs with names like 'formal semantics' or 'formal syntax' or 'formal pragmatics'.[1] 'Formal' here often means the tools and methods of formal logic are used (though other areas of mathematics have played important roles as well). The success of applying logical methods to natural language has led some to see the connection between the two as extremely close. To put the idea somewhat roughly, logic studies various languages, and the only special feature of the study of natural language is its focus on the languages humans happen to speak.

This idea, I shall argue, is too much of a good thing. To make my point, I shall focus on consequence relations. Though they hardly constitute the full range of issues, tools, or techniques studied in logic, a consequence relation is the core feature of a logic. Thus, seeing how consequence relations relate to natural language is a good way to measure how closely related logic and natural language are. I shall argue here that what we find in natural language is not really logical consequence. In particular, I shall argue that studying the semantics of a natural language is not to study a genuinely logical consequence relation. There is indeed a lot we can glean about logic from looking at our languages, and at our inferential practices. But, I shall argue here, we only get to logic *proper* by a significant process of *identification*, *abstraction*, and *idealization*. We first have to identify what in a

[1] For instance, Portner and Partee (2002), Sag et al. (2003), and Kadmon (2001).

language we will count as logical constants. After we do, we still need to abstract away from the meanings of non-logical expressions, and idealize away from a great many features of languages to isolate a consequence relation. This process takes us well beyond what we find in a natural language and its semantics. We can study logic by thinking about natural language, but this sort of process shows that we will need some substantial extra-linguistic guidance—some substantial idea of what we think logic is supposed to be—to do so. We do not get logic from natural language all by itself.

This is not a skeptical thesis, about logic or about language. It accepts that there are substantial facts about what makes a relation a logical consequence relation, and it accepts there are substantial facts about the semantics of natural language. Nor is it a brief against any particular methods in the study of language. Logical methods, as I mentioned, have proved their worth in studying language many times over. It is, rather, an autonomy thesis: the two sets of facts are fundamentally autonomous, though the processes of identification, abstraction, and idealization can forge some connections between them.

There is one large proviso to the conclusions I just advertised. Defending them in their strongest form will require assuming a fairly restrictive view of what logical consequence relations can be like, that distinguishes them from other related notions. This assumption is entirely consonant with a long tradition in logic. It is certainly the way logic was thought about in the work of Frege and Russell, and others at the turn of the twentieth century when the foundations of modern logic were being laid. But in spite of its lofty pedigree, this view is not universally shared, and indeed there is an equally vaunted tradition in logic that rejects it. So, we should not take such a view for granted. Even so, my goal here is not to defend any particular view of the nature of logic, but rather to see how natural language relates to logic given our views on logic's nature. So, I shall begin by assuming a highly restrictive view of the nature of logical consequence, and defend my conclusions about natural language and logical consequence accordingly. I shall then reconsider my conclusions, in light of a more permissive view. I shall argue that the conclusions still hold in the main part, but that extremely permissive views might have some ways to avoid them. I shall suggest, however, that such extreme views run the risk of buying a close connection between logic and natural language at the cost of making it uninteresting, or even trivial.

My discussion in this chapter will proceed in five sections. In section 3.2, I shall motivate the idea that logic and language are closely connected, and spell out my contrary thesis that they are not. I shall also briefly discuss some ideas about logic and related notions I shall rely on throughout the chapter. In section 3.3, I shall offer my main argument that logic and natural language are not so closely connected. In doing so, I shall articulate a little of what I think the semantics of a natural language is like, and show how entailments, but not logical consequence relations, are found in natural language. The arguments of section 3.3 will presuppose a restrictive view of logical consequence. In section 3.4, I shall reconsider those arguments from a permissive view of logical consequence. I shall argue that my main conclusions still hold, though perhaps in weakened form, and some important qualifications are in order. I shall then show in section 3.5 how one can move from natural language to logical consequence by the three-fold process of identification, abstraction, and idealization. Finally, I shall offer some concluding remarks in section 3.6.

3.2 Preliminaries and Refinements

It is very common, at least in some circles, to speak of a logic and a language in the same breath. This perhaps makes the idea that logical consequence and semantics of natural language are closely related an inviting one. The principal thesis of this paper is that they are not so closely related. However, this thesis stands in need of some refinement, and some important qualifications.

To provide them, I shall begin by briefly articulating the perspective that sees logic and language as closely connected. I shall then discuss the notion of logical consequence itself, and distinguish more permissive and more restrictive views on the nature of logical consequence. That will allow us to distinguish (restrictive) logical consequence from some related notions. With these preliminaries in hand, I shall be able to formulate a more refined version of my main thesis. I shall conclude this section with some discussion of what the thesis implies about the application of formal methods to natural language semantics.

3.2.1 Logics and languages

For many logicians, it is languages (i.e. formal languages of particular sorts) that are the primary objects of study. A fairly typical example is the textbook of Beall and van Fraassen (2003), which studies formal languages comprised of a formal syntax, a space of valuations of sentences of the syntax, and a relation of satisfaction between sentences and valuations. Logical consequence is preservation of satisfaction (in most cases, preservation of designated value). This particular textbook focuses on sentential logic, so the valuations are typically determined by assignments of truth values to atomic sentences, but they could very well be determined by models of the right sorts, and for quantificational languages, they will be.

Making such languages the basic elements of logic is especially convenient for the study of a variety of logics, as it gives a natural unit of study that can vary. We ask about different languages, and explore and compare their logical properties. But the connection might well go deeper, and in some ways, it must. Genuine logical relations are interesting not just for their abstract properties, but what they tell us about connections between meaningful sentences, which express what we think. Conversely, the basis for a consequence relation is often thought to be found in the meanings of sentences. For instance, Beall and van Fraassen (2003, 3) write:

> Logic pertains to language. In a logical system we attempt to catalog the valid arguments and distinguish them from the ones that are invalid. Arguments are products of reasoning expressed in language. But there are many languages and different languages may have a different structure, which is then reflected in the appropriate logical principles.

When we look for valid arguments, we look at meaningful language as the medium of expressing them. But moreover, we think of the meanings of sentences as crucial for determining whether or not an argument is valid. Thus, we tend to think of the languages that provide consequence relations as genuine languages, capable of expressing thoughts in much the ways our natural languages do.

Combining these ideas, we can formulate two theses. The first is the *logics in formal languages thesis*:

> Logical consequence relations are determined by formal languages, with syntactic and semantic structures appropriate to isolate those relations.

The logics in formal languages thesis is not entirely trivial, but it is not particularly controversial either. Working in terms of formal languages is one theoretical choice among many on how to develop important logical notions. Like any theoretical choice, it has various consequences that might be weighed differently, but there is little room to say it could be outright wrong. Perhaps more controversial is the idea that there are multiple logics associated with multiple languages. This might be challenged, and strong ideas about logical pluralism are indeed controversial. I shall return to some issues surrounding this point in a moment. But for now, I shall simply observe that the logics in formal languages thesis is optional, but by itself not very contentious. I formulate it to help frame the next thesis, which will be genuinely contentious.

The key idea of the logics in formal languages thesis is that formal languages are the unit of study for logic, and so, formal languages must determine consequence relations. A formal language is a bundle of elements, usually containing something like a syntax, a space of valuations or models, and a relation of satisfaction. A consequence relation is definable from these, typically as preservation of designated value. Variation in ways of presenting formal languages will allow for variation in how consequence relations are defined on them, but the important idea is that formal languages contain enough elements that consequence relations can be defined on them alone, just as we see in the typical case. Thus, the logics in formal languages theses holds that consequence relations are *in* formal languages, in the sense that they are definable from them. I shall likewise sometimes talk about languages *containing* consequence relations.[2]

As we just discussed, it is inviting to think that these formal languages are importantly like natural language. How alike? I suggest the key idea is that a natural language shares important logical features with formal languages. Most importantly, they share the feature of containing logical consequence relations. Thus, studying a range of formal languages may expand our horizons, but it shows us more of what we already can find in natural language. This leads to our second thesis, the *logic in natural language thesis*:

> A natural language, as a structure with a syntax and a semantics, thereby determines a logical consequence relation.

The syntax of natural languages differs from that of our favorite formal languages, and in some ways, perhaps, their semantics does too. But regardless, according to the logic in natural language thesis, they are just more languages, and determine logics just like formal languages do.[3]

Whereas the logics in formal languages thesis is relatively banal, the logic in natural language thesis is one I shall argue against. But, just how strong the thesis is, and what it takes to argue against it, depends on how we view consequence relations. Before we can get an accurate statement of the thesis, and an adequate appreciation of what rejecting it amounts to, we must consider what counts as a logical consequence relation.

3.2.2 *Logical consequence*

In this section, I shall briefly review some ideas about logical consequence. My goal here is limited. I shall not try to defend a view of the nature of logical consequence;

[2] Thus, the claim that a language contains a logic is much stronger than the claim that we can talk about a logic in the language. We can talk about all sorts of things in languages that are not definable on the language itself. We can talk about physics in a language too, but I doubt any formal or natural language contains physics.

[3] This attitude is perhaps most strongly expressed by Montague (1970, 222), who writes "There is in my opinion no important theoretical difference between natural languages and the artificial languages of logicians; indeed, I consider it possible to comprehend the syntax and semantics of both kinds of languages within a single natural and mathematically precise theory." At least, so long as we take the semantics of artificial languages to include a consequence relation, then Montague's view includes, and goes well beyond, the logic in natural language thesis.

rather, I shall try to survey enough options to better frame the logic in natural language thesis.

At its core, logic is the study of valid arguments, as Beall and van Fraassen say above. Of course, this is not all that logicians study, as the many topics in such areas as set theory, model theory, and recursion theory make clear. But this is the core feature that makes logic logic. Logical consequence is the relation that makes arguments valid: it is the relation that holds between a set of sentences and a sentence when the first set comprises the premises and the second sentence the conclusion of a valid argument.[4] But then the important question about the nature of logical consequence is simply: what makes an argument valid? There are a few key ideas that I shall suppose for argument's sake. As I mentioned, my aim here is not to provide a definitive analysis of the notion of logical consequence—that would be far too hard a task. Rather, I simply want to say enough to distinguish logical consequence from some of its neighbors.

Perhaps the main feature of logical consequence is what is sometimes called *necessity*: if S is a consequence of a set X of sentences, then the truth of the members of X necessitates the truth of S, or equivalently, it is impossible that each element of X be true and S be false. What makes an argument valid, necessity notes, is in part that the conclusion cannot be false while the premises are true. This is imprecise, as the notion of necessity at work is not spelled out. Even so, it is sufficient to distinguish logical consequence from, for instance, inductive support. Even a sentence which enjoys strong inductive support on the basis of some assumptions can fail to be true while the assumptions are true.[5]

Necessity is one of the main features of any relation we would call a logical consequence relation. Another, especially in the tradition of Tarski (1936) or Quine (e.g. Quine 1959; 1986), is what is sometimes called *formality*. Formality is hard to spell out in full generality, but the main idea is that logical consequence is somehow a 'formal' relation, holding in virtue of the forms of the sentences in question. This condition is typically taken to rule out implications like 'John is Bill's mother's brother's son, therefore, John is Bill's cousin' as not genuinely logical. It is not, the reasoning goes, because it relies for its validity on the specific meaning of 'cousin' rather than the formal properties of the sentence. Of course, how formality is implemented depends on just what we take the relevant 'formal' structure of a sentence to be. Whereas necessity is likely to be recognized as a requirement on

[4] Of course, things can be complicated here in various ways, but we will not reach a level of detail where such complications would matter in this discussion.

[5] My discussion of logical consequence relies heavily on those of Beall and Restall (2009) and MacFarlane (2009).

consequence relations across the board, formality is more contentious, and how far it will be accepted will depend on how it is spelled out.

One leading approach to spelling out formality, since Tarski if not earlier, has been to identify special logical structure, and propose that logical consequence must hold in virtue of only that structure. It is thus formal in that it holds in virtue of specified structure or form in sentences. In the post-Tarskian tradition, we often capture this by insisting that there are privileged *logical constants* in sentences, and logical consequence holds in virtue of their properties. Spelled out in a model-theoretic way, the idea is that logical consequence holds solely in virtue of the meanings of the logical constants, and hence, we hold those meanings fixed, but allow all other meanings to vary, as we work out model-theoretic consequence relations.

Necessity and formality may well interact. For instance, they do in the standard post-Tarskian model-theoretic view of consequence. This view relies on a range of models to characterize the consequence relation. The models thus provide the possibilities which give substance to necessity. They also implement formality, by allowing the meanings of all the non-logical terms of a language to vary freely, and thereby single out logical constants whose meanings underlie validity. They thus offer a combined version of both formality and necessity. The effect of this combination is enough to rule out familiar analytic entailments as not logical consequences. The 'cousin' inference is an example, and indeed, so is 'Max is a bachelor, therefore, Max is unmarried', which also fails to be a logical consequence on many views.[6]

When implemented this way, necessity and formality are often understood as conspiring to make logical consequence a very narrow notion. We see this, for instance, in the way they rule out analytic entailments as not logical. Behind this conclusion is a very general attitude towards logic, which holds there are substantial constraints on what makes for genuine logical consequence, and finds that only such a narrow notion meets those constraints. This attitude is a long-standing one in the philosophy of logic. It is most prominent in the strand of logic starting with late nineteenth-century developments in the foundations of mathematics in the work of Frege and Russell (e.g. Frege 1879; Russell 1903; Whitehead and Russell 1925–7), going through such figures as Gödel (1930) and Skolem (1922), and then moving on to Tarski (1936) and Quine (1986), to more recent work by Etchemendy

[6] There is considerable historical debate over just what Tarski's view of logical consequence was, sparked, in large part, by Etchemendy (1988). For this reason, I have talked about the 'post-Tarskian view', which is embodied in standard contemporary model theory. This view no doubt stems from work of Tarski, whether it is Tarski's original view or not. For a review of some of the historical work on this and related issues, see Mancosu (2010).

(1990), Shapiro (1991), and Sher (1991), among many others. This tradition is not uniform in its views of what grounds logical consequence (for instance, Frege and Quine and Tarski would see the formality constraint very differently). But it is uniform in thinking that there is some important underlying notion of logic, and that logic plays particular roles and has a particular status. Its special epistemological status was especially important in discussions of logicism, and its metaphysical status in discussions of ontological commitment. Both features are apparent in the discussions of the role of logic in the foundations of mathematics as first-order logic emerged in the early twentieth century. This tradition sees substantial constraints on what makes a relation logical consequence, which flow from the underlying nature of logic. These constraints substantially restrict what can count as logic. This tradition is thus, as I shall call it, *restrictive*, in that it provides highly restrictive constraints answering to a substantial underlying notion.

There is another, more *permissive*, tradition in logic, going hand-in-hand with work on 'non-classical' or 'non-standard' logic. This tradition also has a vaunted pedigree. The range of ideas and issues that it encompasses is very wide, but some of them can be traced back to Aristotle, and were important in the lively medieval logical literature.[7] The permissive tradition does not necessarily abandon the idea that there is some underlying notion of logical consequence, but it interprets the idea much more broadly, and sees much more variety in how the constraints on logical consequence might be applied. The result is a willingness to consider a range of logics as at least candidates for showing us logical consequence relations. It is perhaps natural to think of the permissive approach as going hand-in-hand with *logical pluralism* (e.g. Beall and Restall, 2006). If it does, then it will think that more than one logic genuinely is logic. But permissive views certainly need not be completely indiscriminate. For instance, many logicians in the relevance tradition have doubts about whether classical logic really is logic, often driven by the so-called paradoxes of implication.

Permissive views need not reject the ideas of necessity and formality, and notably, Beall and Restall's logical pluralism does not. But they will interpret these notions more expansively than many classically oriented restrictive views do. One way they can do so is to expand or contract the range of possibilities which inform the necessity constraint, or modify the way truth in a circumstance is characterized. They can also vary the range of logical constants which give content to formality. Both have been done, many times for many different reasons. Often some combination of both proves fruitful.

[7] I am not qualified to review this history, so I shall defer to experts such as Kneale and Kneale (1962), Kretzmann et al. (1982), Read (2012), Smith (1995), and the many references they cite. For a review of more contemporary developments, see Priest (2008) and the many references therein.

Another axis on which we might compare views of logical consequence is according to how broad or narrow a notion of consequence they accept. Classical logic, for instance, is narrower than a consequence relation that includes analytic entailments like the 'bachelor' entailment. These issues are substantially independent of those of permissive versus restrictive views. How broad or narrow the notion(s) of consequence you accept are is determined by just which constraints you impose, not the general issue of permissive versus restrictive approaches. Even so, we may expect permissive views to more readily entertain a range of broader consequence relations than restrictive views do. Likewise, in introducing the restrictive view, I noted that it tends to indicate narrow notions of logical consequence. It will thus simplify our discussion to assume that restrictive views are committed to only a narrow notion of consequence, while permissive views can entertain broad ones as well. This reflects a trend in thinking about consequence, but it is a simplification. For one reason, it leaves out that permissive views might well entertain more narrow consequence relations than restrictive views do. But it will be a useful simplification to make.

For discussion purposes, I shall usually assume that the restrictive view is going to opt for something like classical logical consequence, as in fact the tradition I associated with the restrictive view did opt for; while permissive views can consider consequence relations that reflect other sorts of entailment. I shall not worry here about differences within classical consequence relations, like first-versus second-order logics.[8] In light of the logics in formal languages thesis, and the way we have glossed the constraints of necessity and formality, it will be natural to assume a model-theoretic account of consequence relations, both classical and otherwise. This assumption will set up the most likely route to the logic in natural language thesis, so it is a harmless one to make here.[9]

3.2.3 *Implications and entailments*

We now have at least roughly sketched some ideas about logical consequence, and distinguished permissive from restrictive views of consequence. In what follows, I shall argue against the logic in natural language thesis on the basis of those ideas, especially on the basis of restrictive views of consequence. But when it comes to

[8] Type theories are used in a great deal of work in semantics, but as we will see in section 3.3, not in a way that directly indicates a consequence relation.

[9] I thus have relatively little to say about proof-theoretic accounts of logical consequence, for instance, as explored in work of Dummett (e.g. Dummett 1991) and Prawitz (e.g. Prawitz 1974), all building on seminal work of Gentzen (1935). For a recent survey, see Prawitz (2005). In that survey, Prawitz explicitly endorses the general constraints of necessity and formality, though of course, not the model-theoretic gloss on them I typically employ here.

natural language, it will be important to distinguish logical consequence from potentially broader related notions, such as we have already seen with analytic entailments. In this section, I shall review some of these notions.

To fix terminology, let us start with *implication*. I shall take this term to be very broad, covering many relations between sentences including not only logical consequences and subspecies of them, but looser connections like those captured by defeasible inference. Following the philosophy of language tradition, we might also see implications between utterances of sentences, often defeasible, such as the one discussed by Grice (1975) that typically obtains between 'There is a gas station around the corner' and 'It is open'.

By restrictive lights, implication is much broader than logical consequence, but it is such a wide and loose notion that it is not clear if it is apt for formalization even by a very broad notion of consequence. Permissive approaches have done substantial work on some species of it, notably defeasible inference.[10] Some attempts have been made to make rigorous the computations that might support implicatures, but they tend to focus more on computation than consequence per se.[11]

Within the broad category of implications, two specific notions will be important. One, narrow logical consequence (i.e. classical first-order consequence or something thereabouts), we have already seen. The other is *entailment*. I shall understand entailment as a truth-conditional connection: P entails Q if the truth conditions of P are a subset of the truth conditions of Q. The usual modification for multiple premises holds. We have already seen enough to know that entailment is a wider notion than narrow logical consequence. Analytic entailments like the 'cousin' implication above are entailments by the current definition, but not narrow logical consequences. By many lights, entailments go beyond analytic entailments. They will, for instance, if truth conditions are metaphysically possible worlds. If so, and assuming Kripke–Putnam views of natural kind terms, then 'x is water, therefore x is H_2O' is an entailment. We will encounter more entailments as we proceed. What we need now is simply that entailment is a markedly broader notion than narrow logical consequence, though narrower than some notions of implication, including implicatures like the 'gas station' one.

We have now seen three related ideas: a very broad notion of implication, a very narrow notion of logical consequence, and an intermediate notion of entailment. Permissive views of logic have done extensive work to capture notions of entailment as broad consequence relations. Indeed, such work has identified

[10] See, for instance, Antonelli (2005). See Horty (2001) for a survey of related ideas.
[11] See, for instance, Asher and Lascarides (2003) and Hirschberg (1985).

a number of distinctions within the category of entailment.[12] As I noted above, permissive views have also attempted to capture some aspects of the wide notion of implication. So, we should not think these relations to be beyond the range of permissive approaches to logical consequence, so long as they are broad enough in their notions of consequence. They are, however, clearly beyond the range of the restrictive view, and fail to offer narrow consequence relations. We will see in section 3.3.2 that natural language presents us with a striking variety of entailments, but I shall argue, not narrow logical consequence.

3.2.4 *The refined thesis*

Now that we have some preliminaries about logic and related notions out of the way, we can return to the main claims of this chapter. Above I formulated a thesis of logics in formal languages that claims that consequence relations are determined by formal languages. (Both the permissive and restrictive approaches are compatible with this thesis.) But the important thesis for this chapter is the more contentious logic in natural language thesis, which holds that a natural language, construed as including a syntax and a semantics, determines a logical consequence relation. The main contention of this chapter is that the logic in natural language thesis is false. I shall argue that it is clearly false if we adopt the restrictive view of logical consequence. There are no doubt entailment relations in natural language, determined by the semantics of a language, and there are many other implication relations as well. But the semantics and syntax of a natural language does not determine what the restrictive view of consequence takes to be a logical consequence relation.

If we adopt the permissive view, this claim becomes rather more nuanced. I shall argue that we still have good reason to think the logic in natural language thesis is false, even if we adopt a permissive view. However, I shall also grant that there are some, perhaps extreme, permissive views that might support some forms of the thesis. Even so, I shall argue, they run the risk of stretching the notion of logic too far, and thereby undercutting the interest and importance of the thesis.

Though I shall argue against the logic in natural language thesis here, I shall not claim there is no connection between logic and natural language. We can glean some insight into logical consequence, and indeed even narrow logical consequence, by studying natural language. The reason is that the entailments and other implications we do find in our languages, and our wider inferential practices, provide a rich range of examples around which we can structure our thinking about logical consequence. But to do so correctly, we must get away

[12] See, for instance, Anderson and Belnap (1975) and Anderson et al. (1992).

from the entailments and implications of a human language and human inferential practice, and isolate genuine logical consequence. I shall argue that the strategy of identification, abstraction, and idealization I mentioned above is a useful one for taking this step, given the kinds of information that natural languages really do provide for us. I shall argue that even the permissive view needs to take these steps, to get theoretically substantial logics out of natural language. It may be that for the permissive view, some of the steps may be shorter than those the restrictive view needs to take, but the same kinds of steps must be taken by both.

3.2.5 No logic in semantics?

I am proposing that in a narrow sense, natural language has no logic, and I thereby echo Strawson's famous quip (Strawson, 1950). Though I echo some of the letter of Strawson, I do not follow the spirit. To make this vivid, let me spell out several things I am not claiming. First and foremost, if we adopt a restrictive view and accept only narrow consequence as consequence, saying we do not find logical consequence in natural language does not by any means say that natural language is immune to study by formal or mathematical methods.[13]

In fact, we will see a number of places where we find that natural language semantics makes productive use of tools and techniques from logic. I shall explain as we proceed how this can happen without indicating a narrow consequence relation in natural language, and indeed, seeing how this happens will provide good reasons for rejecting the logic in natural language thesis. Once we see how we really use logic in the study of natural language semantics, the thesis loses its intuitive appeal.

The applicability of logical methods to the study of language, in spite of the failure of the logic in natural language thesis, should not itself be surprising. In singling out the notion of logical consequence as the core of logic, we should not be blind to the impressively wide range of applications of logical (and more generally, formal) methods, which goes well beyond the study of logical consequence per se. Logical structure, in the rough sense of what is tractable via the methods of logic, can be found in many places. Computer science finds it in the organization of data and computation, linguistics finds it not only in some aspects of semantics, but in the syntactic organization of parts of sentences. Sociologists find it in the

[13] The original from Strawson (1950, 27) reads, "Neither Aristotelian nor Russellian rules give the exact logic of any expression of ordinary language; for ordinary language has no exact logic." I am not sure exactly what Strawson has in mind here, but it is common to read him as advocating the view I reject, that natural language is immune to study by formal methods. As I like many of the arguments of Strawson's paper, I often tell my students to pay close attention to every part of it except the last line.

organization of social networks. Indeed, some of the core underlying structures of logic, like Boolean algebras, seem to be found practically anywhere you look for them.

From a restrictive point of view, many of these applications go beyond the study of logical consequence. For instance, the syntactic structure of human language seems clearly not to be a matter of logical consequence according to a restrictive approach, even if we can represent substantial portions of it with the formalisms of logic via the Lambek calculus (Lambek 1958; van Benthem 1991; Moortgat 1997). From a very permissive point of view, say, one which is happy to talk about a logic of syntax via the Lambek calculus, things may look somewhat different. But as I said above, I shall argue they are not all that different. We do not get to clearly articulated broad consequence relations in studying language without departing from the on-the-ground study of language, even if our methods are formal ones.

Denying the logic in natural language thesis in no way argues we should put aside logic when we come to study natural language. Rather, it argues that we should see logical consequence proper and the semantics of natural language as substantially autonomous, but linked by such processes as abstraction and idealization.

3.3 The Logic in Natural Language Thesis from the Restrictive Point of View

We now have a more careful articulation of my main claims, that the logic in natural language thesis fails, but that natural language and logic can be connected by the three-fold process of identification, abstraction, and idealization. My defense of these claims will come in three parts. First, in this section, I shall argue against the logic in natural language thesis assuming a restrictive view of logical consequence. In the next section, 3.4, I shall reconsider those arguments from a permissive point of view. I shall then turn to how to bridge the gap between natural language and logic in section 3.5.

My discussion of the logic in natural language thesis here will present three arguments. The first will argue against the thesis directly, by showing that the semantics of natural language does not provide a consequence relation. The second and third will show that there is no way around this conclusion. The second argument will show that the implications natural language does provide to us are not generally logical consequences. The third will show that natural language does not distinguish logical constants, and so formality cannot be read off the structure of natural language. Thus, we find no consequence relation in the semantics of natural language proper, and cannot find one encoded in natural language by more indirect means. Throughout this section, I shall assume a restrictive view of consequence without further comment.[14]

3.3.1 Semantics, model theory, and consequence relations

In this section, I shall argue that consequence relations are not provided by the semantics of natural language in the way they are provided by formal languages (assuming the logics in formal languages thesis). This is so, I shall argue, even assuming a truth-conditional approach to semantics, and even assuming its model-theoretic variant. The model theory we do for semantics is not the sort of model theory that provides model-theoretic consequence relations (and might be better called something other than 'model theory'). To argue this, I shall first lay out a little of what I think the semantics of a natural language is like. This will show that a viable natural language semantics—in particular, a viable truth-conditional semantics—cannot provide a model-theoretic consequence relation. I shall then explain why this is the case in spite of the apparent use of model theory in so-called 'model-theoretic semantics', and discuss some aspects of how model-theoretic techniques can shed light on natural language without providing consequence relations.

3.3.1.1 ABSOLUTE AND RELATIVE SEMANTICS

Many logicians, lured by the siren song of natural language, have found themselves thinking of model theory as applying to fragments of natural language. This trend got a huge boost in the 1960s and 1970s with developments in the model theory of intensional logics, which made applications to natural language easy to find (e.g. Montague 1968; Scott 1970). Eventually, Montague (e.g. Montague 1970; 1973) in effect proposed that to do the semantics of natural language includes doing its model theory (and proposed that it could really be done). More might be required, such as Montagovian meaning postulates. But along the way to giving a semantics in the Montagovian mold, you will provide a whole space of models, and determine truth in those models for sentences, and so, you will have done your model theory. If this was right, then the logic in natural language thesis would be sustained (though we would want to check that the result lived up to restrictive

[14] To come clean, I am inclined to take a restrictive view, and claim that something like classical logical consequence is the right notion (though I am not really decided on issues like those of first- versus second-order logic, the status of intensional logic, etc.). But it is not my goal to defend any such views here.

standards). Indeed, this is the main reason the logic in natural language thesis might be thought to be correct.[15]

I shall argue that this is a mistake. This will show that the Montagovian route to the logic in natural language thesis fails. Moreover, it will show that the semantics of natural language does not build in a consequence relation in anything like the way the Montagovian route supposes. I take this to be a good reason to reject the logic in natural language thesis, though I shall provide two other reasons below. My argument goes by way of a reconsideration of an old debate about model-theoretic semantics for natural language. The debate, with neo-Davidsonian advocates of so-called 'absolute' truth-conditional semantics, challenged the idea that model-theoretic techniques could be used to study natural language at all.

The debate takes place within a program of truth-conditional semantics for natural language. The goal of this program is to provide an account of a key aspect of speakers' linguistic competence. In particular, semantics seeks to provide an account of what a speaker understands when they grasp the meanings of their sentences: that is, what they know when they know what their sentences mean.

I shall take it for granted that truth conditions provide a central aspect of this knowledge. A key part of what a speaker knows when they know the meaning of 'Snow is white' is precisely that it is true if and only if snow is white. This is a non-trivial assumption, rejected, for instance, by a number of conceptual role or cognitive approaches to semantics. I think it is correct, but I shall not defend it here. Rather, I shall take it as a starting point for asking about how logic, and particularly consequence relations, might find their way into natural language.[16]

[15] For more recent presentations of Montague Grammar, see for instance Dowty et al. (1981) or Gamut (1991). Other important early contributions along similar lines include Cresswell (1973) and Lewis (1970). Of course, these works did not appear in a vacuum, and earlier works of Frege (e.g. Frege 1891), Tarski (e.g. Tarski 1935), Carnap (e.g. Carnap 1947), Church (e.g. Church 1940), Kripke (e.g. Kripke 1963), and others stand as important predecessors.

Though Montague and a number of these other authors link logic and natural language, this marks a departure from the main trend of early work in the restrictive tradition that I discussed in section 3.2. Frege and Russell, for instance, saw only distant connections between logic and natural language.

[16] For discussion of this role for truth conditions, see for instance Higginbotham (1986; 1989b), Larson and Segal (1995), and Partee (1979).

The important question, of course, is whether this assumption is correct; and on that point, I do not have much to add to the current literature, and this is certainly not the place to pursue the issue in depth. But one might wonder if making this assumption makes my whole argument somewhat parochial. I do not think it does. At least in empirically minded work in semantics and related work in philosophy, truth-conditional semantics is a well-established research program. It might well be the dominant one (I think it is), though it is not my intention to dismiss work in cognitive semantics, which has been important in the literature on lexical semantics. Whether it is the dominant research program or not, truth-conditional semantics certainly enjoys a sufficiently important place in research in semantics to make asking about its connections to logic important. If, as I believe, assuming a truth-conditional perspective is correct, then all the more so. In the less empirically minded literature in philosophy of language, there has been more discussion of conceptual role

From this starting point, it is no surprise that logicians have sometimes found a very close connection between semantics and model theory, and through that, a connection with logical consequence, just as the logic in natural language thesis would have it.[17] Models, after all, seem just right to play the role of individual conditions under which the sentences of a language are true. So, to specify the truth conditions of a sentence, we might suppose, is precisely to specify the class of models in which it is true. This is one of the key components of Montague's own approach to the semantics of natural language (e.g. Montague 1973).[18]

On this view, a good semantic theory will, among other things, assign sets of models to sentences, which are taken to capture their truth conditions. Of course, we want more than that; not least of which, we want our theory to correctly derive the truth conditions of sentences from the meanings of their parts (we presumably want some form of *compositionality* to hold). Putting these two together, a reasonable requirement for a theory which will assign sets of models as truth conditions is that it can derive results like:

(1) For any model $\mathfrak{M}$, 'Ernie is happy' is true in $\mathfrak{M} \iff \text{Ernie}^{\mathfrak{M}} \in \text{happy}^{\mathfrak{M}}$.

($\text{Ernie}^{\mathfrak{M}}$ is the extension (or other appropriate value) of 'Ernie' in $\mathfrak{M}$.) If we can do this for a large fragment of a human language, compositionally, the view holds, we have thereby elaborated a good semantic theory.

theories of meaning, inferentialist theories of meaning, and use theories of meaning. But these have never really gotten off the ground as empirical theories. Of course, all these deserve more discussion, but from the empirically minded perspective I am adopting here, they do not really provide well-developed alternatives to truth-conditional semantics.

[17] Though in many cases, logicians are more interested in the pure mathematics than the psychology. This was certainly Montague's attitude. As Thomason (1974, 2) puts it, "Many linguists may not realize at first glance how fundamentally Montague's work differs from current linguistic conceptions. Before turning to syntactic theory, it may therefore be helpful to make a methodological point. According to Montague the syntax, semantics, and pragmatics of natural languages are branches of mathematics, not of psychology." (For further discussion, see again Partee (1979), or Zimmermann (1999).) Perhaps this was not realized by many, and it is perhaps no surprise that authors such as Higginbotham identified with the neo-Davidsonian tradition in semantics. But whether intended by Montague or not, some aspects of his apparatus have been incorporated into an approach to semantics within the broader tradition of generative linguistics, as is witnessed by the textbooks of Chierchia and McConnell-Ginet (2000) and Heim and Kratzer (1998). I shall briefly discuss what of Montague's apparatus got so incorporated below, but a more full discussion shall have to wait for other work in progress. Finally, I should note that the general idea that semantics has something to do with what we know when we know a language can be completely divorced from the Chomskian perspective of Higginbotham or Larson and Segal, as in the work of Dummett (e.g. Dummett 1991).

[18] In sketching Montague's ideas, I am suppressing a great deal of detail that is not important for the issue at hand. For instance, Montague's approach also relies on other elements, including categorial grammar, intensional type theory, and meaning postulates, but these do not change the basic place of models in the theory.

In many applications, the models involved are modal, and contain a set of possible worlds, as in the classic Montague (1973) or Lewis (1970). But the assignment of semantic values is still done relative to a model, as well as to a world, and sometimes a time. We will not be worried here about intensional constructions, so we can just think about assigning reference and truth in a model. In the long run, the use of intensional model theory might well affect what consequence relations are at issue, but it will not affect the basic route from truth conditions to models to consequence, so we can safely ignore this issue here.

The model-theoretic or Montagovian approach to truth-conditional semantics is one of two classic approaches. The other, following Davidson (1967) (who himself in some ways follows Tarski 1935), emphasizes deriving clauses like:

(2) 'Ernie is happy' is true $\Longleftrightarrow$ Ernie is happy.

Davidson, following Quine (e.g. Quine 1960), emphasized the extensional nature of such a theory, and also its non-model-theoretic pedigree. No model is mentioned, and clauses like these are typically derived from statements of reference and satisfaction properties, like that 'Ernie' refers to Ernie. But what is most important for us is that the resulting T-sentences or disquotational statements state the truth conditions of sentences.[19]

It may look like these two approaches do essentially the same thing. One is more proof-theoretic, emphasizing theories that can derive canonical statements of truth conditions. The other is more model-theoretic, explicitly referring to models. Yet both seem to be in the business of working out how the truth conditions of a sentence are determined by the reference and satisfaction properties of its parts. Both thereby hope, in light of the assumptions we have made about meaning, to represent some important aspects of a speaker's knowledge of meaning: that is, their semantic competence.

In spite of this, it is often thought that the two approaches to semantics are very different. In fact, it has been argued, notably by Lepore (1983), that model-theoretic semantics is somehow defective, or at least less satisfactory than absolute semantics (cf. Higginbotham 1988). Far from being a variant on the same basic idea, Lepore argues, the model-theoretic approach has built-in failings that make it inappropriate for doing semantics at all, and it is hardly equivalent to a neo-Davidsonian semantics.

Though I do not think the morals to be drawn from this argument are what many neo-Davidsonians think, there is something importantly right about the

[19] Davidson himself did not ascribe to some of the assumptions about linguistic competence I made above, but some neo-Davidsonians do, including Higginbotham and Larson and Segal cited above.

argument, and it will reveal something important about the connections between logical consequence and truth-conditional semantics. This will lay the groundwork for rejecting the logic in natural language thesis.

Lepore's main point is that model-theoretic semantics can only provide *relative* truth conditions: that is, conditions for truth in or relative to a model. And you can know those and not know what the sentence means. You can know that for any model, 'Snow is white' is true in that model if the extension of 'white' in that model includes the referent of 'snow', without having any idea what the sentence means. You could have no idea that it talks about snow, or whiteness. It is no better than knowing that 'The mome raths outgrabe' is true in a model if the extension of 'raths outgrabe' in the model includes the referent of 'the mome' in the model. We know that, but (I at least) have no idea what this sentence means.[20]

Davidsonian or *absolute* statements of truth conditions, of the kind we get from techniques stemming from Tarski's work on truth (Tarski 1935), do tell you much more. They tell you the sentence 'Snow is white' is true if and only if snow is white, which is what we wanted. As you do in fact understand what 'snow' and 'white' mean, this tells you much more than that in any model, some element falls in some extension. It tells you that this stuff—snow—has this color—white. Similarly, we cannot, in a language we understand, write down a T-sentence for the Lewis Carroll 'The mome raths outgrabe'. Hence, Lepore argues, the model-theoretic approach to semantics fails to characterize enough of what speakers know about the meanings of their words and sentences, while the absolute or neo-Davidsonian approach does much better.

I think the conclusion of this argument is correct. It follows that we do not provide a consequence relation as part of the semantics of natural language. In particular, it follows that we do not provide a model-theoretically defined consequence relation, by providing a space of models and a truth in a model relation, in the course of natural language semantics. Thus, natural language semantics does not do what is taken for granted by the logic in natural language thesis, and so, the thesis looks doubtful. There are some remaining issues, of course, such as whether we can press truth conditions into service to provide a consequence relation in some other, less direct way. I shall return to these below. But we can see from Lepore's argument that a model-theoretic consequence relation is not built into the basic apparatus of semantic theory.

To further support this conclusion, I shall explore a little more where model-theoretic techniques do fit into natural language semantics, and show how they do so without providing consequence relations.

[20] Philosophers of language have thus long been indebted to Lewis Carroll (Carroll 1960).

3.3.1.2 MODEL THEORY IN CURRENT SEMANTIC THEORY

Nearly all current work in semantics, including that work done in the 'model-theoretic' or Montagovian tradition, is in fact really doing absolute semantics. In this section, I shall illustrate this point by discussing a few of the important features of current model-theoretic semantics, and showing that they do not lead to consequence relations. Indeed, we will see that what is characteristic of such approaches to semantics these days is a reliance on *type theory*, not model theory in the sense needed to get logical consequence. Thus, we will see that the conclusion of the previous section, that the basic idea of truth-conditional semantics does not lead to a model-theoretic consequence relation, is not specific to a neo-Davidsonian view of semantics. It follows just as much on a model-theoretic view. We will return in section 3.3.3 to further questions of how model theory might find its way into absolute semantics.[21]

What is characteristic of most work in the model-theoretic tradition is the assignment of semantic values to all constituents of a sentence, usually by relying on an apparatus of types (cf. Chierchia and McConnell-Ginet 2000; Heim and Kratzer 1998). Thus, we find in model-theoretic semantics clauses like:[22]

(3) a. $[\![\text{Ann}]\!] = \text{Ann}$
b. $[\![\text{smokes}]\!] = \lambda x \in D_e .\ x \text{ smokes}$

We also find rules like function application (Heim and Kratzer 1998; Klein and Sag 1985):

(4) If α is a branching node and β, γ its daughters, then $[\![\alpha]\!] = [\![\beta]\!]([\![\gamma]\!])$ or vice versa.

These are not things a neo-Davidsonian theory (one using traditional Tarskian apparatus) is going to have.

Even though clauses like this look different from those preferred by neo-Davidsonians, they provide absolute statements of facts about truth and reference. They just put those facts in terms of functions and arguments (as Frege would have as well!). We see that the value of 'Ann' is Ann, not relative to any model. The value of 'smokes' is a function, but one that selects the things that smoke, again, not relative to any model. Semantics needs to be absolute, but both model-theoretic

[21] This section is a brief discussion, focused on the issues at stake for the status of logical consequence in natural language. There are a great many more questions that are raised by the place of model-theoretic or other mathematical techniques in semantics. I address more of them in work in progress.

[22] In common notation, $[\![\alpha]\!]$ is the semantic value of α. I write $\lambda x \in D_e .\ \phi(x)$ for the function from the domain D_e of individuals to the domain of values of sentences (usually truth values).

and neo-Davidsonian semantic theories provide absolute truth conditions. These days, semantics in either tradition is absolute.[23]

Perhaps the most obvious of the distinctive features of the model-theoretic approaches is that it uses the typed λ-calculus to assign semantic values to all constituents. Compare the neo-Davidsonian (5a) (e.g. Larson and Segal 1995) with its model-theoretic variant (5b):

(5) a. $\mathtt{Val}(x, \text{smokes}) \iff x \text{ smokes}$
b. $[\![\text{smokes}]\!] = \lambda x \in D_e.\, x \text{ smokes}$

Though section (5b) provides an object to be the semantic value where section (5a) states that a relation holds, as far as truth conditions goes, these do pretty much the same thing, in the same way. That one posits a function is a difference in the apparatus the theories use (and so an ontological commitment in the long run), but not in the explanations they provide for the basic semantic property of this word. The speaker is not being attributed understanding of the theory of functions and relations, either for semantic values or for $\mathtt{Val}$ relations. These are used to attribute to the speaker knowledge that 'smokes' applies to things that smoke.

Thus, the absolute nature of semantics, and the lack of appeal to a space of models that constitutes a consequence relation, is not specific to neo-Davidsonian semantic theories. It holds just as much for what is called the model-theoretic approach. That being said, there are some genuinely significant differences between model-theoretic and neo-Davidsonian theories, and the use of λs is important for them. It is not my goal here to explore them deeply, but let me mention one, just to make clear that there are some, and they are important. The two sorts of theories disagree on the nature of semantic composition, and the use of λs, really the use of a wide-ranging apparatus of functions, allows model-theoretic semantics to see semantic composition in terms of functions and arguments. Neo-Davidsonian theories see composition much differently—in many cases, in terms of conjunction (e.g. Pietroski 2005). This does lead to some real empirical differences between model-theoretic and neo-Davidsonian analyses of various phenomena and, especially, has potentially far-reaching implications for issues of logical form in natural language. But this can happen while the two theories agree on the fundamental idea of specifying truth conditions as a way to capture speakers' linguistic competence.

[23] Montague's original work (Montague 1973) and subsequent presentations like Dowty et al. (1981) did officially rely on a notion of truth in a model. But even so, they usually drop reference to models when the linguistic analysis starts to get interesting.

I have dwelt at length on an internal dispute among semanticists about what sort of semantic theory works best, though in the end I have suggested that the dispute has been resolved, and everyone has opted for absolute semantics, in either model-theoretic or neo-Davidsonian guises. Current model-theoretic semantics does so using type theory, and so, where we have grown accustomed to saying 'model-theoretic semantics' it might be better to say 'type-theoretic semantics'. Names aside, the apparatus neither of Tarski-style truth theories nor of type theory itself provides any sort of logical consequence relation, and in their uses in semantics they cannot. One way or another, semantics must be absolute, and so not relative to a model.

I suggest this makes vivid how the enterprise of truth-conditional semantics is distinct from that of studying logical consequence. Relative truth conditions, in Lepore's terminology, are just what anyone studying logical consequence should want. If we want to understand the logical properties of a sentence of a language, we look at how the values of the sentence can vary across models. This is just what the logics in formal languages thesis builds on. But semantics of natural language—the study of speakers' semantic competence—cannot look at that, and still capture what speakers understand. To capture what the speakers understand, semantics must be absolute, and so blind to what happens to a sentence across any non-trivial range of models. Thus, we cannot find the basic resources for studying logical consequence in natural language semantics, even in its truth-conditional form. We cannot take the step from the logics in formal languages thesis to the logic in natural language thesis. To give the argument of this section a name, let us call it the *argument from absolute semantics* against the logic in natural language thesis. The argument from absolute semantics, I submit, gives us good reason to reject the logic in natural language thesis.

3.3.2 Entailments and consequences

The argument from absolute semantics shows that the semantics of natural language and the model theory of consequence relations are different things. But it might be objected that there are other ways to find consequence relations in natural language. Moreover, there is an obvious place to look to find such relations. Whether or not semantics is absolute, it must endow natural language with some implication properties. Indeed, implications are among the main data that empirical semantic theory builds on, and I doubt we would find any viable approach to semantics that failed to account for them. Even if the simple idea of capturing speakers' understanding of truth conditions does not hand us a consequence relation, it might be that the implications that are built into natural language do. If so, we would have an alternative route to defending the logic in natural language thesis. I shall argue in this section there is no such route. On the restrictive view of logical consequence we are now assuming, what we find in natural language are entailments, but not logical consequences.

As we saw above, sentences of natural language do present us with obvious implications, in that in certain cases competent speakers consistently judge that the truth of one sentence follows from the truth of another. Analytic entailments like the 'cousin' inference of section 3.2.3 are good examples. And truth-conditional semantics is ready-made to capture some of these implications. It endows natural languages with *entailment* properties. Each sentence is associated with a set of truth conditions, and so truth-conditional containment properties are fixed. A sentence S entails a sentence T if the truth conditions of S are a subset of the truth conditions of T.[24]

We have already seen that in general, such entailment relations are not narrow logical consequence. Natural language provides entailments of other sorts. So, as I noted, advocates of the restrictive view of consequence will not find their preferred narrow notion in natural language entailments.[25] But if the only issue was that we find something more like the patterns of strict entailment in natural language, rather than narrow logical consequence, it would not impress anyone with even modestly permissive views. After all, the logic of such entailments, and many related notions, have been studied extensively.[26] At core, what would be needed is a somewhat different approach to the *necessity* constraint than classical logic uses, which would make possibilities more like metaphysically possible worlds than like classical models. Thus, it might be proposed, what we find in natural language may not be the most narrow of consequence relations, but it is close enough to what a restrictive view of consequence is after to offer an interesting version of the logic in natural language thesis.

I shall argue here that this sort of response is not adequate. The reason is that natural language puts pressure on the *formality* constraint on logical consequence. Absolute semantic theory itself has no room for any such notion of formality. It specifies absolute meanings for each expression of a language, and sees no distinction between the ones whose meanings determine consequence relations

[24] We might worry about whether extensional and intensional theories provide exactly the same entailment relations, and whether extensional theories really provide for strict entailment in the sense of Lewis and Langford (1932). But I shall not dwell on that here, as both provide reasonable notions of entailment, and both turn out not to be logical consequence.

[25] As was observed also by Cresswell (1978).

[26] For some indications of the scope of this research, see among many sources Anderson and Belnap (1975), Anderson et al. (1992), Priest (2008), and Restall (2000).

and those which do not. And, I shall show, the facts about implication in natural language reflect this. Natural language is filled with rather idiosyncratic *lexical entailments*, driven by the meanings of a huge variety of lexical items. These depart too far from formality to satisfy restrictive views of logical consequence. In section 3.4, I shall argue they depart too far for most permissive views as well.

We have already noted that natural language provides us with non-logical implications, like analytic entailments. These, as the common wisdom goes, are determined by the meanings of non-logical terms like 'bachelor', not by the meanings of the logical constants. But natural language is permeated by entailments which strike us as evidently non-logical (by restrictive lights). Here is another case, much discussed by semanticists (see Anderson 1971; Fillmore 1968; Levin and Rappaport Hovav 1995):

(6) a. We loaded the truck with hay.
ENTAILS
We loaded hay on the truck.

b. We loaded hay on the truck.
DOES NOT ENTAIL
We loaded the truck with hay.

This is a report of semantic fact, revealed by judgments of speakers, both about truth values for the sentences, and about entailments themselves. It indicates something about the meaning of the word 'load' and how it combines with its arguments. More or less, the 'with' variant means we loaded the truck *completely* with hay, while the 'on' variant does not.[27]

To take one more much-discussed example, we see (Hale and Keyser 1987; Higginbotham 1989a):

(7) John cut the bread.
ENTAILS
The bread was cut with an instrument.

The meaning of 'cut', as opposed to 'tear', for example, requires an instrument, as Hale and Keyser famously noted.[28]

Entailments like these are often called *lexical entailments*, as they are determined by the meanings of specific lexical items. As a technical matter, it is not easy to

[27] There is some debate about whether the connection of the 'with' variant to being completely loaded is entailment or implicature, but I believe the judgments on the (a/b) contrast above are stable enough to indicate an entailment.

[28] Hale and Keyser (1987) gloss the meaning of 'cut' as 'a linear separation of the material integrity of something by an agent using an instrument'.

decide on the exact source of these entailments: that is, whether they simply come from the atomic meanings of the verbs, the compositional semantics of verbs and arguments, or some form of predicate decomposition, presumably at work inside the lexicon. But this is an internal issue for natural language semantics (a very interesting one!). What is important for us is that any viable semantics must account for these sorts of entailment patterns. Though we have only looked at a couple of examples, I believe they make plausible the empirical claim that natural language is rife with such lexical entailments. Any semantic theory, of either the model-theoretic or neo-Davidsonian stripe, must capture them.

It is clear that these lexical entailments will not count as narrow logical consequences. The reason is not that we are looking at a wide space of classical models, rather than a smaller space like that of metaphysically possible worlds, or even some smaller space. Rather, the reason is that these entailments are fixed by aspects of the meanings of words like 'load' and 'cut'. If we start with any consequence relation which does not treat these as logical constants, we will not get the right entailments without violating the formality constraint.

From the restrictive point of view that we are adopting here, this shows that we will not find a logical consequence relation in the lexical entailments that natural language provides. But, anticipating discussion of more permissive approaches to come, it is worth asking how far we would have to go to accommodate such lexical entailments in a consequence relation. Assuming we can take care of necessity, could we satisfy formality with an appropriately permissive notion of consequence which captures these cases? I shall offer two reasons why we cannot. One is that we will find lexical entailments permeate language too far to generate what even modestly restrictive views would recognize as a logic. The other is that we will run afoul of another important condition on logical constants.[29]

Let us take up the first of these points first. Even if we were willing to consider taking, for example, 'load' and 'cut' to be logical constants, we would not have the 'logic of lexical entailments'. We obviously would not, as many more words than these trigger lexical entailments. Practically every one does! We cannot just take these two. We would have to make nearly every word a logical constant. This would render the formality constraint virtually trivial. The result would not be accepted as logic by restrictive views, but in trivializing the formality constraint, I doubt it would be accepted by most permissive views either.

There are a couple of reasons why the situation might not be so dire, but I do not think they are enough to recover a viable consequence relation, according to

[29] Related points are made by Lycan (1989), though he emphasizes the difference in degree between logical consequences and lexical entailments more than I do.

restrictive or even moderately permissive views. It might not be that we have to take every individual word as a constant, as the patterns we are discussing occur across families of expressions. For instance, the pattern we see in section (6) is one that reappears across a family of verbs, including 'spray', 'brush', 'hang', etc.[30] But as far as we know, we will still wind up with a very large and very quirky collection of families. Looking at the list of verb classes in Levin (1993), for instance, we find differences between classes of verbs of sound emission and light emission, and differences between classes of verbs of bodily processes and verbs of gestures involving body parts. Taking each such class to indicate a logical constant would still give us a huge and quirky list of constants, which would still undercut the formality constraint. Even if it does not completely trivialize formality, it will still undercut formality well beyond what restrictive views can accept. I believe it would undercut formality enough to be unacceptable to modestly permissive views as well.

The only hope for narrowing down our domain of logical constants would be that underlying the many lexical entailments we find might be a small group of factors that generate them. In fact, it is a theoretically contentious issue just what generates the sorts of entailment we have been looking at, and what might group various words together into natural classes. But there is an optimistic idea that what appear to be idiosyncratic lexical entailments—entailments determined by the idiosyncratic properties of words' meanings—are really entailments determined by some hidden structure of classes of lexical items. But I suggest that, as far as we know, we will still wind up with a group of 'logical constants' that are too quirky and too heterogeneous to satisfy formality in any substantial way. We will identify as formal a range of selected features driven by the quirky structure of natural language, not by the forms of valid arguments. For instance, it is a persistent thought that behind the entailments in section (6) is something like a requirement for the 'with' variant that a container or surface be completely filled or covered.[31] (This constraint might have something to do with kinds of arguments of the verb, or just be coded up into the right meanings.) So our logic will have to be in part a logic of filling containers or covering surfaces. Restrictive views will reject this. And this is just the start. It will also have to be a logic of instruments for cutting to account for the entailments in section (7). The result would be a formal listing of a quirky range of entailment properties, not the kind of logic the

[30] See Levin (1993) for many more examples. See Levin and Rappaport Hovav (2005) for a survey of some theories that seek to explain the sorts of entailment I am discussing here.

[31] See Dowty (1991) and Pinker (1989) for discussion and numerous references.

restrictive view is after. Again, it will not be the kind of logic modestly permissive views are after either, as I shall discuss more in section 3.4.

This reminds us that even though both logic and semantics are concerned with implications in general, they are concerned with different ones. Semantics is absolute, and interested in the specific meanings of all the terms in a given language. It is thus interested in the entailments that arise from those meanings. These are typically not logical consequence relations. The idiosyncrasy of natural languages makes this striking, as the 'load' case shows. We find in natural language all kinds of unexpected and odd entailments, driven by the quirky meanings of various words. When we study these, we are not doing logic; we are certainly not looking at restrictive logical consequence.

There is a second reason for restrictive views to reject the idea of modifying logic to take in lexical entailments. By restrictive lights, the expressions we would have to count as 'logical constants' will not meet the standards for being logical constants. This will apply equally to the words like 'load', or to any possible underlying features that generate families of lexical entailments.

It is a common idea that to satisfy formality, the logical constants have to meet certain standards. The main standard is being 'topic-neutral', or not specifically about the particular things in the world, but only its general or 'formal' structure. This is indeed a further constraint than we have so far considered, but it is a natural way to fill out the idea of formality. In technical terms, it is often spelled out as requiring that logical constants be invariant under permutations of the elements of the universe, or under the right isomorphisms.[32]

If we impose a constraint like this on logical constants, it is very unlikely that anything responsible for the kinds of lexical entailment we have been considering would meet it. Words like 'load' and 'cut' do not. Might some underlying features that might account for lexical entailments meet it? It is perhaps possible, but unlikely. One reason for skepticism is that it is well known that practically no predicates and relations satisfy permutation invariance. For a domain M, the only predicates satisfying it are $\emptyset$ and M, and the only binary relations satisfying it are $\emptyset$, $M \times M$, identity, and non-identity (Peters and Westerståhl 2006; Westerståhl 1985). If we require normal permutation invariance as a constraint on logical constants, we will not find enough in the domain of predicates to capture the kinds of lexical entailment we are here considering. It might be that a more permissive view could appeal to a different range of isomorphisms to recover a notion of logical constant applying to some of the terms we have considered here, as I shall discuss more in

[32] This idea is discussed by Mautner (1946) and Tarski (1986). See also van Benthem (1986) and Sher (1991).

section 3.4. But nonetheless, this fact about permutation invariance shows how far from a restrictive view we would have to go to get a logical consequence relation from the lexical entailments of a natural language.[33]

I have argued in this section that we will not find logical consequence relations in the lexical entailments of natural language. As we will need to refer back to this argument, let us call it the *argument from lexical entailments*. This argument supplements the argument from absolute semantics. We will not find logical consequence relations in the basic apparatus of truth-conditional semantics, and we will not find it in the entailments present in natural language either. We thus have two reasons to reject the logic in natural language thesis.

3.3.3 Logical constants in natural language

There is one more route I shall consider to saving the logic in natural language thesis. It might be objected that in spite of the points I made in the last two sections, there are some genuinely logical expressions in natural language, and their properties might somehow provide a consequence relation for natural language even if truth-conditional semantics and lexical entailments do not. In this section, I shall argue this is not so either. First, following up the discussion of model-theoretic semantics in section 3.3.1, I shall show that the (model-theoretic) semantics of logical terms in natural language does not provide a genuine consequence relation. Second, I shall point out that natural language does not really come pre-equipped with a distinguished class of logical constants. If we wish to find logical constants in natural language, we have to identify them in ways that go beyond the semantics of natural language itself. This will point the way towards our discussion of how one can move from natural language to logic proper in section 3.5.

First, let us look at how the semantics of uncontroversially logical terms will be handled in an absolute semantics. As an example, I shall focus on the quantifiers, or more property the determiners: words like 'every', 'most', 'some', etc. Some of these are clearly logical constants by even the most restrictive lights, and aside from a few contentious cases, they satisfy the permutation invariance constraint we discussed in section 3.3.2.[34] We will see that these logical expressions of natural language get an entirely standard model-theoretic semantic treatment, putting aside some structural differences between the most common logical formalisms

[33] There has been some interesting recent work on conditions related to permutation invariance (e.g. Bonnay 2008; Feferman 1999), but if anything, this work suggests even more restrictive criteria for logical constants. For a survey of related ideas about logical constants, see MacFarlane (2009).

[34] See Sher (1991) for a defense of the idea that generalized quantifiers are genuinely logical constants by restrictive measures.

and natural language; but they do so in a way that preserves the absolute nature of the semantics of natural language.

Methods of logic, particularly of model theory, have proved extremely fruitful in the study of the determiners, as classic work of Barwise and Cooper (1981), Higginbotham and May (1981), and Keenan and Stavi (1986) has shown. Indeed, this led Higginbotham (1988) to declare that in natural language, model theory is the lexicography of the logical constants.

But model theory does this in a specific way. Consider, for example, a common way of representing the meaning of a determiner like 'most':

(8) $[\![\text{most}]\!](A, B) \iff |A \cap B| > |A \setminus B|$

The meanings of determiners, on this view, are relations between sets expressing cardinality properties. This definition is drawn from the theory of generalized quantifiers. This is a rich mathematical field, and it has led to an impressive number of non-trivial empirical predictions and generalizations in semantics.[35]

The application of generalized quantifier theory to the semantics of determiners is one of the most well-explored applications of model theory to natural language, and a key example of the success of model-theoretic semantics. But it remains an example of *absolute* semantics, as the arguments of section 3.3.1 show it must. There is no tacit quantification over a domain of models in the semantics of determiners. The sets A and B in the above definition (8) must be drawn from whatever fixed domain of individuals is involved in the rest of the semantics. There is no tacit 'for any model $\mathfrak{M}$'. In this way, we can do what is often labeled 'model theory', while still doing absolute semantics, and still not generate a consequence relation.

Actually, this point is already enshrined in the theory of generalized quantifiers, as the distinction between *local* and *global* generalized quantifiers.

(9) a. Local: $[\![\text{most}]\!]_M = \{\langle A, B\rangle \subseteq M^2 : |A \cap B| > |A \setminus B|\}$
b. Global: function from M to $[\![\text{most}]\!]_M$

The direct application of generalized quantifiers to semantic theory uses *local* generalized quantifiers, as an absolute semantics should.

On the other hand, if we are to embed generalized quantifier theory in our theory of logical consequence, it is global quantifiers that we need. As has been much discussed, to capture logical consequence relations with quantifiers, the

[35] The mathematics of generalized quantifiers was first studied by Lindström (1966) and Mostowski (1957). See Barwise and Feferman (1985) and Westerståhl (1989) for surveys. Among the most striking empirical predictions is the *conservativity* constraint articulated by Barwise and Cooper (1981) and Higginbotham and May (1981). This may well be a semantic universal.

domain must be allowed to vary. That is just what global generalized quantifiers do. (We will return to this issue in section 3.5.) For the study of semantics of natural language—absolute semantics—local generalized quantifiers are the basic notion; while for the study of logical consequence, global ones are.

Attending to the local versus global distinction, we can reconcile two facts that might have seemed in tension. First, familiar determiners in natural language have more or less the semantics that logical theory says they should. Though there is some interesting linguistics subtlety (a little of which I shall mention below), 'every' is pretty much a universal quantifier as we come to learn about in logic. And yet, the semantics of this expression is absolute, and does not make reference to the range of models essential to the logic of quantification. But the reason is simply that semantics of natural language only uses local properties of quantifiers in spelling out the semantics of determiners. These are readily available for absolute semantics.

I thus conclude that the presence of recognizably logical expressions in natural language does not help support the logic in natural language thesis. We can find terms which we recognize as logical, and give them essentially the semantics that logic should make us expect, while keeping semantics entirely absolute, and not involving any true consequence relations.

Now, this does not mean we can never look at the global notion of quantifier in thinking about natural language. The basic idea for giving absolute truth conditions is the local one, and in fact, sometimes we can get interesting further results out of local definitions.[36] But on occasion, we learn something by abstracting away from absolute truth conditions, by looking at global generalized quantifiers. A example is the idea that natural language determiners express restricted quantification. This is captured in two ways:

(10) a. CONSERV (local): For every $A, B \subseteq M$, $Q_M(A, B) \Longleftrightarrow Q_M(A, B \cap A)$
b. UNIV (global): For each M and $A, B \subseteq M$, $Q_M(A, B) \Longleftrightarrow Q_A(A, A \cap B)$

UNIV is a generally stronger principle (Westerståhl, 1985), and captures an interesting way in which natural language determiners really quantify only over a restricted domain. If we refused to look at global properties of generalized quantifiers, we would not see it.

[36] For instance, counting and classifying available denotations over a fixed domain can be interesting. One example is the 'finite effability theorem' of Keenan and Stavi (1986), which shows that over a fixed finite universe, every conservative type $\langle 1, 1 \rangle$ generalized quantifier is the denotation of some possibly complex English determiner expression.

In looking at this sort of global property, we are not simply spelling out the semantics of a language. Rather, we are abstracting away from the semantics proper—the specification of contributions to truth conditions—to look at a more abstract property of an expression. It turns out, in this case, that abstracting away from the universe of discourse is the right thing to do. Particularly when asking about logical or more generally mathematical properties of expressions, this sort of abstraction can be of great interest. And, we can prove that typical natural language determiners satisfy UNIV, invoking a little bit of mathematics, even if it goes beyond the semantics of any language per se.

This sort of possibility shows how we might take the step from semantics proper to logic, as I shall discuss more in section 3.5. But for the moment, I shall pause to note one feature of how the application of the model theory of generalized quantifiers to natural language works. Rather than indicating anything like a genuine consequence relation in natural language, it illustrates how the application of model-theoretic techniques to natural language semantics really turns out to be an instance of the general application of mathematical techniques. In applying local generalized quantifiers, what we are really applying is a little bit of set theory (deeply embedded in standard model theory too), to represented cardinality-comparing operations expressed in language, and studying their properties. We can do so over an arbitrary or a fixed domain, depending on the question under investigation. But especially when we are stating absolute truth conditions, it is the mathematics of cardinality comparison over a fixed domain that we really invoke. Though this can reasonably be counted as logic, as it is the subject-matter of local generalized quantifiers, it is not *logic* in the core sense of logical consequence relations. When we use logic in the study of natural language semantics, we are typically using logic in the broad sense in which logic can be found in many domains, not the narrow one of the study of logical consequence relations.[37]

The case of determiners shows how we can grant that there are logical expressions in natural language, which get more or less standard logical analyses, and still reject the logic in natural language thesis. But one might still wonder if natural language supports the logic in natural language thesis in another way, by presenting a distinguished class of logical terms. Even if their semantics is absolute, this might be a partial vindication of the thesis.

We have already seen some reason to be skeptical of whether language does the job of identifying the logical constants for us. We have seen that the mere fact

[37] The application of mathematical techniques to natural language semantics is not specific to techniques found in model theory, nor is it specific to the logical expressions in natural language. I discuss more ways that mathematics applies to absolute semantics in work in progress.

that some expressions are well analyzed by techniques like those of generalized quantifier theory only shows that they are amenable to a kind of mathematical analysis. In fact, the class of expressions that are at least partially analyzable in mathematical terms is very wide, and contains clearly non-logical expressions, as well as plausibly logical ones. To cite one example, a little bit of the mathematical structure of orderings (a tiny bit of topology) has proved useful in analyzing certain adjectives (see Kennedy and McNally 2005). As we discussed in section 3.3.2, there are good reasons to doubt expressions like these are logical, and they are certainly not by restrictive lights. So, natural language will not hand us a category of logical constants identified by having a certain sort of mathematically specifiable semantics.

Is there anything else about a language—anything about its grammar, semantics, etc.—that would distinguish the logical constants from other expressions? No. Generally, expressions we are inclined to accept as logical constants by restrictive lights, like quantifiers, conjunction, negation, etc., group with a much larger family of elements of a natural language. They are what are known as 'functional' categories, which include also complementizers ('that', 'which'), tense, mood, and other inflectional elements, morphological elements like degree terms ('-er'), etc. By strict restrictive standards, it is doubtful that all of these will be counted as logical constants.[38] Some of these elements, like tense, have been points of dispute among restrictive views over what counts as logical. But most restrictive views will refuse to count the complementizer 'that' or the comparative morpheme '-er' as logical constants.

This claim relies on the highly restrictive tendencies of traditional restrictive views, which simply find no place in their logics for such terms. Whether or not they fail to meet some criterion, like permutation invariance, is a rather more delicate matter, and I will not try to decide it here. One of the typical features of functional categories is that their semantics is not the familiar semantics of predicates and terms; rather, their semantics tends to be 'structural', involving times or worlds for inflectional elements, degrees for comparatives, etc.[39] We cannot easily apply permutation invariance to these, and instead, we will have to ask if they satisfy the right property of invariance under isomorphism. Whether they do will depend on just what isomorphisms are at issue.[40] This will become more of an issue when we turn to permissive views in section 3.4. But as with

[38] As we will see in section 3.5.3, even the elements we are inclined to accept as logical do not behave exactly as the expressions in formal languages do.

[39] Many linguists will gloss this as their not having a 'theta-involving semantics', meaning it does not involve regular predicates and their linguistically provided arguments.

[40] Hence, we do find, for instance, logics of comparison, as in Casari (1987).

other abstract properties of expressions, it is not something that is specifically marked by the grammar. Natural languages do not tag certain expressions as invariant under isomorphisms—absolute semantics cannot do this! Rather, as we saw with UNIV, they provide absolute semantics for expressions, and we can work out mathematically what would happen if we take a global perspective and look at invariance properties. Hence, if we are to conclude that some or all functional categories meet some invariance conditions, we must go beyond what the grammars of our languages tell us.

Linguistically, there are lots of distinguishing marks of the class of functional categories. Functional categories are closed classes. Unlike nouns, verbs, and adjectives, it is impossible to add to these classes (except at the glacial pace of language change). You cannot simply add a new determiner or a new tense to your language the way you can easily add a new verb or noun. Functional categories also play special grammatical roles. In many languages some determiner, even a semantically minimal one, is needed to make a noun phrase an argument of a predicate. Generally, functional categories act as 'grammatical glue' that binds sentences together. This sort of grammatical role is clearly on display with complementizers. In a more theoretical vein, unlike nouns, verbs, and adjectives, functional categories do not assign argument structure.[41]

But these sorts of features group all the functional categories together. They do not distinguish the logical sub-class from the wider one, as they do not really pertain to logicality at all. Indeed, I do not know of any linguistic property that does distinguish logical elements from other functional ones (assuming the restrictive view that they are not all logical). So, I conclude, natural language does not sort expressions into logical and non-logical. It contains expressions that will count as logical—just how many depends on your standards for logical terms—but it does not itself distinguish them. That is something we can do when we look at a language from a more abstract perspective, as we will discuss more in section 3.5.[42]

We thus see one more way that the logic in natural language thesis fails. Let us call this one the *argument from logical constants*. It joins the argument from absolute semantics and the argument from lexical entailments in showing what is wrong with the logic in natural language thesis. In particular, like the argument from lexical entailments, it shows that a route to finding significant logical structure in natural language is not open.

[41] Many recent syntax texts will review the notion of functional category. See also Abney (1987), Fukui (1995), Grimshaw (2005), and Speas (1990).

[42] This conclusion is not so far off from the claim of Evans (1976) that what he calls 'semantic structure' is distinct from 'logical form', and does not determine logical consequence.

This concludes my main discussion of the logic in natural language thesis. But there is one loose end to tie up, and one further issue to address. Both return to points we have seen in passing in this section. First, the conclusions I have reached have been under the assumption of a restrictive view of logical consequence, and we need to see how they fare with a permissive view. We will do this in the next section, 3.4. Second, we have already seen ways that we can go beyond semantics proper to explore logical properties of natural language expressions. This reminds us that we can take the step from natural language to logic, even if the logic in natural language thesis is false. We will explore how this step may be taken in section 3.5.

3.4 The Logic in Natural Language Thesis from the Permissive Point of View

Above I offered three arguments against the logic in natural language thesis, but generally assumed a restrictive view of logical consequence. In this section, I shall examine how those arguments fare if we adopt a permissive view, as I did at a few points above as well. I shall argue that we still have some good reasons to reject the logic in natural language thesis. Even so, permissive views will have more opportunities to find logical properties in natural language. I shall suggest that this will not really support the logic in natural language thesis, without also weakening it to a point where it threatens to become uninteresting. But at the same time, as we will discuss more in section 3.5, it will show how permissive views might see logic and natural language as more closely linked than restrictive views do.

The three arguments I offered against the logic in natural language thesis in section 3.3 were the argument from absolute semantics, the argument from lexical entailments, and the argument from logical constants. The argument from absolute semantics was the main argument, which showed that a natural language does not have a consequence relation in virtue of having a semantics. The other two arguments served a supporting role, by showing that we will not find restrictive logical consequence in natural language by other means.

The argument from absolute semantics in fact made no use of the premise of restrictive logical consequence. It made no assumptions about logical consequence beyond that it will require a non-trivial space of models (with more than one element). That is not something even the permissive views in question will deny. So, I conclude, the argument from absolute semantics is sustained on a permissive view. As this was my main reason for rejecting the logic in natural language thesis, I believe that thesis looks doubtful even on a permissive view of logical consequence.

We have already seen in sections 3.3.2 and 3.3.3 that matters are somewhat more complicated with the other two arguments. Let us consider the argument from lexical entailments first. The idea was that one might find a logical consequence relation in the entailments presented in natural language, and use that as a route to the logic in natural language thesis. We saw in section 3.3.2 that those entailments are lexical entailments, and fail to be logical consequences according to restrictive requirements. But we also saw that there might be ways to count some of them as logical by more permissive lights. I argued in section 3.3.2 that we could not do so without undercutting the formality condition. I offered two reasons for this. One reason was that we would have to accept logical constants which would not meet the permutation invariance condition. The other was that the needed logical constants would permeate language so extensively that we could only preserve the logic in natural language thesis at the cost of weakening formality too much. In the worst case, preserving the logic in natural language thesis threatened to trivialize formality.

Of course, extremely permissive views can reject both these claims. The question is what the results of doing so would be. Let us consider permutation invariance first. Though this could be rejected out of hand, it might make the formality constraint unacceptably weak if there is are no restrictions on what counts as a logical constant. So, the more likely avenue for a permissive approach, as we saw above, is to find some alternative constraints. The most obvious way, as we also saw, is to look for some appropriate structure for which a candidate expression will be invariant under isomorphism. Returning to the 'load' example (6), for instance, we might wonder if we can find an appropriate structure for filling containers or covering surfaces, which would render it invariant. A hint of how that might be done is its similarity to ideas from mereology. So, though I am not sure just how it would go, there is no reason why a permissive view might not be able to find a logic of filling and covering, and capture the 'load' entailments through it. More generally, case by case, a permissive view might be able to see each lexical entailment as derived from some sort of logical constant.[43]

For each individual case, this will be plausible enough by permissive lights. After all, permissive views are interested in finding many different candidate logics, and will not find it odd to look for a logic of, for example, filling or covering. There is some risk of losing any sense of what makes a logical constant a logical constant, but presumably good permissive views will find some way to keep this notion reasonably constrained. The more pressing worry is about how pervasive

[43] As I mentioned in n. 33, there has been some exploration of alternatives to isomorphism invariance, but it has tended to offer more, not fewer, restrictive conditions.

the lexical entailments of natural language are, and what accounting for all of them as logical would do to formality.

At the extreme, as we saw in section 3.3.2, to sustain the logic in natural language thesis we might need to count nearly every word in a language as a logical constant, as nearly every expression will generate some lexical entailments. Even if we are willing to grant that many of them can be viewed as permissive logical constants, making nearly all of them constants will radically weaken formality. If nearly every expression is a logical constant, then there is little left of the idea that inferences are valid based on distinguished formal structure. That might rescue the logic in natural language thesis, but at the cost of trivializing it.

The more hopeful option for the permissive defense of the logic in natural language thesis is that there might be a smaller set of features that generates the lexical entailments. As I discussed in section 3.3.2, it is not clear whether this is so, and if it is, it is not clear just how large or how heterogeneous the features involved will be. Again, a permissive view will probably be willing to grant that each such feature individually can be a logical constant, and have a logic for the entailments associated with it. The question, again, is what we get if we put them all together. Here, if indeed the set is small enough and homogeneous enough, there could possibly be a defensible permissive view. But I am doubtful that will work. If the set is too large, it will trivialize the notion of formality again. If it is too heterogeneous, it will undermine the logic in natural language thesis in other ways. If our group of logical constants is too heterogeneous, it is not clear if we will find any single consequence relation which makes sense of all of them in one logic, as the logic in natural language thesis demands. Even if we can, it is not clear if the result will wind up being any better than a brute force way to capture the logic of natural language by enumerating the lexical entailments. Technically, that would not falsify the logic in natural language thesis, but it would undercut its interest. The thesis would be interesting if we had some, even permissive, idea of what counts as logic, and thought we could find this logic in natural language. A brute force coding of lexical entailments would not provide that. Of course, finding any tractable way to enumerate all the lexical entailments in natural language would be of huge interest! The problem is that it may not produce a natural or interesting consequence relation, which is what the logic in natural language thesis looks for.

I thus grant that the argument from lexical entailments might not work on a permissive view (or at least, an extreme permissive view). Even so, I still register a great deal of skepticism over whether the argument can be bypassed in a way that leaves the logic in natural language thesis substantial and interesting. As the argument from absolute semantics still holds, I think we should be dubious of the logic in natural language thesis by permissive lights, and I count my skepticism about ways permissive views might bypass the argument from lexical entailments as more reason to doubt the logic in natural language thesis. Yet all the same, we should grant that by permissive lights, the relation of lexical entailment to logic is not so clear-cut.

Finally, we need to reconsider the argument from logical constants from section 3.3.3. This argument shows that natural language does not present us with a group of distinguished logical constants, thus blocking one other potential route to the logic in natural language thesis. Part of this argument applies for permissive views. As the semantics of all terms, logical ones included, must be absolute, we will not find any terms of a natural language distinguished within the language by having any particular global or non-absolute logical properties. The semantics and the rest of the grammar of a natural language do not do that. This part of the argument does not rely on restrictive assumptions, and works equally well for permissive views.

But there is another part of the argument from logical constants that is likely to fail by permissive lights. I argued that the only grammatically distinguished class of terms that contains the uncontroversially logical ones is the class of functional expressions. By restrictive standards, this proved too large to count as the class of logical constants. But it remains open to permissive views simply to accept that all functional expressions are logical constants. As we discussed in section 3.3.3, it will be easier to sustain this claim than the corresponding one for lexical entailments, as functional categories do have features that may make them amenable to logical analysis. Permissive views hunting for new interesting logics might well find the functional categories of natural language fertile ground.

Some of the worries I just raised surrounding the argument from lexical entailments will apply here as well. It is not clear if the result of counting all functional expressions as logical constants will be a single coherent consequence relation, or a brute force coding of multiple logics corresponding to multiple classes of functional elements. Hence, as with lexical entailments, I remain skeptical about whether permissive views of logical constants will offer a viable route to the logic in natural language thesis, even if they are able to undermine some parts of the argument from logical constants of section 3.3.3. All the same, I grant, this is one of the more promising possibilities for permissive views of logical consequence to consider.

I conclude that for permissive views, the logic in natural language thesis still looks doubtful. The argument from absolute semantics still holds, and there is good reason to be skeptical of the possibility of avoiding the arguments from lexical entailments and logical constants. Both these latter two arguments still have some force, even if they are weakened on permissive views. Yet the relative weakness of the arguments from lexical entailments and logical constants does indicate possibilities for permissive views to explore, which might lead to something related to the logic in natural language thesis, I suspect in a much-weakened form.

It is clear that permissive views of logic can readily find properties of natural language that lead to interesting logics (as they can in many other domains as well). This does not by itself sustain the logic in natural language thesis, but it reminds us that both permissive and restrictive views can find logically relevant material in natural language. Not surprisingly, permissive views do so more easily, but both can. How this can happen, without the logic in natural language thesis, will be the topic of the next section.

3.5 From Natural Language to Logic

Let us suppose the logic in natural language thesis is false. This claim may be more secure by restrictive than permissive lights, but I have given some reasons to think so for both views. Regardless, we have also seen throughout the discussion above that we can often find things of interest to logic in natural language. We have seen that we can find expressions in natural language which turn out to have interesting logical properties, or even turn out to be logical constants, and we can find entailments which might prove of interest to permissive views of logic. This should not be a great surprise, as it has been clear all along that natural language does present us with a range of implication phenomena, and a range of expressions which might be of interest to logic. We can get from language to logic somehow, and this is a fact that logicians of both permissive and restrictive varieties have often exploited.

The question I shall explore in this section is how the step from natural language to logic can be taken. I shall argue that the space between natural language and logic can be bridged in a fairly familiar way, by a process of *identification* of logical constants and *abstraction*. But I shall highlight that we also need a third component of *idealization*, and argue that this component is more substantial than many logicians might have assumed. Together, these processes allow us to move from the absolute semantics of natural language proper to a logical consequence relation. Isolating them will also help us to see how much is required for permissive or restrictive views to make the jump from natural language to logic.

A metaphor might help illustrate what the processes do. The richness of natural language shows us logical consequence 'in the wild', in our discursive and inferential abilities and practices. Sometimes we want to take it back to the lab, magnify some part of it, dissect it, modify it, and purify it. What we get is a cousin of the wild type, just as implications can be distant cousins of narrow logical consequence. How distant a cousin, of course, depends on what your view of consequence was to begin with. Frege did something like this with quantifiers. The problem of polyadic quantification theory that Frege solved came up in part because it came up in natural language (and in mathematics as well). But neither Frege's solution to the problem, nor many model-theoretic variants, amount to doing natural language semantics. Rather, they require identification of the relevant logical features of polyadic quantification, abstraction from the meanings of non-logical expressions, and idealization from features of natural language grammar not relevant to the core logical issues. That is what we generally do when we move from natural language to logic.

I shall address each of these processes in turn. Actually, I shall start out of order by discussing abstraction, as it is the main process by which we move from the absolute semantics of natural language to model-theoretic consequence relations. I shall then discuss the role played by identification, especially when it comes to logical constants, even though identification typically has to happen before abstraction. Finally, I shall discuss the important role for idealization in getting us the kinds of logic that have proved so useful.

3.5.1 Abstraction from absolute semantics

I have argued that the semantics of natural language is absolute, and so does not provide a consequence relation. But there is a way to move from absolute semantics to the kinds of model we need to build model-theoretic consequence relations. Once we have them, we will be well positioned to start working out logics, by looking at (the right sort of) preservation of truth or designated value across the range of models we produced. Presumably the resulting consequence relations will be genuinely logical, and also reflect some aspects of the absolute semantics we started with. That would be a major step towards building a laboratory-refined logic out of a wild-type natural language.

Actually, the way to do this is well known from discussions of logical consequence and model theory. All we need to do is abstract away from the absolute features of our semantics. Absolute semantics will typically provide extensions for predicates, and referents for terms, etc. It will provide more than that, as it will provide semantic values for many more sorts of natural language expressions. It might also provide structures for tense, mood, etc., which might enrich the structures we are working with. But for now, let us simply focus on familiar extensions for predicates and terms.

How can we get from absolute extensions to a space of models? As has been much discussed, we need somehow to abstract away from the specific meanings of the expressions of a language. The obvious way to do this, if we are thinking of extensions as sets of individuals, is simply to allow these sets to vary. How much they are allowed to vary will determine properties of the logic that results, but the basic idea is that they should vary freely. We might also approach abstraction by allowing substitutions of different expressions with different meanings (as discussed, for example, by Quine 1959, 1986, following Bolzano), but with enough set theory around, simply allowing the sets to vary is the natural way to abstract from the specific meanings of terms, and it avoids some problems that are well known for substitutional approaches. Doing this will get us a non-trivial space of many models.

It is well known, however, that just varying the extensions of predicates and terms is not enough. Only doing this will make our consequence relation dependent on what actually exists in the world (or in the domain of our absolute semantics). Many philosophers have concluded this is not enough to satisfy necessity, and not enough to satisfy the topic-neutrality idea behind formality. But it is also well known what is needed to do better: we need to let the domain from which we draw individuals (the 'universe of discourse') vary as well, and draw our extensions for predicates and referents of terms from those varying domains.[44] Varying the domain, and the extensions of predicates and terms, produces a wider space of models. Indeed, as described, it produces the space of models that generates classical logic. But the main idea here is quite general, and can be applied to generate different sorts of consequence relations, depending on what structure is involved in the models, and how variation is understood. The case of classical logic gives us good reason to think that this process of varying domains and extensions is a successful one for getting us from absolute semantics to the kinds of spaces of models that serve logical consequence.

This is the process I shall call *abstraction*, as it involves abstraction from the meanings of terms. Abstraction gets us from absolute semantics to a space of models, and so, gets us the essentials of a logical consequence relation. I understand abstraction as a *kind* of process, which can be implemented in different ways, resulting in different logics.

[44] This is the received view in the tradition stemming from Tarski. There are some historical issues about just where and how domain variation was first recognized, but I shall not pursue them here. See the references mentioned in n. 6.

3.5.2 *The identification of logical constants*

The use of abstraction to move from absolute semantics to logical consequence is standard, at least in the post-Tarskian tradition. But, as is very well known, it also brings us back to the issue of logical constants. We cannot abstract away from the meaning of every expression of a language and get back an interesting consequence relation, or one that will count as a consequence relation by restrictive lights. We have to keep fixed the meanings of the logical terms, and that means identifying the logical constants. Indeed, we have to identify the logical constants before we can abstract away from the meanings of the non-logical terms. The formality constraint requires that we identify the right formal structure for valid inferences, and the logical constants provide that structure.

We discussed the issues of how to get logical constants from natural language in section 3.3.3 and in section 3.4. That discussion reminds us that when we abstract away from absolute semantics, we will need to make sure our logical constants get suitable *global* interpretations in our space of models, or are properly handled when we define satisfaction in a model. But the more important result of those sections, for our current concern, is that natural language does not do the job of identifying the logical constants for us. At least, if we are not so permissive as to count virtually every term as logical, or perhaps every functional category, then natural language will not distinguish the logical constants for us.

Thus, if we are to carry out the process of abstraction to get a consequence relation, we will also have to carry out a process of *identification* to identify the logical constants. I argued above that natural language does not do the identification for us, and we will have to do it ourselves. The discussion of section 3.3.3 gave a few indications of how we might proceed. By restrictive lights, at least, we might appeal to some property like permutation invariance, or some global property of isomorphism invariance. (Such global properties might have to be articulated together with the abstraction process that would give us a space of models in which isomorphism could be defined.) If we are very permissive, we might find the process easier, and perhaps more a matter of identifying terms of interest than terms that meet some restrictive standard. But nonetheless, aside from the extreme case, we will have to do the identification, be it easy or hard to do. Doing so goes beyond what natural language itself tells us. We will have to go beyond observations of natural language 'in the wild', to our laboratory setting, and do the work of identifying the logical constants ourselves.

Abstraction and identification work together, and in doing so, they characterize a common post-Tarskian understanding of model-theoretic consequence relations. Isolating them helps to make clear how we can get to such consequence relations starting with absolute semantics of natural language. They also allow us to measure different ways that more or less permissive views might carry out the processes differently. We have seen that some permissive views might find the process of identification easier. Some related views might find the process of abstraction harder, as, for instance, constraints on metaphysical possibility or the structure of strict entailment might be required. Especially when it comes to identification, we will have to rely on some prior grasp of what logic is about to carry out the process. Perhaps, as we have been assuming, notions like formality or permutation invariance will be our guide. Regardless, both processes mark ways in which we depart from natural language when we build consequence relations.

3.5.3 The idealization problem

The two processes of identification of the logical constants and abstraction from lexical meaning give us the tools to get from natural language to logic, and their use in tandem is quite familiar. But, I shall argue here, they are not sufficient to get us something we would want to call a logic. One more step, *idealization*, is needed. The reason is that even after we have performed abstraction, we are still going to be stuck with some idiosyncratic and quirky features of natural language grammar that we will not want to contaminate our logic. Even after we have abstracted away from absolute lexical meaning, we still have not finished our laboratory work, and further steps of purification will be in order.

To illustrate the kinds of features of natural language we will want to idealize away from, let us return to the behavior of quantifiers in natural language. Even when we have recognizably logical expressions, like 'every', their behavior in natural language will be determined by a wide range of features of the grammar. This will not undercut the expressions having the kinds of semantics logic leads us to expect (in local form), but it can, and does, produce some unexpected behavior that makes them not work in quite the way we expect a logical constant figuring in a consequence relation to work.

We see this, for instance, in the scoping behavior of the determiner 'every'. 'Every' is a universal quantifier, no doubt, and behaves like one. In fact, it is a *distributive* universal quantifier, in a way that 'all' is not in English. As we have seen, it gets essentially the semantics of universal quantification, in local form. But natural language quantifier scoping behavior is well known to be complex and delicate (or if you like, very quirky), compared to the essentially uniform scope potentials nearly every formal language builds in. To start with, there are well-known subject/object asymmetries in scoping which affect 'every'. They are brought out most strikingly if we also throw in a negation for good measure. Observe that the following, as judged by native speakers, is *not* ambiguous:

(11) John did not read every book.

This sentence does not have a reading where 'every' scopes over the negation: that is, it cannot mean $\forall x(B(x) \rightarrow \neg R(j, x))$. On the other hand, if we substitute in an existential (i.e. an 'indefinite'), then the reading is available. The following has just the scope ambiguity that standard logic would lead you to expect:

(12) John did not read a book.

In fact, this just scratches the surface of the maddening behavior of quantifier-like constructions in natural language. Even so, it is enough to show how unlike the nice uniform behavior of quantifiers in formal languages the behavior of quantifiers in natural language can be.[45]

Existential quantifiers in natural language show other sorts of behavior that we might not expect, if logic were to be our guide. Not least of which, existential and universal quantifiers do not share the same distribution in some puzzling environments, such as:

(13) a. i. Max is a friend of mine.
ii. *Max is every friend of mine.
b. i. There is a book on the table.
ii. *There is every book on the table.

Indeed, if we thought 'there is' expressed existential quantification, or 'heralded existence' in the much-noted phrase of Quine (1958), we might already be puzzled as to why it can pair with a quantifier at all. But even putting that aside, nothing in our familiar logical formalisms tells us to expect that some quantifiers occur in some positions, and some in others, or that existential (and related) quantifiers show special behavior.[46]

These are two illustrations, specific to quantifiers, of how the behavior of logical constants in natural language can be quirky. More specific to our concerns is that presumably we do not want our notion of logical consequence to include such quirks. I take it we do not want a notion of logical consequence captured in a language that builds in significant distributional differences between quantifiers, or builds in different scope behavior for existential and universal quantifiers.

[45] The literature on these issues is large, but for some starting points, see Aoun and Li (1993) on subject/object asymmetries, Beghelli and Stowell (1997) on 'every', and Fodor and Sag (1982) on indefinites. The latter has spawned an especially huge literature.

[46] Again, the literature on these issues is large, but the phenomenon illustrated in (13) was first discussed by Milsark (1974).

Of course, when we build formal languages of the usual sorts, we smooth out this sort of quirky behavior. Standard syntax for formal languages provides the same distribution for elements of each category, so all quantifiers enjoy the same distribution. We thus will not have the kind of behavior we see in (13). When it comes to scope, standard formal languages mark scope explicitly, by such devices as parentheses. Thus, the kinds of questions about the possible scopes of quantifiers in surface forms we encounter in (11) and (12) will not arise once we fix the syntax of scope in a formal language. My point is that when we set up formal languages this way, we are not simply reflecting the grammar of the logical constants in the natural languages with which we start. We make substantial departures from the structure of natural language when we set up familiar sorts of formal language.

We do so for good reason. We want our formal languages to display uniform grammatical properties in important logical categories. We want our quantifiers to show the important logical properties of quantifiers, for instance, not the quirky properties they pick up from the grammar of natural language. I am not going so far as to claim that any formal structure infused with such natural-language peculiarities would fail to be a logic (by permissive lights!). At least, those with permissive views might be interested in such formal languages, and some logical apparatus has been used to address, for example, quantifier scope in natural language.[47] But when we think about the core idea of logic as the study of valid arguments, it is just hard to see why these sorts of quirk should be relevant. Good reasoning about 'every' versus 'some' just does not seem to be different in the ways grammar sometimes makes these words behave differently. So, in building logical consequence relations that reflect this core idea, we should *idealize* away from these quirks of grammar, even if we see them in natural languages. It is clear that restrictive views will have to idealize this way. In fact, even very permissive approaches to logic usually idealize in just the same way, when they are exploring core logical consequence. We see this in the standard formalisms of non-classical logics (e.g. Priest 2008). Both permissive and restrictive views will typically idealize substantially in moving from natural language to logic.[48]

Idealization, as it figures here, is a familiar kind of idealization in scientific theorizing that builds idealized models. One way to build idealized models is to remove irrelevant features of some phenomenon, and replace them with uniform or simplified features. A model of a planetary system is such an idealized model: it

[47] For instance, see the textbook discussion of Carpenter (1997), or the very interesting work of Barker (2002).

[48] Quine (1960) strikes a similar note in his discussion of the role of 'simplification' in regimentation.

ignores thermodynamic properties, ignores the presence of comets and asteroids, and treats planets as ideal spheres (cf. Frigg and Hartmann 2009). When we build a logic from a natural language, I suggest, we do just this. We ignore irrelevant features of grammar, and replace them with uniform and simplified logical categories. We do so for particular purposes. If we want to think of logic as the general, and highly idealized, study of valid arguments, such an idealized tool will serve our purposes well. But different purposes will require different tools. If we want to use logic as a tool for analyzing the structure of natural language in detail, as I mentioned, different idealizations will be in order. For the kinds of concern with core logical consequence we have been focusing on here, we will want to idealize away from the quirks of natural language grammar; and regardless, some idealization will be in order for any purpose.

Thus, we need to add a process of idealization to those of abstraction and identification. We need all three to get from natural language to logic. We only get to logic—something that serves our purposes in analyzing valid reasoning, and is recognizably like what logicians work with—when we include idealization.

3.6 Conclusion

I have argued for two main claims in this chapter. First, the logic in natural language thesis is false: we do not find logical consequence relations in our natural languages. Though the logics in formal languages thesis might have made the logic in natural language thesis seem appealing, it should still be rejected. Part of my argument for this claim relied on a restrictive view of logical consequence; but only part did, and we saw several reasons to doubt the logic in natural language thesis on permissive views as well.

Second, I have tried to show how natural language can serve as a useful object of study for those of us interested in logic, even if the logic in natural language thesis is false. The history of logic tells us that this must be possible, as many advances in logic have come from at least glancing at our own languages. As the quote from Beall and van Fraassen above reminds us, arguments are presented in language, so we have little chance of producing an analysis of valid arguments which paid no attention to language at all. Indeed, as we considered the semantics and grammar of natural language above, we found many things of great interest to logic. To explain this, in light of the failure of the logic in natural language thesis, I sketched a three-fold process that allows us to get from a natural language to a logical consequence relation. The process involves identifying logical constants, abstracting away from other features of meaning, and idealizing away from quirks in the structure of human languages. The relation between logic and natural language is thus less close than advocates of the logic in natural language thesis would have it, while the three-fold process allows that there is some connection.

As I said at the outset, this is an autonomy thesis. Natural language semantics per se does not include logic. Logic, in its core, does not include the quirks or specific contents of natural language semantics. Moreover, to get from natural language to logic, you will have to rely on some fairly robust, if perhaps general, ideas of what makes something count as logical. You cannot find logical consequence just by looking at natural language, any more than you can find natural language semantics by looking at your favorite logic. Logic and natural language are autonomous. All the same, if you already have some prior, perhaps rough or general, grip on logic, you can find lots of interesting facts in natural language, which you might use in further theorizing. This could even lead you to extract a full-blown logical consequence relation from a natural language, but you have to start with some logic if you want to get some logic.[49]

References

Abney, Steven. 1987. "The English noun phrase in its sentential aspect." Ph.D. dissertation, MIT.

Anderson, Alan Ross and Belnap, Nuel D. 1975. *Entailment: The Logic of Relevance and Necessity*, volume I. Princeton, NJ: Princeton University Press.

Anderson, Alan Ross, Belnap, Nuel D., and Dunn, J. Michael. 1992. *Entailment: The Logic of Relevance and Necessity*, volume II. Princeton, NJ: Princeton University Press.

Anderson, Stephen R. 1971. "On the role of deep structure in semantic interpretation." *Foundations of Language* 7:387–96.

Antonelli, G. Aldo. 2005. *Grounded Consequence for Defeasible Logic*. Cambridge: Cambridge University Press.

Aoun, Joseph and Li, Yen-hui Audrey. 1993. *Syntax of Scope*. Cambridge, MA: MIT Press.

Asher, Nicholas and Lascarides, Alex. 2003. *Logics of Conversation*. Cambridge: Cambridge University Press.

Barker, Chris. 2002. "Continuations and the nature of quantification." *Natural Language Semantics* 10:211–42.

Barwise, Jon and Cooper, Robin. 1981. "Generalized quantifiers and natural language." *Linguistics and Philosophy* 4:159–219.

[49] Thanks to Jc Beall for valuable discussions of the topics explored in this chapter. Versions of this material were presented at the Conference on the Foundations of Logical Consequence sponsored by Arché, St Andrews, 2010, and at Tel Aviv University, 2011. Thanks to all the participants there for many helpful questions and comments, and especially to Colin Caret, Josh Dever, Christopher Gauker, Ole Hjortland, Graham Priest, Agustín Rayo, Stephen Read, Ofra Rechter, Stewart Shapiro, and Brian Weatherson. I am especially grateful to my commentator at St Andrews, Derek Ball, and to an anonymous referee, for incisive comments.

Barwise, Jon and Feferman, Solomon (eds.). 1985. *Model-Theoretic Logics*. New York: Springer.

Beall, Jc and Restall, Greg. 2006. *Logical Pluralism*. Oxford: Oxford University Press.

Beall, Jc and Restall, Greg. 2009. "Logical consequence." In Edward N. Zalta (ed.), *The Stanford Encyclopedia of Philosophy*. CSLI, fall 2009 edition. <http://plato.stanford.edu/archives/fall2009/entries/logical-consequence/>.

Beall, Jc and van Fraassen, Bas. 2003. *Possibilities and Paradox: An Introduction to Modal and Many-Valued Logic*. Oxford: Oxford University Press.

Beghelli, Filippo and Stowell, Tim. 1997. "Distributivity and negation: the syntax of *each* and *every*." In A. Szabolcsi (ed.), *Ways of Scope Taking*, 71–107. Dordrecht: Kluwer.

Bonnay, Denis. 2008. "Logicality and invariance." *Bulletin of Symbolic Logic* 14:29–68.

Carnap, Rudolf. 1947. *Meaning and Necessity*. Chicago: University of Chicago Press.

Carpenter, Bob. 1997. *Type-Logical Semantics*. Cambridge, MA: MIT Press.

Carroll, Lewis. 1960. *The Annotated Alice*. Introduction and Notes by Martin Gardner. New York: Meridian.

Casari, Ettore. 1987. "Comparative logics." *Synthese* 73:421–49.

Chierchia, G. and McConnell-Ginet, S. 2000. *Meaning and Grammar: An Introduction to Semantics*. Cambridge, MA: MIT Press, 2nd edition.

Church, Alonzo. 1940. "A formulation of the simple theory of types." *Journal of Symbolic Logic* 5:56–68.

Cresswell, Max J. 1973. *Logics and Languages*. London: Methuen.

Cresswell, Max J. 1978. "Semantics and logic." *Theoretical Linguistics* 5:19–30.

Davidson, Donald. 1967. "Truth and meaning." *Synthese* 17:304–23. Reprinted in Davidson (1984).

Davidson, Donald. 1984. *Inquiries into Truth and Interpretation*. Oxford: Oxford University Press.

Dowty, David R. 1991. "Thematic proto-roles and argument selection." *Language* 67: 547–619.

Dowty, David R., Wall, Robert E., and Peters, Stanley. 1981. *Introduction to Montague Semantics*. Dordrecht: Reidel.

Dummett, Michael. 1991. *The Logical Basis of Metaphysics*. Cambridge, MA: Harvard University Press.

Etchemendy, John. 1988. "Tarski on truth and logical consequence." *Journal of Symbolic Logic* 53:51–79.

Etchemendy, John. 1990. *The Concept of Logical Consequence*. Cambridge, MA: Harvard University Press.

Evans, Gareth. 1976. "Semantic structure and logical form." In G. Evans and J. McDowell (eds.), *Truth and Meaning*, 199–222. Oxford: Oxford University Press.

Feferman, Solomon. 1999. "Logic, logics, and logicism." *Notre Dame Journal of Formal Logic* 40:31–54.

Fillmore, Charles J. 1968. "The case for case." In E. Bach and R. T. Harms (eds.), *Universals in Linguistic Theory*, 1–88. New York: Holt, Rinehart, and Winston.

Fodor, Janet Dean and Sag, Ivan A. 1982. "Referential and quantificational indefinites." *Linguistics and Philosophy* 5:355–98.

Frege, Gottlob. 1879. *Begriffsschrift, eine der arithmetischen nachgebildete Formelsprache des reinen Denkens*. Halle: Nebert. References are to the translation as "Begriffsschrift, a Formal Language, Modeled upon that of Arithmetic, for Pure Thought" by S. Bauer-Mengelberg in van Heijenoort (1967).

Frege, Gottlob. 1891. *Function und Begriff*. Jena: Pohle. References are to the translation as "Function and Concept" by P. Geach in Frege (1984).

Frege, Gottlob. 1984. *Collected Papers on Mathematics, Logic, and Philosophy*. Edited by B. McGuinness. Oxford: Basil Blackwell.

Frigg, Roman and Hartmann, Stephan. 2009. "Models in science." In Edward N. Zalta (ed.), *The Stanford Encyclopedia of Philosophy*. CSLI, summer 2009 edition. <http://plato.stanford.edu/archives/sum2009/entries/models-science/>.

Fukui, Naoki. 1995. *Theory of Projection in Syntax*. Stanford, CA: CSLI Publications.

Gamut, L. T. F. 1991. *Logic, Language, and Meaning*, volume 2: *Intensional Logic and Grammar*. Chicago: University of Chicago Press. 'Gamut' is a pseudonym for van Benthem, Groenendijk, de Jongh, Stokhof, and Verkuyl.

Gentzen, Gerhard. 1935. "Untersuchungen über das logische Schliessen." *Mathematische Zeitschrift* 39:175–210, 405–31. English translation "Investigations into logical deduction" in Szabo (1969, 68-131).

Gödel, Kurt. 1930. "Die Vollständigkeit der Axiome des logischen Funktionenkalküls." *Monatshefte für Mathematik und Physik* 37:349–60. References are to the translation as "The completeness of the axioms of the functional calculus of logic" by S. Bauer-Mengelberg in Gödel (1986).

Gödel, Kurt. 1986. *Collected Works*, volume I. Oxford: Oxford University Press. Edited by S. Feferman, J. W. Dawson Jr, S. C. Kleene, G. H. Moore, R. M. Solovay, and J. van Heijenoort.

Grice, Paul. 1975. "Logic and conversation." In Peter Cole and Jerry L. Morgan (eds.), *Speech Acts*, volume 3 of *Syntax and Semantics*, 41–58. New York: Academic Press. Reprinted in Grice (1989).

Grice, Paul. 1989. *Studies in the Way of Words*. Cambridge, MA: Harvard University Press.

Grimshaw, Jane. 2005. "Extended projection." In *Words and Structure*, 1–73. Stanford, CA: CSLI Publications.

Hale, Kenneth and Keyser, Samuel J. 1987. "A view from the middle." Lexicon Project Working Papers 10, Center for Cognitive Science, MIT.

Heim, Irene and Kratzer, Angelika. 1998. *Semantics in Generative Grammar*. Oxford: Blackwell.

Higginbotham, James. 1986. "Linguistic theory and Davidson's program in semantics." In E. Lepore (ed.), *Truth and Interpretation: Perspectives on the Philosophy of Donald Davidson*, 29–48. Oxford: Blackwell.

Higginbotham, James. 1988. "Contexts, models, and meanings: a note on the data of semantics." In R. Kempson (ed.), *Mental Representations: The Interface between Language and Reality*, 29–48. Cambridge: Cambridge University Press.

Higginbotham, James. 1989a. "Elucidations of meaning." *Linguistics and Philosophy* 12: 465–517.

Higginbotham, James. 1989b. "Knowledge of reference." In A. George (ed.), *Reflections on Chomsky*, 153–74. Oxford: Basil Blackwell.

Higginbotham, James and May, Robert. 1981. "Questions, quantifiers and crossing." *Linguistic Review* 1:41–79.

Hirschberg, Julia. 1985. "A theory of scalar implicature. Ph.D." dissertation, University of Pennsylvania.

Horty, John F. 2001. "Nonmonotonic logic." In L. Goble (ed.), *The Blackwell Guide to Philosophical Logic*, 336–61. Oxford: Blackwell.

Kadmon, Nirit. 2001. *Formal Pragmatics*. Oxford: Blackwell.

Keenan, Edward L. and Stavi, Jonathan. 1986. "A semantic characterization of natural language determiners." *Linguistics and Philosophy* 9:253–326. Versions of this paper were circulated in the early 1980s.

Kennedy, Christopher and McNally, Louise. 2005. "Scale structure, degree modification, and the semantics of gradable predicates." *Language* 81:345–81.

Klein, Ewan and Sag, Ivan A. 1985. "Type-driven translation." *Linguistics and Philosophy* 8:163–201.

Kneale, William and Kneale, Martha. 1962. *The Development of Logic*. Oxford: Clarendon.

Kretzmann, Norman, Kenny, Anthony, and Pinborg, Jan (eds.). 1982. *The Cambridge History of Later Medieval Philosophy*. Cambridge: Cambridge University Press.

Kripke, Saul. 1963. "Semantical considerations on modal logic." *Acta Philosophica Fennica* 16:83–94.

Lambek, Joachim. 1958. "The mathematics of sentence structure." *American Mathematical Monthly* 65:154–70.

Larson, Richard and Segal, Gabriel. 1995. *Knowledge of Meaning*. Cambridge, MA: MIT Press.

Lepore, Ernie. 1983. "What model-theoretic semantics cannot do." *Synthese* 54:167–87.

Levin, Beth. 1993. *English Verb Classes and Alternations*. Chicago: University of Chicago Press.

Levin, Beth and Rappaport Hovav, Malka. 1995. *Unaccusativity: At the Syntax-Lexical Semantics Interface*. Cambridge, MA: MIT Press.

Levin, Beth and Rappaport Hovav, Malka. 2005. *Argument Realization*. Cambridge: Cambridge University Press.

Lewis, Clarence Irving and Langford, Cooper Harold. 1932. *Symbolic Logic*. New York: Century.

Lewis, David K. 1970. "General semantics." *Synthese* 22:18–67. Reprinted in David K. Lewis, *Philosophical Papers*, volume I, Oxford: Oxford University Press, 1983.

Lindström, Per. 1966. "First order predicate logic with generalized quantifiers." *Theoria* 32:186–95.

Lycan, William G. 1989. "Logical constants and the glory of truth-conditional semantics." *Notre Dame Journal of Formal Logic* 30:390–400.

MacFarlane, John. 2009. "Logical constants." In Edward N. Zalta (ed.), *The Stanford Encyclopedia of Philosophy*. CSLI, fall 2009 edition. <http://plato.stanford.edu/archives/fall2009/entries/logical-constants/>.

Mancosu, Paolo. 2010. "Fixed- versus variable-domain interpretations of Tarski's account of logical consequence." *Philosophy Compass* 5:745–59.

Mautner, F. I. 1946. "An extension of Klein's Erlanger program: logic as invariant-theory." *American Journal of Mathematics* 68:345–84.

Milsark, Gary. 1974. "Existential sentences in English." Ph.D. dissertation, MIT. Published by Garland, New York, 1979.

Montague, Richard. 1968. "Pragmatics." In Raymond Klibansky (ed.), *Contemporary Philosophy: A Survey*, 102–22. Florence: La Nuova Italia Editrice. Reprinted in Montague (1974).

Montague, Richard. 1970. "Universal grammar." *Theoria* 36:373–98. Reprinted in Montague (1974).

Montague, Richard. 1973. "The proper treatment of quantification in ordinary English." In J. Hintikka, J. Moravcsik, and P. Suppes (eds.), *Approaches to Natural Language*, 221–42. Dordrecht: Reidel. Reprinted in Montague (1974).

Montague, Richard. 1974. *Formal Philosophy: Selected Papers of Richard Montague*. Edited by R. H. Thomason. New Haven, CT: Yale University Press.

Moortgat, Michael. 1997. "Categorial type logics." In J. van Benthem and A. ter Meulen (eds.), *Handbook of Logic and Language*, 93–177. Amsterdam: Elsevier.

Mostowski, A. 1957. "On a generalization of quantifiers." *Fundamenta Mathematicae* 44: 12–36.

Partee, Barbara H. 1979. "Semantics—mathematics or psychology?" In A. Bäuerle, U. Egli, and A. von Stechow (eds.), *Semantics from Different Points of View*, 1–14. New York: Springer.

Peters, Stanley and Westerståhl, Dag. 2006. *Quantifiers in Language and Logic*. Oxford: Clarendon.

Pietroski, Paul M. 2005. *Events and Semantic Architecture*. Oxford: Oxford University Press.

Pinker, Steven. 1989. *Learnability and Cognition*. Cambridge, MA: MIT Press.

Portner, Paul and Partee, Barbara H. (eds.). 2002. *Formal Semantics: The Essential Readings*. Oxford: Blackwell.

Prawitz, Dag. 1974. "On the idea of a general proof theory." *Synthese* 27:63–77.

Prawitz, Dag. 2005. "Logical consequence from a constructivist point of view." In Stewart Shapiro (ed.), *The Oxford Handbook of Philosophy of Mathematics and Logic*, 671–95. Oxford: Oxford University Press.

Priest, Graham. 2008. *An Introduction to Non-classical Logic*. Cambridge: Cambridge University Press, 2nd edition.

Quine, W.V. 1958. "Speaking of objects." *Proceedings and Addresses of the American Philosophical Association* 31:5–22. Reprinted in Quine (1969).

Quine, W.V. 1959. *Methods of Logic*. New York: Holt, revised edition.

Quine, W.V. 1960. *Word and Object*. Cambridge, MA: MIT Press.

Quine, W.V. 1969. *Ontological Relativity and Other Essays*. New York: Columbia University Press.

Quine, W.V. 1986. *Philosophy of Logic*. Cambridge, MA: Harvard University Press, 2nd edition.

Read, Stephen. 2012. "The medieval theory of consequence." *Synthese* 187:899–912.

Restall, Greg. 2000. *An Introduction to Substructural Logics*. London: Routledge.

Russell, Bertrand. 1903. *Principles of Mathematics*. Cambridge: Cambridge University Press.

Sag, Ivan A., Wasow, Thomas, and Bender, Emily M. 2003. *Syntactic Theory: A Formal Introduction*. Stanford, CA: CSLI Publications, 2nd edition.

Scott, Dana. 1970. "Advice on modal logic." In K. Lambert (ed.), *Philosophical Problems in Logic*, 143–73. Dordrecht: Reidel.

Shapiro, Stewart. 1991. *Foundations without Foundationalism: A Case for Second-order Logic*. Oxford: Oxford University Press.

Sher, Gila. 1991. *The Bounds of Logic*. Cambridge, MA: MIT Press.

Skolem, Thoralf. 1922. "Einige Bemerkungen zur axiomatischen Begründung der Mengenlehre." In *Matematikerkongressen i Helsingfors den 4–7 Juli 1922, Den femte skandinaviska matematikerkongressen, Redogörelse*, 217–32. Helsinki: Akademiska Bokhandeln. References are to the translation as "Some remarks on axiomatized set theory" by S. Bauer-Mengelberg in van Heijenoort (1967).

Smith, Robin. 1995. "Logic." In J. Barnes (ed.), *The Cambridge Companion to Aristotle*, 27–65. Cambridge: Cambridge University Press.

Speas, Margaret. 1990. *Phrase Structure in Natural Language*. Dordrecht: Kluwer.

Strawson, P.F. 1950. "On referring." *Mind* 59:320–44. Reprinted in Strawson (1971).

Strawson, P.F. 1971. *Logico-linguistic Papers*. London: Methuen.

Szabo, M.E. (ed.). 1969. *The Collected Papers of Gerhard Gentzen*. Amsterdam and London: North-Holland.

Tarski, Alfred. 1935. "Der Wahrheitsbegriff in den formalisierten Sprachen." *Studia Philosophica* 1:261–405. References are to the translation as "The concept of truth in formalized languages" by J.H. Woodger in Tarski (1983). Revised version of the Polish original published in 1933.

Tarski, Alfred. 1936. "O pojciu wynikania logicznego." *Przegląd Filozoficzny* 39:58–68. References are to the translation as "On the concept of logical consequence" by J.H. Woodger in Tarski (1983).

Tarski, Alfred. 1983. *Logic, Semantics, Metamathematics*. Edited by J. Corcoran with translations by J.H. Woodger. Indianapolis: Hackett, 2nd edition.

Tarski, Alfred. 1986. "What are logical notions? Text of a 1966 lecture edited by J. Corcoran." *History and Philosophy of Logic* 7:143–54.

Thomason, Richmond H. 1974. "Introduction." In R. H. Thomason (ed.), *Formal Philosophy: Selected Papers of Richard Montague*, 1–69. New Haven, CT: Yale University Press.

van Benthem, Johan. 1986. *Essays in Logical Semantics*. Dordrecht: Reidel.

van Benthem, Johan. 1991. *Language in Action*. Amsterdam: North-Holland.

van Heijenoort, Jean (ed.). 1967. *From Frege to Gödel: A Source Book in Mathematical Logic, 1879–1931*. Cambridge, MA: Harvard University Press.

Westerståhl, Dag. 1985. "Logical constants in quantifier languages." *Linguistics and Philosophy* 8:387–413.

Westerståhl, Dag. 1989. "Quantifiers in formal and natural languages." In D.M. Gabbay and F. Guenthner (eds.), *Handbook of Philosophical Logic*, volume IV, 1–131. Dordrecht: Kluwer.

Whitehead, Alfred North and Russell, Bertrand. 1925–7. *Principia Mathematica*. Cambridge: Cambridge University Press, 2nd edition.

Zimmermann, Thomas Ede. 1999. "Meaning postulates and the model-theoretic approach to natural language semantics." *Linguistics and Philosophy* 22:529–61.

4

Is the Ternary R Depraved?

Graham Priest

4.1 Introduction: Routley–Meyer Semantics

When modal logic was reinvented by C. I. Lewis early in the twentieth century, it was formulated simply as a number of axiom systems. There were certainly intuitions which drove thoughts about what should be an axiom, and what should not. But the axiom systems had no formal semantics. In modern logic, a system of proof with no such semantics appears distinctly naked. In some sense, without a semantics, we would not seem to know what the system is *about*. It was therefore a happy event when world-semantics for modal logics were invented by Kripke (and others).

In a very similar way, relevant logics were also invented (some years later) as purely axiomatic systems. It was therefore a happy event when Routley and Meyer (and others) produced a world-semantics for them. As they put it in the first of a ground-breaking series of papers:[1]

> Word that Anderson and Belnap had made a logic without semantics leaked out. Some thought it wondrous and rejoiced, that the One True Logic should make its appearance among us in the Form of Pure Syntax, unencumbered by all that set-theoretic garbage. Others said that relevant logics were Mere Syntax. Surveying the situation Routley... found an explication of the key concept of relevant implication. Building on [this], and with help from our friends... we use these insights to present here a formal semantics for the system R of relevant implication.

But critics were not impressed. The world-semantics of modal logic were certainly contentious, but no one could deny that the notion of a possible world, and of a binary relation of relative possibility, clearly had intuitive and relevant content.

[1] Routley and Meyer (1973, 199).

The Routley–Meyer semantics, by contrast, said the critics, employed devices of a purely technical nature. As one put it:[2]

> [T]he Routley–Meyer semantics . . . fails to satisfy those requirements which distinguish an illuminating and philosophically significant semantics from a merely formal model theory.

Or, as another critic put it more bluntly, commenting on the promises held out by those who took this kind of semantics to explain why contradictions do not entail everything: 'What else can one do but ask for one's money back?'[3]

The devices in the semantics which drew the ire of critics were two:[4] a function, * (the Routley Star), from worlds to worlds, employed in giving the truth conditions of negation:[5]

$$\nu_w(\neg\alpha) = 1 \text{ iff } \nu_{w^*}(\alpha) = 0$$

and a ternary relation, R, on worlds, employed in giving the truth conditions of the conditional:

$$\nu_w(\alpha \to \beta) = 1 \text{ iff for all } x, y, \text{ such that } Rwxy, \text{ when } \nu_x(\alpha) = 1, \nu_y(\beta) = 1$$

The first of these is not now in such bad shape. A plausible understanding of the Routley * may be given in terms of a primitive notion of incompatibility.[6] Matters with the ternary relation are in a less happy state. Some interpretations have certainly been put forward. The most successful so far, I think, is in terms of the notion of information flow, suggested by a connection between relevant semantics and situation semantics.[7] But even this appears somewhat tenuous.[8] The following question is still, therefore, a pressing one: what, exactly, does the ternary relation mean, and why is it reasonable to employ it in giving the truth conditions of a conditional? In what follows, I will attempt an answer to the question.

4.2 Validity

Before I turn to this matter, however, it will be useful to take a step back, and put the matter in perspective, so that we can see what is being asked of the semantics and why. Let's start at the beginning.

[2] Copeland (1979, 400). He repeats the charge in Copeland (1983), citing others who have made the similar claims: Scott, van Bentham, Hintikka, Keilkopf, and (David) Lewis.

[3] Smiley (1993, 19).

[4] It it worth noting that there are semantics for relevant logics which avoid both of these techniques. See Priest (2008, chs. 8, 9).

[5] $\nu_w(\alpha) = 1\,[0]$ means that the value of α at world w is true[false].

[6] As in Restall (1999).

[7] See, for example, Restall (1995), Mares (1997).

[8] See Priest (2008, 10.6.)

When we reason, we deploy premises and conclusions. Successful reasoning requires that the premises really do support that conclusion: that is, that the argument is valid. Logic is essentially the study of validity. We need a logical theory to tell us which inferences are valid and which are not. But the theory should do more than just give us two washing lists: logic, like, arguably, any science, should explain the *why* of things. Without understanding why things are valid, we are in no position, for example, to evaluate inferences that do not, as yet, appear on either list, or to adjudicate disputes about whether something should be on one or other of the lists.

Now, again, when we reason, we reason in a natural language. The language may be augmented by technical notions, such as those of physics or mathematics; but it is a natural language nonetheless. The notion of validity which logic investigates must apply to such arguments. However, in the methodology of modern logic, an account of validity is given, not for natural languages, but for various formal languages. Of course, there must be a connection. Some, such as Montague, have taken English itself to be a formal language. This is somewhat implausible. No one ever used a formal language to write poetry or make jokes. And even if English is a formal language, it is not one of those which is in standard use in logic (e.g. the first-order predicate calculus). Perhaps more plausibly, the formal languages we use can be thought of as providing reliable models (in the scientists' sense) of certain aspects of natural language. Thus, a correlation is made between certain formal symbols and certain worlds of natural language. Standardly, '$\wedge$' is paired with 'and', '$\rightarrow$' with 'if', '$\forall$' with 'for all', and so on. No one would suppose that the formal symbol behaves exactly as does its natural-language counterpart. For example, 'and' is much more versatile, often in an idiosyncratic way, than is '$\wedge$'. Nonetheless the behaviour of '$\wedge$' provides a model of certain key aspects of the behavior of 'and'; and so on for the other paired symbols. How exactly to understand this is a hard matter, which, fortunately, we may bypass here. The important point is that once we have an account of the validity of inferences in the formal language, the correlation will provide, transitively, an account for the natural-language inferences.

We are not at the why of the matter yet, though. We still want for an account of why the inferences of the squiggle language are valid or invalid. In particular, any appeal to what is valid or invalid in natural language will not provide the required explanation of why certain inferences in the squiggle language are valid or invalid; for understanding validity in the formal language was meant to deliver an understanding of natural language validity, not the other way around.

So how is this to be provided? In modern logical methodology, two very different strategies for solving the problem are commonly espoused: one is proof-theoretic; the other is model-theoretic. The proof-theoretic strategy applies, as far as I am aware, only to deductive validity. The thought is that an inference is valid in virtue of the meanings of some of the symbols involved. The meanings are constituted, in turn, by the rules of inference governing these constituents. There are many obstacles to pursuing this strategy successfully. However, this approach is not my concern here, and I mention it only to set it aside. Our focus is on the other strategy.[9]

4.3 Pure and Applied Semantics

The model-theoretic strategy is quite different, and can be applied to both deductive and inductive inference.[10] We define certain set-theoretic structures called *interpretations*, and also what it means for a sentence of our formal language to *hold* in an interpretation. A valid inference is one the conclusion of which holds in every interpretation in which all the premises hold.[11] So much is easy. But we hardly have something which gives us the why of validly yet. An arbitrary semantics of this kind will not provide it. Thus, for example, we can give a semantics for intuitionist logic in which interpretations have as a component a topological space; sentences are assigned open subsets of the space, and the sentences that hold in the interpretation are the ones that get assigned the whole space. There is no reason whatever (at least without a *very* much longer story) as to why the fact that an inference preserves taking as a value the whole of a topological space should explain how it is that the premises of a valid argument provide any rational ground for the conclusion.

If this is not clear, just recall that given any set of rules that is closed under uniform substitution, we can construct a many-valued semantics in a purely formulaic way. The values are the formulas themselves; the designated values are the theorems, and the sentences which hold in the interpretation are those which get designated values. Given some fairly minimal conditions, it can be shown that an inference is vouchsafed by the rules iff it is valid in this model theory.[12] A semantic construction which can be made to fit virtually any set of inferences whatever, does not have the discrimination required to justify any one of them.

We are forced to distinguish, then, between a model theory with explanatory grunt, and one without. The distinction is a well-acknowledged one, though

[9] In Priest (2006, ch. 11), the proof-theoretic strategy is discussed at length, and I argue that it cannot, in the end, be made to work.

[10] See Priest (2006, ch. 11).

[11] Or every interpretation of a certain kind, in the inductive case.

[12] See Priest (2008, 7.10.9–7.10.10).

terminology and a precise characterization vary. One fairly standard account of it is given by Haack, who calls the distinction one between a formal semantics and an applied semantics—or, following Plantinga (1979), a pure and a depraved semantics. For her, any construction of the kind I have described is a formal (pure) semantics. An applied (depraved) semantics is a pure semantics which has a suitable interpretation. She describes the matter as follows:[13]

> I distinguished ... four aspects relevant to one's understanding of ordinary, non-modal sentence logic; the distinction applies, equally, to modal logic. One has:
>
> (i) the syntax of the formal language
> (ii) informal readings of (i)
> (iii) formal semantics for (i) (pure semantics)
> (iv) informal account of (iii) ('depraved semantics')
>
> In the case of the sentence calculus, the formal semantics (iii) supplies a mathematical construction in which one of t, f is assigned to wffs of the calculus, and in terms of which (semantic) validity is defined and consistency and completeness results proved. For all the formal semantics tells one, however, the calculus could be a notation representing electrical circuits, with 't' standing for 'on', and 'f' for 'off' ... But the claim of the calculus to be a sentence logic, to represent arguments the validity of which depends upon their molecular sentential structure, depends on one's understanding the formal semantics in such a way that 't' represents truth and 'f' falsehood; it depends, in other words, on the informal account of the formal semantics—level (iv).

Dummett characterizes the distinction as one between a semantic notion of logical consequence, properly so called, and a merely algebraic one. For him, the difference is that, in the former, the notions involved must themselves be semantic ones, having an appropriate connection to meaning:[14]

> We have examples of purely algebraic [interpretations]. For instance, the topological interpretations of intuitionist logic were developed before any connection was made between them and the intended meanings of the intuitionistic logical constants. Thus, intuitionistic sentential and predicate logic is complete with respect to the usual topology on the real line, under a suitable interpretation, relative to that topology, of the sentential operators and quantifiers. No one would think of this as in any sense giving the meanings of the intuitionistic logical constants, because we have no idea what it would mean to assign an actual statement, framed within first-order logic, a 'value' consisting of an open subset of the real line.
>
> Semantic [interpretations] are framed in terms of concepts which are taken to have a direct relation to the use which is made of the sentences of a language; to take the most obvious example, the concepts of truth and falsity. It is for this reason that the definition of the semantic valuation of the formula under a given interpretation of its schematic letters is

[13] Haack (1978, 188f). [14] Dummett (1975, 293 of reprint).

thought of as giving the meanings of the logical constants. Corresponding algebraic notions define a valuation as a purely mathematical object—an open set, or a natural number—which has no intrinsic connection with the uses of sentences.

Though Haack and Dummett characterize the distinction differently, the difference between them is, I think, somewhat superficial. Both agree on the fundamental point: to have a model-theory with philosophical grunt, the notions used in the model-theoretic construction must be ones which either are, or may be interpreted as, intrinsically semantic ones—something, to put it in Fregean terms, to do with sense and reference. That Routley–Meyer semantics are not of this kind is essentially the critics' complaint. Thus Copeland again:[15]

> The key semantical function of [Routley and Meyer's] theory [truth at a point in an interpretation] receives no more than a bare formal treatment in their writings, and we are offered no explanation of how the formal account of the logical constants given in the theory is to be related to the body of linguistic practices within which the logical constants receive their meaning. The Routley–Meyer 'semantics' as it stands, then, is merely a formal model-theoretic characterisation of the set of sentences provable in NR, no connection being exhibited between the assessment of validity and the intended meanings of the logical constants . . . it is totally unclear what account of the meanings of the logical constants is given in the Routley–Meyer 'semantics'.

4.4 Model-Theoretic Validity

None of this is to disparage the usefulness of pure semantics. Clearly, such semantics are very useful in proving various metatheoretic results concerning independence, and so on. But for a semantics to give us an account of the why of validity, its notions must be (interpretable as) semantic in an appropriate way. But what way?

When we reason, we reason about all sorts of situations: actual, merely possible, and maybe impossible as well. Deductive reasoning is useful because a valid argument is one which gives us a guarantee that whatever situation we are reasoning about, if the premises are true of that situation, so is the conclusion. This is exactly what an applied semantics is all about. Its interpretations are not literally situations—at least, certainly not in the case of actual situations: real situations are

[15] Copeland (1979, 406). Copeland also requires that the semantics deliver the classical meanings of the connectives, and especially negation. This is a much more contentious point, and he is taken to task over the claim in Routley et al. (1982b). He replies in Copeland (1983), but I think that this particular criticism sticks. The very issue here is whether classical semantics or relevant semantics get the meaning of negation in the vernacular right. See Priest (2006, ch. 10, esp. 10.9).

not sets. But the set-theoretic constructions represent situations. They do this by sharing with them a certain structure, in virtue of which a sentence holding in an interpretation faithfully represents being true in the situation represented.[16]

It should be noted, as an aside, that it would seem to be necessary on this conception of model-theoretic validity that every situation about which we reason has an appropriate set-theoretic representation, or we have no reason to suppose that valid arguments will do the job—or better, if they do, we still lack an explanation of why they do. This is not a toothless requirement. In fact, it is not even satisfied by standard model theory, couched in ZF set theory. One situation about which we reason—indeed, about which we reason when we do model theory—is set theory. And there is no interpretation the domain of which contains all sets. This is not a problem that arises if we use a set theory not so limited, such as a naive set theory.[17] But it is a problem which must be faced if we wish to conduct our model theory in ZF. To discuss strategies for doing this and their adequacies,[18] would, however, take us away from the matter at hand.

Part of what is normally involved in an account of truth in an interpretation is the provision of recursive truth-in-a-model conditions for the logical operators. The thought here is that these spell out the meanings of the operators. That the meaning of a sentence is provided by its truth conditions is one which is widely subscribed to in logic—though people may disagree about which notion of truth should be deployed here: for example, whether or not it should be epistemically constrained. The meanings of the logical operators are then naturally thought of as being their contributions to the truth conditions of sentences in which they occur. And this, in turn, is naturally thought of as provided by their recursive truth conditions. Model theory adds a twist to this picture. In model theory, we are not dealing with truth *simpliciter*, but with truth-in-an-interpretation—or, at one remove, truth in a situation. The extension is a natural one, however. To know the meaning of a sentence is not just to know what it is for it to be true or false of some particular situation. The understanding has to be one which can be ported to any situation with which one is presented. If someone knows what it would be for a sentence of the form $\alpha \wedge \beta$ to be true in some situations but not others, we would be disinclined to say that they knew what '$\wedge$' meant. Thus,

[16] Thus, to reason about a situation in which there is a desk and two books (with various properties), take an interpretation where the domain is the set containing those three objects, the extension of one monadic predicate, P, is the set containing the desk, the extension of another, Q, is the set containing the books, etc. With this interpretation, sentences of the language express things about the situation, and may be used to reason about it.

[17] See Priest (2006, 11.5, 11.6).

[18] Such as that in Kreisel (1967).

the truth-in-an-interpretation conditions of a logical operator can be thought of as specifying the meaning of that operator, and so transitively, of the vernacular notion with which is coordinated. Of course, given any particular formal language with its semantics, there is an issue of how good a model it is. There can be legitimate philosophical disagreement about this. Thus, both intuitionists and dialetheists, for example, will insist that the account of negation given in classical semantics provides a bad model for the meaning of ordinary language negation. Nonetheless, when the model is right, the truth conditions of the operator give its meaning.

But what is it for a model to be right? Hard issues in the philosophy of language lie here, such as those that divide realists and Dummettian anti-realists. But all can agree, as Haack and Dummett insist, that for the model to be an adequate one, the notions deployed in stating the recursive truth conditions must be ones which are plausibly thought of as connected with the meaning of the natural language correlate of the formal operator, and are deployed in an appropriate fashion. Presumably, there can be no objection to stating the truth conditions "homophonically", as one would normally do for conjunction. In any interpretation:

$$\nu_w(\alpha \wedge \beta) = 1 \text{ iff } \nu_w(\alpha) = 1 \text{ and } \nu_w(\beta) = 1$$

That the connection is met here is patent. (Though it should be remembered that the natural language 'and' is already to be understood as regimented in a certain way.) But appropriate truth conditions are not necessarily homophonic. Thus, in the world-semantics for modal logic, the truth conditions for $\Box$ are the familiar:

$$\nu_w(\Box\alpha) = 1 \text{ iff for all } w' \text{ such that } wRw', \nu_{w'}(\alpha) = 1$$

Indeed, such non-homophonic truth conditions may well be highly desirable; for example, if our grasp of the behavior of the operator for which truth conditions are being given is insecure, and the truth conditions are given in terms of notions our grasp of which is more secure. However, if we are giving truth conditions for some form of necessity, the following must be considered failures:

$$\nu_w(\Box\alpha) = 1 \text{ iff } f(\alpha, w) = 37$$

where f is some function from sentences and worlds to numbers, and:

$$\nu_w(\Box\alpha) = 1 \text{ iff } \nu_w(\alpha) = 1$$

The one does not even get to first base because, even if the definition gets the extension of $\Box\alpha$ correct, having value 37 is not a semantic notion at all. (Just being 37 has no connection with meaning.) In the other, truth (having value 1) is certainly a semantically relevant notion, but the posited connection between it and necessity is all wrong.

What we have seen, in summary, then, is this: applied 'semantics do not come free. The notions deployed must be intelligible [as semantic notions] in their own right, and their deployment in the framing of a semantics similarly so'.[19] We have seen some of the things this means. There is surely more to be said about matters. But we at least have enough to turn, at last, to Routley–Meyer semantics.

4.5 The Ternary Relation

If Routley–Meyer semantics are to be more that merely algebraic, the ternary relation employed must have an intuitive meaning, and one, moreover, that is plausibly connected with conditionality. How is this to be done? I claim no originality for the answer I will offer. In some sense, it is part of the folklore of relevant logic.[20] If what follows has any originality, it is in dragging the idea from the subconscious of relevant logicians into the full light of day. In the rest of this section I will describe the main idea. In the next, we will see how it is implemented.

The idea that a proposition is a function is a familiar one in modern logic. For example, in intensional logics one can take a proposition as a function from worlds to truth values. One can think of this as something like the sense of the sentence: given a world/situation, it takes us to its truth value there. The idea that the propositional content of a conditional is a particular sort of function is also familiar. In intuitionist logic, the semantic content of a conditional, $\alpha \to \beta$, is a construction which applies to any proof of α to give a proof of β. This construction is obviously a function. I want to suggest that the conditional in relevant logic is also best thought of as a function. Clearly, a conditional is something which, in some sense, takes you from its antecedent to its consequent. It is therefore natural to think of the proposition expressed by the conditional $\alpha \to \beta$ as a function which, when applied to the proposition expressed by α, gives the proposition expressed by β. The ternary relation can be understood in these terms. Let us look at the details.[21]

[19] Priest (2008, 585).

[20] For example, the thought that applying a conditional is something like functional application is found in the motivating remarks of Slaney (1990); the idea that $Rxyz$ means that $x(y) = z$ can be found in Fine (1974); and Restall (2000, 12–13 and 246–8), notes the connection between relevant conditionals and functions.

[21] After this paper was written, Beall et al. (2012) was written. This offers three understandings of the ternary relation. The one offered in this paper is the second of these.

4.6 Interpreting Routley–Meyer Semantics

In what follows, I take it that we are dealing with a positive propositional relevant logic, so avoiding issues to do with the Routley Star. Let us consider the most fundamental of these, the relevant logic B^+. I will discuss its extensions in the next section.

A Routley–Meyer interpretation for this[22] is a tuple $\langle @, N, W, R, \nu \rangle$. W is a set of worlds (situations); N (the normal worlds) is a non-empty subset of W; @ is a distinguished member of W; R is a ternary relation on W; ν is a function which assigns a truth value (1 or 0), $\nu_w(p)$, to every parameter at each world, w.

For $x, y \in W$, the relation $x \leqslant y$ is defined as follows: $\exists n \in N, Rnxy$. The worlds of a structure must satisfy the following conditions:

R0 $@ \in N$
R1 $x \leqslant x$
R2 If $x \leqslant y$ and $Ryzw$ then $Rxzw$
R3 If $x \leqslant y$ and $\nu_x(p){=}1$ then $\nu_y(p){=}1$

where, in R3, p is any propositional parameter. R3 is called the *heredity condition*, and, employing the truth conditions of the connectives and R2, it can be shown to extend to all formulas, not just propositional parameters.

The truth conditions for the logical constants of the language are as follows:

T1 $\nu_w(\alpha \wedge \beta) = 1$ iff $\nu_w(\alpha) = 1$ and $\nu_w(\beta) = 1$
T2 $\nu_w(\alpha \vee \beta) = 1$ iff $\nu_w(\alpha) = 1$ or $\nu_w(\beta) = 1$
T3 $\nu_w(\alpha \to \beta) = 1$ iff for all x, y, such that $Rwxy$, when $\nu_x(\alpha) = 1$, $\nu_y(\beta) = 1$

α holds in an interpretation if $\nu_@(\alpha) = 1$, and an inference is valid if in every interpretation in which all the premises hold, the conclusion holds.[23]

We may understand the meanings of the various notions as follows. Sentences express propositions. We do not need to worry too much about what these are; they are just whatever it is that sentences express. I write a, b, etc. for propositions. If α and β express the propositions a and b, I write the propositions expressed by $\alpha \wedge \beta$, $\alpha \to \beta$, as $a \wedge b$, $a \to b$, etc. We do not need to worry too much, either, about what, exactly, worlds are. It will suffice that they are the sort of thing

[22] As found, for example, in Routley et al. (1982a, 4.1–4.6). They do not include a base world, @. Including one makes no difference to what is valid in the semantics, and brings the semantics into line with the abstract characterization I gave.

[23] Note, then, that the situation about which we reason is (represented by) @, which is to be thought of as coming with its own raft of alternative worlds.

characterized by a set of propositions. In fact, as a matter of convenience, we may simply identify a world with a set of propositions. Each world is closed under under conjunction, and is prime (that is, whenever a disjunction is a member, so is at least one disjunct). In particular, we have, for all $w \in W$:

P1 $\mathfrak{a} \wedge \mathfrak{b} \in w$ iff $\mathfrak{a} \in w$ and $\mathfrak{b} \in w$
P2 $\mathfrak{a} \vee \mathfrak{b} \in w$ iff $\mathfrak{a} \in w$ or $\mathfrak{b} \in w$

Further, say that $\mathfrak{a}$ *entails* $\mathfrak{b}$ just if every world that $\mathfrak{a}$ is in, $\mathfrak{b}$ is in. It follows that each world is also closed under entailment.

The propositions expressed by conditionals are functions. Specifically, the proposition $\mathfrak{a} \to \mathfrak{b}$ is a function which maps $\mathfrak{a}$ to $\mathfrak{b}$. One can think of the function as a procedure which takes certain propositions into others: one which, when applied to the proposition expressed by α, gives one expressed by β. Given the conceptual connection between conditions and inference, it is natural to take this procedure to be one grounded in inference. Thus, it might take things of a logical form of α into things of a logical form of β.[24]

Now, if $x, y \in W$, let $x[y]$ be:

$$\{\mathfrak{b}: \text{for some } \mathfrak{a} \in y,\ \mathfrak{a} \to \mathfrak{b} \in x\}$$

Thus, $x[y]$ is the result of taking any $\mathfrak{a}$ in y, and applying any function of the form $\mathfrak{a} \to \mathfrak{b}$ in x. Note that $x[y]$ may not be a world. For example, there is no reason to suppose it to be prime. However, we can use it to define the relationship R on worlds as follows:

$$Rxyz \text{ is } x[y] \subseteq z$$

In other words, $Rxyz$ iff whenever the result of applying any function, $\mathfrak{a} \to \mathfrak{b}$, in x to a proposition, $\mathfrak{a}$, in y is in z.[25] Given that a function and its application are involved, why there should be a three-place relation is obvious: one place is for the function; one is for its argument; and one is for its value. The third place in the relation simply records the propositions one gets by applying the relevant functions to the relevant arguments.

It remains to say what @, ν, and N are. @ is (represents) the situation about which we are reasoning. $\nu_w(\alpha) = 1$ means that α is true at a w: that is, if α

[24] There is a well-known connection between the Routley–Meyer semantics and the λ-calculus (and combinatory logic)—see, for example, Dunn and Meyer (1997). In the λ-calculus *all* objects are functions. This suggests that an investigation of the present proposal in that context might be fruitful.

[25] The semantics given here are the non-simplified semantics. In the simplified semantics, to give truth conditions for $\to$ uniformly in terms of R, we need the condition: $Rnyz$ iff $y = z$ (where n is a normal world). See Priest (2008, 10.2). For this condition to hold on the present account, we would need: $n[y] \subseteq z$ iff $y = z$. This clearly fails, since we can have distinct z_1 and z_2 for which $n[y] \subseteq z_1$ and $n[y] \subseteq z_2$.

expresses the proposition a, then $a \in w$. The members of N are exactly those worlds, n, such that for any a and b, $a \to b \in n$ iff a entails b. Thus, since worlds are closed under entailment, we have, for any w:

P3 If $a \to b \in n$ then if $a \in w$, $b \in w$

Given these explanations of the semantic notions, both the conditions R0–R3, and the truth conditions of the connectives make perfectly good sense. That is, they are justified by these understandings.

R0 says that, in the situation about which we are reasoning, $\to$ really represents the entailment relation. That is, we may interpret sentences of the form $\alpha \to \beta$ as saying that α entails β. In other words, $\to$ gets the right meaning. (This is not a vacuous constraint, since at non-normal worlds, $\to$ may represent a different relation.)

For R1: We need to show that, for some $n \in N$, $Rnxx$: that is, $n[x] \subseteq x$. Choose any $n \in N$, and let $b \in n[x]$. Then for some $a \in x$, there is an $a \to b \in n$. By P3, $b \in x$.

For R3: Suppose that $x \leqslant y$. Then for some $n \in N$, $Rnxy$: that is, $n[x] \subseteq y$. If $a \in x$ then, since $a \to a \in n$, $a \in n[x]$. So $a \in y$. That is, $x \subseteq y$. Now suppose that $\nu_x(p) = 1$. Then if a is the proposition expressed by p, $a \in x$. Hence, $a \in y$: that is, $\nu_y(p) = 1$.

For R2: Suppose that $x \leqslant y$. Then, as we have just seen, $x \subseteq y$. It follows that $x[z] \subseteq y[z]$. (For if $b \in x[z]$, then for some $a \in z$, $a \to b \in x$. Since $x \subseteq y$, $a \to b \in y$, so $b \in y[z]$.) Thus, if $Ryzw$, that is, $y[z] \subseteq w$, it follows that $x[z] \subseteq w$: that is, $Rxzw$.

Turning to the truth conditions: P1 and P2 obviously deliver T1 and T2. For T3: Suppose that the function $a \to b \in w$, and that $Rwxy$: that is, $w[x] \subseteq y$. Then if $a \in x$, $b \in y$. Conversely, if the function $a \to b \notin w$, then it is natural to suppose that there are worlds, x and y, such that $Rwxy$, with $a \in x$ and $b \notin y$. For take a world, x, which contains just a and what it entails; we can form y by applying all the functions $c \to d \in w$ to it. Because $a \to b \notin w$, b will not be in y. (More precisely, if $b \in y$, then there is something, c, entailed by a, such that $c \to b \in w$. But in that case, $a \to b$ would be in w, since $c \to b$ entails $a \to b$.)[26]

4.7 Extensions

Positive logics stronger than B^+ are obtained, in a standard way, by adding further constraints on the ternary relation. For example, the relevant logic R^+—

[26] Essentially, this is the heuristic which is implemented in constructing the canonical model in the completeness proof for relevant logics. See Routley et al. (1982a, 4.6).

the strongest standard-relevant logic—is obtained by imposing the following constraints:

R4 If $Rxyz$ then $Ryxz$
R5 If $Rxyz$ then $\exists w(Rxyw$ and $Rwyz)$
R6 If $\exists w(Rxyw$ and $Rwuv)$ then $\exists w(Rxuw$ and $Rywv)$

The functional interpretation of R does not, in itself, make these constraints plausible. Indeed, it makes them implausible. Consider R4. This says that for any x, y, z, if $x[y] \subseteq z$ then $y[x] \subseteq z$. $x[y]$ is obtained by applying functions in x to arguments in y; $y[x]$ is obtained by applying functions in y to arguments in x. These are not, in general, the same—functional application is not commutative! Similarly, R5 tells us that for all x, y, z, if $x[y] \subseteq z$ then, for some w, $x[y] \subseteq w$ and $w[y] \subseteq z$. But the things guaranteed to be in w are the results of applying functions in x to arguments in y. There is no reason to suppose that, if these are functions, applying them to arguments in y *again* will give the same things: the application of a second function could take us anywhere. Similar comments apply to R6.

In the canonical model construction, used in the completeness proofs for these stronger logics, one invokes the appropriate axiom to show that the corresponding constraint holds in the model. (Thus, in the case of R4, one invokes the axiom $\alpha \to (\beta \to \gamma) \vdash \beta \to (\alpha \to \gamma)$). One might therefore appeal to the plausibility of such inferences about conditionality to justify the corresponding constraint. However, in this case, the semantics cannot be used to justify the properties of the conditional, on pain of circularity. In other words, they fail to explain the *why* of things, as is required for a genuine applied semantics.[27] (See the discussion in section 4.3 of appealing to the behaviour of vernacular notions.)

The interpretation of the semantics we have been looking at cannot, therefore, be used to justify the stronger relevant logics—or if they can, this requires a much more complicated version of the story than the one told here. Some might see this as a vice. Personally, I see it as a virtue. One of the embarrassments of relevant logics is their multiplicity. A challenge has always been to single out one of the multiple as the correct relevant logic. From this perspective, the discriminating nature of the interpretation of the Routley–Meyer semantics which I have offered is an advantage. It is true that it does not justify the stronger relevant logics, like R, which have been the favourites of North American relevant logicians. Instead, it justifies weaker (depth-relevant) logics, like B. These have always been preferred

[27] In a similar way, one cannot invoke the plausibility of the inference $\Box\alpha \vdash \alpha$ to justify the reflexivity of the binary R in modal logic. If justification is to be forthcoming, this has to be in terms of the *meaning* of the binary R.

by Australian relevant logicians (such as myself), because of their applications to naive truth theory and set theory.

4.8 Conclusion

Let me summarize. As we saw, for a semantics to provide a satisfactory model-theoretic account of validity, it must be possible to understand it as an applied (depraved) semantics. Routley–Meyer semantics, and especially its ternary relation, have always had a problem being seen in this light. We now see that they can be. Specifically:

1. It is perfectly natural to understand the meaning of a conditional as a function.
2. If one does this, then an intelligible meaning for the semantic ternary relation is straightforward. Essentially, it records the results of applying the function which the conditional expresses.
3. Understanding the meaning of the conditional in this way motivates the relevant logic B^+, though not stronger relevant logics, like R^+.

Critics of the semantics of relevant logic have had a tendency to see them as perverted—a turning away from true semantics. In particular, the ternary relation has been taken to be depraved: that is, debased, corrupt (*OED*). What we have seen is that it is not: its depravity is not of the vicious kind, but of the virtuous.[28]

References

Beall, Jc, Brady, Ross, Dunn, J. Michael, Hazen, A.P., Mares, Edwin, Meyer, Robert K., Priest, Graham, Restall, Greg, Ripley, David, Slaney, John, and Sylvan, Richard. 2012. "On the ternary relation and conditionality." *Journal of Philosophical Logic* 41:595–612.

Copeland, B.J. 1979. "On when a semantics is not a semantics: some reasons for disliking the Routley–Meyer semantics for relevance logic." *Journal of Philosophical Logic* 8:399–413.

Copeland, B.J. 1983. "Pure and applied semantics." *Topoi* 2:197–204.

Dummett, Michael. 1975. "The justification of deduction." *Proceedings of the British Academy* 59:201–32. Reprinted as ch. 17 of *Truth and Other Enigmas*, London: Duckworth, 1978.

[28] A version of this paper was given at a workshop of the Foundations of Logical Consequence at the University of St Andrews, January 2009. It was also given to the meeting of the Australasian Association for Logic, Melbourne University, July 2009. I am grateful to the audiences for their comments, and particularly to Jc Beall, Colin Caret, Ira Kiourti, Stephen Read, Greg Restall, Crispin Wright, and Elia Zardini.

Dunn, J. Michael and Meyer, Robert K. 1997. "Combinators in structurally free logic." *Logic Journal of the IGPL* 5:505–37.

Fine, Kit. 1974. "Models for entailment." *Journal of Philosophical Logic* 3:347–72.

Haack, Susan. 1978. *Philosophy of Logics*. Cambridge: Cambridge University Press.

Kreisel, Georg. 1967. "Informal rigour and completeness proofs." In I. Lakatos (ed.), *Problems in the Philosophy of Mathematics*, 138–86. Amsterdam and London: North-Holland.

Mares, Edwin. 1997. "Relevant logic and the theory of information." *Synthese* 109:345–60.

Plantinga, Alvin. 1979. *The Nature of Necessity*. Oxford: Oxford University Press.

Priest, Graham. 2006. *Doubt Truth to Be a Liar*. Oxford: Oxford University Press.

Priest, Graham. 2008. *An Introduction to Non-classical Logic: From Ifs to Is*. Cambridge: Cambridge University Press, 2nd edition.

Restall, Greg. 1995. "Information flow and relevant logics." In J. Seligman and D. Westerståhl (eds.), *Logic, Language, and Computation: The 1994 Moraga Proceedings*, 463–77. Stanford, CA: CSLI Publications.

Restall, Greg. 1999. "Negation in relevant logics: how I stopped worrying and learned to love the Routley star." In D. Gabbay and H. Wansing (eds.), *What is Negation?* 53–76. Dordrecht: Kluwer.

Restall, Greg. 2000. *An Introduction to Substructural Logics*. London: Routledge.

Routley, Richard and Meyer, Robert K. 1973. "The semantics of entailment." In H. LeBlanc (ed.), *Truth, Syntax, and Modality*, 199–243. Amsterdam and London: North-Holland.

Routley, Richard, Plumwood, Val, Meyer, Robert K., and Brady, Ross. 1982a. *Relevant Logics and Their Rivals*, volume I. Atascadero, CA: Ridgeview.

Routley, Richard, Routley, Valerie, Meyer, Robert K., and Martin, Erroll. 1982b. "On the philosophical bases of relevant logic." *Journal of Non-classical Logic* 1:71–102.

Slaney, John K. 1990. "A general logic." *Australasian Journal of Philosophy* 68:74–88.

Smiley, Timothy J. 1993. "Can contradictions be true?, I." *Aristotelian Society Supplementary Volume* 68:17–33.

5

Proof-Theoretic Validity

Stephen Read

5.1 Introduction

Could analytic statements be false? An analytic statement, or analytic truth, is often characterized as one which is true in virtue of the meaning of the words. So it seems to follow that analytic statements are *ipso facto* true. But must they be true? That question is many-ways ambiguous. To begin with, there are, of course, analytic falsehoods, statements which are false simply in virtue of their meaning. But that aside, analytic truths are often thought to be necessarily true. Yet even necessary truths could have been false, on some accounts of necessity. In **S4**, what is necessary, is necessarily so; but in other modal systems, like **T**, that does not follow, and some necessary truths might not have been necessary. In this sense, the question is whether analytic truths, though in fact true, might have been false.

But these are not the issue I want to address. For the question whether analytic statements must be true has a further sense. Consider Tarski's discussion of semantically closed languages, for example. Tarski (1933, 164–5) claimed that natural languages are universal and semantically closed, and that semantic closure leads to inconsistency: that is, that natural language is inconsistent. This seems at first rather puzzling. How can a language, a set of sentences or statements, be inconsistent? Is it not theories, based on languages, that are consistent or inconsistent? The fact which Tarski was invoking is that natural languages are interpreted languages—they consist not just in a set of sentences, but in a set of interpreted, meaningful sentences.[1] In virtue of that meaning, some of those sentences may force themselves on us: for example, the Liar sentence, 'This

[1] Tarski (1933, 166). Note that even his formalized languages are interpreted: "we are not interested here in 'formal' languages and sciences in one special sense of the word 'formal', namely sciences to the signs and expressions of which no material sense is attached."

sentence is not true' (call it $\mathcal{L}$). Why is $\mathcal{L}$ true in virtue of its meaning: that is, of what it says? Because we can prove it true. Suppose $\mathcal{L}$ were not true. Then things would be as it says they are. So it would be true. Hence, by *reductio* (if it weren't true it would be true), $\mathcal{L}$ is true. But if $\mathcal{L}$ is true then $\mathcal{L}$ is also not true (since that is what it says), and so the meanings of certain terms in natural language—in particular, the word 'true'—commit their users to assenting to contradictory sentences. Yet the meaning of 'not' also commits its users to refusing to assent to contradictory sentences. Hence, simply in virtue of the meanings of words, users of such a language are subject to contradictory and inconsistent demands. The language itself is inconsistent. In this sense, certain statements may, although analytic, be false. '$\mathcal{L}$ is true' and '$\mathcal{L}$ is not true' are both analytic, but they cannot both be true, so one of them is false. Analytic statements can be false.

Boghossian (1997, 334) makes a useful distinction between metaphysical and epistemic analyticity:

> On [one] understanding, 'analyticity' is an overtly *epistemological* notion: a statement is 'true by virtue of its meaning' provided that grasp of its meaning alone suffices for justified belief in its truth. Another, far more metaphysical, reading of the phrase 'true by virtue of meaning' is also available, however, according to which a statement is analytic provided that, in some appropriate sense, it *owes its truth-value completely to its meaning*, and not at all to 'the facts'.

Boghossian believes that the metaphysical conception was undermined by Quine's arguments, but that the epistemic conception, grounding a notion of the a priori, survives them. Nonetheless, it seems, *pace* Boghossian, that meaning does not confer knowledge, since some statements to which the speaker is committed a priori are false. Indeed, Boghossian says only that knowledge of meaning leads to a justified belief in the resulting statements. For without the metaphysical underpinning, that belief and its justification cannot make the statement true. Some analytic statements are false.

But if analytic commitments are not guaranteed to be true, wherein lies their truth when they are true? In particular, what makes logical truths true, when they are true, if not the meaning of the logical constants? It is their necessity, their being true come what may. If one does not reject Quine's arguments against metaphysical analyticity, one will be condemned to endorse whatever one's concepts dictate. But those concepts may be confused or otherwise mistaken. That is where metaphysics and epistemology come apart.

Azzouni (2007) and Patterson (2007) have inferred from Tarski's observation that natural language is inconsistent; that it is in fact meaningless, if not trivial. Eklund (2002) and Scharp (2007), however, have drawn the more reasonable conclusion that inconsistent languages can be meaningful and useable, but stand in need of revision. The constitutive principles on which they are based—in particular, those for truth—pick out, for Scharp, inconsistent concepts which need further elucidation, like the pre-relativistic concept of mass. For Eklund, semantic competence "exerts pull" (as he puts it): speakers' semantic competence disposes them to accept the untrue premises or invalid steps in arguments concerning inconsistent concepts. To reject those premises or inferences would indicate lack of competence with the concepts concerned.

Scharp cites another example, Dummett's discussion of the term 'Boche'. Dummett (1973, 454) claims that simply using this term in a language commits its practitioners to inferring that Germans are barbaric and prone to cruelty. Simply as a consequence of using a word with a certain meaning, one is committed, analytically, to assert falsehoods. That someone is German warrants the claim that he is a Boche, and that he is Boche implies that he is barbaric. It follows that Germans are barbaric, by the very meaning rules of the term 'Boche'. But that is false. So analytic statements can be false.

This doctrine may sound reminiscent of dialetheism, the doctrine that there are true contradictions. But it adds a further twist to dialetheism. From the fact that the meanings of certain expressions have the consequence that they commit their users to assenting to contradictory statements, dialetheists infer that those contradictions are true. In contrast, the position I am exploring reverses this inference. Anyone who uses such expressions would indeed be committed to asserting contradictory statements. But since contradictory statements cannot both be true, it is clearly a mistake to use such expressions. Indeed, this was surely Tarski's response. What he effectively claimed was that both '$\mathcal{L}$ is true' and '$\mathcal{L}$ is not true', as formulated in a natural language, are analytic: that is, users of such languages are committed to assenting to them both. That is precisely the respect in which natural languages are, in Tarski's eyes, inconsistent. Consequently, Tarski's response was to "abandon . . . the language of everyday life and restrict [himself] entirely to formalized languages."[2]

There are three positions one might adopt about a concept such as 'true':

1. that of Eklund and others (perhaps including Tarski): that the semantic paradoxes show that the concept encapsulated in natural languages is inconsistent and needs to be revised;
2. that of the dialetheists: that the concept we have is indeed inconsistent, but we must accept this fact and its consequences; and

[2] Tarski (1933, 165).

3. my own position: that the paradoxes reveal that our theory of truth is mistaken, and more work is needed to find out how our concept of truth really works.[3]

My contention is that adopting certain logical conceptions commits one to analytic consequences some of which are mistaken and unacceptable. Only the right conceptions of conjunction, implication and so on yield the right account of analyticity. If we are not careful, our analysis may over-generate,[4] declaring invalid arguments to be valid.

What, then, is the advantage of proof-theoretic validity over model theory? The one is fallible, the other hollow. Proof-theoretic validity cannot guarantee truth preservation, but it does reveal immediately from the inference rules we use what the meaning of the constituent concepts is, and how they commit their user to their consequences. Model theory is hollow in that it either invokes essentially modal concepts, or deflates that modality by reduction to extensional terms. But Etchemendy showed that the reduction is conceptually confused:

> Tarski's analysis [of logical consequence] involves a simple, conceptual mistake: confusing the symptoms of logical consequence with their cause. Once we see this conceptual mistake, the extensional adequacy of the account is not only brought into question—itself a serious problem given the role the semantic definition of consequence is meant to play—but turns out on examination to be at least as problematic as the conceptual adequacy of the analysis.... Suppose we have an argument form all of whose instances preserve truth, just as the reductive account requires, but suppose that the only way to recognize this is, so to speak, serially—by individually ascertaining the truth values of the premises and conclusions of its instances... [T]he premises would provide no justification whatsoever for a belief in the conclusion. For, by hypothesis, knowing the specific truth value of the conclusion in question would be a prerequisite to recognizing the 'validity' of the argument. (Etchemendy 2008, 269)

Alternatively, one might try to accept the modal criterion at face value: an argument is valid if and only if it is impossible that the premises be true and the conclusion false. However, that puts validity beyond our reach: who knows what is possible and what is not? Proof-theoretic validity shows directly what inferences a concept permits. It is fallible, in that it may permit too much or too little. So conceptual revision (by rule revision) may be necessary in the light of (logical) experience. But that only serves to make it human.

[3] See, e.g., Read (2008; 2009).

[4] In Etchemendy's phrase: Etchemendy (1990, 8).

5.2 Analytic Validity

We see, then, that adopting a certain language may not only commit a speaker to believing certain statements; it may also commit a speaker to making certain inferences. Prior in his famous 'Runabout Inference Ticket' (Prior 1960) described such a commitment as "analytic validity". He didn't like it, and argued that validity must instead be based on truth preservation, not on meaning. Others claimed instead that the flaw lay in supposing that arbitrary rules can confer meaning on the relevant expressions. Arguably, 'tonk' had not been given a coherent meaning by Prior's rules. Rather, whatever meaning tonk-introduction had conferred on the neologism 'tonk' was then contradicted by Prior's tonk-elimination rule. But if rules were set down for a term which did properly confer meaning on it, then certain inferences would be "analytic" in virtue of that meaning. The question is: what constraints must rules satisfy in order to confer a coherent meaning on the terms involved?

Dummett introduced the term 'harmony' for this constraint: in order for the rules to confer meaning on a term, two aspects of its use must be in harmony. Those two aspects are the grounds for an assertion as opposed to the consequences we are entitled to draw from such an assertion. Those whom Prior was criticizing, Dummett claimed, committed the "error" of failing to appreciate "the interplay between the different aspects of 'use', and the requirement of harmony between them."[5] All aspects of an expression's use reflect its meaning. But when they are in harmony, each of the aspects contributes in the same way.

Dummett is here following out an idea of Gentzen's, in a famous and much-quoted passage where he says that "the E-inferences are, through certain conditions, *unique* consequences of the respective I-inferences."[6] The idea was that the meaning encapsulated in the introduction- or I-rule serves to justify the conclusions drawn by the elimination- or E-rule. But in fact, Dummett has loftier ambitions than this. He introduces the idea of the proof-theoretic justification of logical laws.[7] Not only does the I-rule justify the E-rule: the I-rule serves to justify itself.

Dummett distinguishes three grades of proof-theoretic justification. The first is familiar, namely, the notion of a derived rule, that one can reduce a complex inference, and so justify it, by articulating it into a succession of simpler, more

[5] Dummett (1973, 396). Cf. Dummett (1978, 221): "If the linguistic system as a whole is to be coherent, there must be a harmony between these two aspects." See also Dummett (1991, 215). He also put constraints on the introduction-rules. See section 5.4 below.

[6] Gentzen (1935, 81). [7] Dummett (1991, ch. 11).

immediate, inference-steps. Then the simpler steps of immediate inference justify and validate the indirect consequence.

The second grade of proof-theoretic validity rests on another familiar notion, namely, that of an inference being admissible. We may not be able to derive the conclusion from the premise or premises, but it may be that if the premises are derivable, so is the conclusion. The classic example of such a justification is cut-elimination, established by Gentzen for his sequent systems **LJ** and **LK** in his famous Hauptsatz. As a theorem about the elimination of Cut, this result takes a calculus which includes the Cut principle:

$$\frac{\Gamma \Rightarrow \Theta, D \quad D, \Delta \Rightarrow \Lambda}{\Gamma, \Delta \Rightarrow \Theta, \Lambda} \text{ Cut}$$

and shows that any derivation can be transformed into a derivation of the same end-sequent in which no application of the Cut rule is made. As a result about the admissibility of Cut, in contrast, one takes a calculus which does not include the Cut rule and shows that if the premises of the Cut rule are derivable in the system, then so is the conclusion.

There is a crucial difference between derived rules and an admissible rule like Cut, which reflects the difference between axiomatic and sequent calculi on the one hand and natural deduction systems on the other. What is purportedly "natural" about natural deduction systems is that they work from assumptions, as Gentzen claimed was the natural method of reasoning in mathematics.[8] Sequent calculi, by contrast, consist of apodeictic assertions, starting with axioms (aka basic sequents) and such that the end-sequent of a derivation is asserted categorically. Thus, whereas natural deduction derivations may be open—that is, contain open assumptions—sequent calculus derivations are always closed. The premises of a Cut inference can be interpreted as saying that their succedent-formulae follow from their antecedent-formulae as assumptions (given the tight connection between sequent calculi and the corresponding natural deduction system), but they say this categorically. In general, an admissible inference is one whose conclusion is provable (from closed assumptions) when its premises are provable. This will be generalized to the case of open assumptions in Dummett's third grade of proof-theoretic justification, but first let us concentrate on the second grade.

When transferring the concept of an admissible inference from sequent calculus (or axiomatic systems) to the natural deduction case, caution is necessary; and at

[8] Gentzen (1935, 74 ff.). Cf. Gentzen (1935, 2 ff.).

its heart is the notorious Fundamental Assumption. The Fundamental Assumption makes its first appearance in Dummett (1991, 252) in this form:

> Whenever we are justified in asserting [a] statement, we *could have* arrived at our entitlement to do so by [direct or canonical] means.

Dummett attributes the Fundamental Assumption to Prawitz in this form:

> Prawitz expressly assumes that, if a statement whose principal operator is one of the logical constants in question can be established at all, it can be established by an argument ending with one of the stipulated introduction rules. (1991, 252)

The reference is to a series of articles on the "foundations of a general proof theory" which Prawitz published in the early 1970s. (Recall that Dummett's book, though not published until 1991, had been originally given as the William James lectures at Harvard in 1976.) Prawitz's idea was to find a characterization of validity of argument independent of model theory, as typified by Tarski's account of logical consequence. Whereas Tarski's analysis deems an inference valid when it preserves truth under all reinterpretations of the non-logical vocabulary, Prawitz, following Gentzen's lead in the passage cited earlier, accounts an argument or derivation valid by virtue of the meaning or definition of the logical constants encapsulated in the introduction-rules. The idea is this: we take the introduction-rules as given. Then any argument (or in the general case, below, any argument-schema) is valid if there is a "justifying operation" (Prawitz 1973, 233) ultimately reducing the argument to the application of introduction-rules to atomic sentences:

> The main idea is this: while the introduction inferences represent the form of proofs of compound formulas by the very meaning of the logical constants when constructively understood and hence preserve validity, other inferences have to be justified by the evidence of operations of a certain kind.[9]

The operations are essentially the reduction steps in the proof of normalization, and depend on the Fundamental Assumption. An argument is valid if either it reduces to a non-logical justification of an atomic sentence, or it reduces to an argument whose last inference is an introduction inference and whose immediate subarguments are valid.[10]

The connection between the second grade of justification and the reduction to an atomic base is usefully spelled out by Humberstone (2011, section 4.13). Take

[9] Prawitz (1973, 234). Prawitz (n.d., 21) justifies the I-rules as follows: "To make an inference is to apply a certain operation to the given grounds for the premisses, and that the inference is valid is now defined just to mean that the result of applying this operation to grounds for the premisses is a ground for the conclusion, and hence it justifies the person in question in holding the conclusion true."

[10] See Prawitz (1973, 236).

a natural deduction system in which no rule discharges assumptions (or closes open variables): for example, with the sole connectives $\wedge$ and $\vee$. Say that an argument from Γ to A is proof-theoretically valid if for any atomic base Π (i.e. set of propositional variables, justified by what Dummett (1991, 204) calls "boundary rules") for which, for each $B \in \Gamma$, B can be derived from Π by I-rules alone, so can A. Then (Humberstone 2011, Theorem 4.13.3) "all and only the provable sequents of [the] natural deduction system for $\wedge$ and $\vee$ are proof-theoretically valid." That is, the E-rules extend what can be proved by the I-rules alone only to infer what can be "introductively derived" from an atomic base of closed assumptions if their premises can be so derived.

Following Prawitz, Dummett's third grade of proof-theoretic justification generalizes this account from closed arguments to open arguments: that is, argument-schemata.[11] Argument-schemata have either open assumptions which will later be closed by rules which discharge those assumptions, or free variables which will later be the eigenvariables of a quantifier rule. In Prawitz's discussion, these two cases are exemplified just by $\to$I and $\forall$I. Such an argument-schema is valid if, besides the earlier cases, the result of closing the assumption (by a closed proof) or the eigenvariable (by substitution of a closed term) is valid.

Let us consider a couple of examples: in this section, implication, and in section 5.3, negation. First, $\to$I:

$$\dfrac{\begin{matrix}[A]\\ \vdots \\ B\end{matrix}}{A \to B}\to I$$

inferring (an assertion of the form) $A \to B$ from (a derivation of) B, permitting the optional discharge of (several occurrences of) A. But A and B may themselves contain occurrences of '$\to$', so we need not only to infer (assert) '$\to$'-wffs, but also, as Dummett noted, to draw consequences from assertions of the form $A \to B$. Consequently, we need a further rule of $\to$E, which will tell us when we may do so. Accordingly, all further inferences should reduce, by the first grade of proof-theoretic justification, to application of $\to$I and $\to$E. The crucial task now is to identify the form of $\to$E which is justified by $\to$I by the second and third grades of proof-theoretic justification, so that all inference involving '$\to$' reduces to and is justified by the meaning conferred on '$\to$' by $\to$I.

Whatever form $\to$E has, there must be the appropriate justificatory operation of which Prawitz spoke. That is, we should be able to infer from an assertion of $A \to B$

[11] Prawitz (2006, 511) calls them "argument skeletons".

no more (and no less) than we could infer from whatever warranted assertion of $A \to B$. We can represent this as follows:[12]

$$\dfrac{A \to B \qquad \begin{matrix}\begin{bmatrix}[A]\\ B\end{bmatrix}\\ \vdots \\ C\end{matrix}}{C}\to E$$

That is, if we can infer C from assuming the existence of a derivation of B from A, we can infer C from $A \to B$. Then by the Fundamental Assumption, assuring us that there is a derivation α of $A \to B$ which terminates in an application of $\to$I, we can apply the justifying operation:

$$\dfrac{\dfrac{\begin{matrix}[A]\\ \alpha \\ B\end{matrix}}{A\to B}\to I \qquad \begin{matrix}\begin{bmatrix}[A]\\ B\end{bmatrix}\\ \vdots\\ C\end{matrix}}{C}\to E \qquad \text{reduces to} \qquad \begin{matrix}\begin{bmatrix}A\\ \alpha\\ B\end{bmatrix}\\ \vdots\\ C\end{matrix}$$

inferring C directly from the grounds (the actual derivation α of B from A) for asserting $A \to B$. Here we have replaced the assumption of a derivation of B from A, marked by $\begin{bmatrix}[A]\\ B\end{bmatrix}$, with the actual derivation α of B from A, written $\begin{bmatrix}A\\ \alpha\\ B\end{bmatrix}$.[13]

So much for the general theory. What, however, is the rule $\to$E which has here been justified? In particular, what does the minor premise actually say? It speaks of whatever (C) can be inferred from the assumption that B can be inferred from A. In other words, if we had a proof of A, we could infer B (and continue to infer whatever, C, from B):[14]

$$\dfrac{A\to B \qquad A \qquad \begin{matrix}[B]\\ \vdots\\ C\end{matrix}}{C}$$

[12] Cf. Schroeder-Heister (1984, 1294).

[13] This notation is inspired by, but extends further than, that introduced in Gentzen (1932), in passages which did not appear in the published version (translated in Gentzen (1935)). See also von Plato (2008). Note that in both proofs, the assumption A is closed, in the first being discharged by the application of $\to$I, in the second by its derivation or discharge within the overall derivation of C.

[14] See Dyckhoff (1988) and Negri and von Plato (2001, 8). For limits to the legitimacy of this procedure, see Schroeder-Heister (2012) and Read (2015).

Other things being equal,[15] we can permute the derivation of C from B with the application of the elimination-rule, to obtain the familiar rule of Modus Ponendo Ponens (MPP):

$$\dfrac{A \to B \qquad A}{B}\ \text{MPP} \\ \vdots \\ C$$

Following Francez and Dyckhoff (2012), I dub the general procedure by which we obtain the →E-rule from the I-rule, "general-elimination harmony". It is the inverse of Prawitz's justifying operation. The crucial idea is distilled by Negri and von Plato (2001, 6):

Inversion Principle: Whatever follows from the direct grounds for deriving a proposition must follow from that proposition.

Given a set of introduction-rules for a connective (in general, there may be several, as in the familiar case of '∨'), the elimination-rules (again, there may be several, as in the case of '∧') which are justified by the meaning so conferred are those which will permit a justifying operation of Prawitz's kind. Each E-rule is harmoniously justified by satisfying the constraint that whenever its premises are provable (by application of one of the I-rules, by the Fundamental Assumption), the conclusion is derivable (by use of the assertion-conditions framed in the I-rule).

Elsewhere (Read, 2010), I have discussed the application of the general-elimination procedure to '∧' and '∨'. The I-rules for these connectives differ from the I-rule for '→' in three ways: no assumptions are discharged, ∧I has two premises, and ∨I has two cases. The general conclusion reached was that, where there are m I-rules each with n_i premises ($0 \leqslant i \leqslant m$), there will be $\prod_{i=1}^{m} n_i$ E-rules. Take a formula $\$\vec{A}$ with main connective '\$' and immediate subformulae $\vec{A}$. Then the E-rules have the form :

$$\dfrac{\$\vec{A} \qquad \begin{matrix}[\pi_{1j_1}] \\ \vdots \\ C\end{matrix} \quad \cdots \quad \begin{matrix}[\pi_{mj_m}] \\ \vdots \\ C\end{matrix}}{C}\ \$E$$

where each minor premise derives C from one of the grounds, π_{ij_i} for asserting $\$\vec{A}$. π_{ij_i} may be a wff (as in ∧I), or a derivation of a wff from certain assumptions (as in →I). Then the justifying operation permits one to infer C from one of the grounds for assertion of $\$\vec{A}$ whenever one can infer C from $\$\vec{A}$ itself:

[15] In a particular case, we need to check that the permutation preserves the conditions on being a proof. For an example where that condition may fail, see section 5.4 below.

$$\dfrac{\dfrac{\begin{matrix}\alpha_1 & \dots & \alpha_n \\ \pi_{i1} & & \pi_{in_i}\end{matrix}}{\$\vec{A}}\$I \quad \begin{matrix}[\pi_{1j_1}] \\ \beta_1 \\ C\end{matrix} \quad \dots \quad \begin{matrix}[\pi_{mj_m}] \\ \beta_m \\ C\end{matrix}}{C}\$E \qquad \text{reduces to} \qquad \begin{matrix}\begin{bmatrix}\alpha_{j_i} \\ \\ \pi_{ij_i}\end{bmatrix} \\ \beta_i \\ C\end{matrix}$$

Having one minor premise in each E-rule drawn from among the premises for each I-rule ensures that, whichever I-rule justified assertion of $\$\vec{A}$ (here it was the i-th), one of its premises can be paired with one of the minor premises to carry out the reduction.[16]

5.3 The Fundamental Assumption

Before we turn to the second example, we need to give further consideration to the Fundamental Assumption. This is the claim that if there is a proof of $\$\vec{A}$, there is a proof of $\$\vec{A}$ in which the final step is an application of \$I. In the case of open arguments, this is false. Given a derivation of $\$\vec{A}$ by means of an open argument (i.e. an argument-schema), with open assumptions or free variables, there is no such guarantee. However, proof-theoretic justification of the third grade reduces to the second grade by closing those open assumptions and free variables. This is reflected in Prawitz's theorem on the form of normal deductions.[17] Define a *thread* as a sequence of successive formulae in a derivation running from an assumption to the end-formula, and let a *branch* be an initial segment of a thread which ends in the minor premise of an application of an E-rule (or in the end-formula if there is no such minor premise). Prawitz's result shows that, given a proof, there is a proof in which each branch is divided into two parts, an E-part and an I-part, separated by a minimum formula. Each formula in the E-part is major premise of an E-rule, and each formula in the I-part (except the last) is premise of an I-rule. Reduction of proof-theoretic justification of the third to the second grade depends on this result. Moreover, this result itself depends on the existence of what Dummett (1977, 112) calls "permutative reductions".

The proof of normalization consists in the removal of maximum formulae from derivations. A maximum formula is a formula which is both the conclusion of an

[16] The point carries over to the quantifier rules, and vindicates what was said in Read (2000, sections 2.5–2.6) about the quantifier rules: the multiplicity of cases of ∀E (one for each term) matches the multiplicity of premises in the real introduction-rule for '∀', and similarly for '∃'. Note that the justifying procedure not only ensures that the E-rule does not permit inference of more than is warranted by the I-rule; it also ensures that it is strong enough, in permitting inference of everything that is so warranted. See Read (2015).

[17] Theorem 3 in ch. III, section 2 of Prawitz (1965). Cf. Gentzen's "Sharpened Hauptsatz" in Gentzen (1935, section IV.2). Schroeder-Heister (2006, 531) calls the Fundamental Assumption a corollary of normalizability, but it is really a lemma essential to its proof.

I-rule and major premise of an E-rule, as was $\$\vec{A}$ above. Dummett, in a graphic expression, calls such a part of a derivation a "local peak" (Dummett 1991, 248), so that normalization is the "levelling of local peaks". However, before Prawitz's "justifying operations" can be applied to local peaks, a preliminary permutative reduction is needed:

> The other reduction steps are auxiliary, being principally concerned to rearrange the order in which the rules are applied, so that a proof in which a sentence is introduced by an introduction rule, and only later removed by means of an elimination rule in which it is the major premiss, can be transformed into one in which the elimination rule is applied immediately after the introduction rule to form a local peak. (Dummett 1991, 250)

What we have in general is called by Prawitz a "maximum segment" (Prawitz 1965, 49), what we might graphically describe as a "local plateau":[18] a succession of occurrences of the same formula as minor premises of applications of E-rules separating its introduction by an I-rule and elimination by the corresponding E-rule. The procedures by which we reduce plateaux to peaks are Dummett's permutative reductions. They correspond to the reductions of rank in the proof of Cut-elimination. By successively permuting the application of an I-rule for a connective with the E-rule for which its conclusion is a minor premise, we can eventually bring the formula in question face to face with the corresponding application of the E-rule for the same connective. Then what Dummett calls the "fundamental reduction step" which we described at the end of section 5.2 can be applied to eliminate the maximum formula entirely.

What this shows, however, is that the Fundamental Assumption is more than just an assumption. It requires proof that the rules as a whole are such that each branch has this hour-glass shape, with a minimum formula separating the branch into an E-part and an I-part. That is a holistic matter, depending on the interaction between the various rules and connectives. It cannot be simply assumed, but must be proved.

The other example which invites exploration is negation. Prawitz makes little mention of negation in his papers on general proof theory. Where it does appear, $\neg A$ is treated by definition as $A \to \bot$, where the "absurdity constant" $\bot$ is governed solely by an elimination-rule, from $\bot$ infer anything:

$$\frac{\bot}{A}\ \bot E$$

As far as I know, every other author follows Prawitz' lead except Dummett (1977, ch. 4); thereby, the harmony of '$\neg$' devolves on the harmony of '$\to$' (and '$\bot$').

[18] Gentzen calls it a "hillock" (*Hügel*): see von Plato (2008, 247).

Gentzen (1935, II 2.21, 5.2) treats negation in two ways in his published paper, as primitive and as defined, but using the absurdity constant $\perp$ in both cases. In the MS previously mentioned, he treated '$\neg$' as primitive without appeal to $\perp$. As introduction-rule, he took *reductio ad absurdum* (dubbed by him 'R') in this form:

$$\dfrac{\begin{matrix}[A] \\ \vdots \\ B\end{matrix} \qquad \begin{matrix}[A] \\ \vdots \\ \neg B\end{matrix}}{\neg A}\ \neg I$$

What elimination-rule does this justify? We can infer from $\neg A$ all and only what we can infer from its grounds. There is one I-rule with two premises ($m = 1, n_1 = 2$), so there will be two E-rules, one for each premise of the I-rule:

$$\dfrac{\neg A \qquad \begin{matrix}\left[\begin{matrix}[A] \\ B\end{matrix}\right] \\ \vdots \\ C\end{matrix}}{C}\ \neg E_1 \qquad \text{and} \qquad \dfrac{\neg A \qquad \begin{matrix}\left[\begin{matrix}[A] \\ \neg B\end{matrix}\right] \\ \vdots \\ C\end{matrix}}{C}\ \neg E_2$$

Applying the simplification as before, where we infer C from assuming the existence of derivations, respectively, of B and of $\neg$B from A, we obtain:

$$\dfrac{\neg A \qquad A \qquad \begin{matrix}[B] \\ \vdots \\ C\end{matrix}}{C} \qquad \text{and so} \qquad \begin{matrix}\dfrac{\neg A \qquad A}{B} \\ \vdots \\ C\end{matrix}$$

and

$$\dfrac{\neg A \qquad A \qquad \begin{matrix}[\neg B] \\ \vdots \\ C\end{matrix}}{C} \qquad \text{and so} \qquad \begin{matrix}\dfrac{\neg A \qquad A}{\neg B} \\ \vdots \\ C\end{matrix}$$

The second of these is simply a special case of the first, and so we have justified, by considerations of ge-harmony, a form of *ex falso quodlibet* (EFQ) as the matching elimination-rule for '$\neg$' (dubbed 'V' by Gentzen):

$$\dfrac{\neg A \qquad A}{B}\ \neg E$$

We need to check, however, that this rule does accord harmoniously with ¬I and permit a justification of Prawitz's kind. So suppose we have an assertion of ¬A justified by ¬I, immediately followed by an application of ¬E:

$$\dfrac{\dfrac{\begin{matrix}[A]\\ \alpha_1\\ \neg B\end{matrix} \qquad \begin{matrix}[A]\\ \alpha_2\\ B\end{matrix}}{\neg A}\,\neg I \qquad \begin{matrix}\beta\\ A\end{matrix}}{C}\,\neg E$$

If we now close the open assumptions of the form A in α_1 and α_2 with the derivation β, we obtain:

$$\dfrac{\begin{matrix}\left[\begin{matrix}\beta\\ A\end{matrix}\right]\\ \alpha_1\\ \neg B\end{matrix} \qquad \begin{matrix}\left[\begin{matrix}\beta\\ A\end{matrix}\right]\\ \alpha_2\\ B\end{matrix}}{C}\,\neg E \tag{†}$$

An obvious worry, and it was Gentzen's worry at this point, is that we still have an occurrence of the wff ¬B, major premise of an application of ¬E and possibly inferred by ¬I. Indeed, since A and B are independent, the degree of ¬B may be greater than that of ¬A. So has a suitable reduction been carried out?[19]

Gentzen's solution, described in the manuscript but not appearing in the published version (where the contradictories B and ¬B were replaced by '⊥'), is first to perform a new kind of permutative reduction on the original derivation of ¬A, so that it concludes in a single application of ¬I. So suppose that the derivation of ¬A concludes in successive applications of ¬I:

$$\dfrac{\dfrac{\begin{matrix}[A, B]\\ \beta_1\\ \neg C\end{matrix} \qquad \begin{matrix}[A, B]\\ \beta_2\\ C\end{matrix}}{\neg B}\,\neg I \qquad \begin{matrix}[A]\\ \alpha_2\\ B\end{matrix}}{\neg A}\,\neg I$$

The detour through ¬B is unnecessary. The derivation can be simplified as follows:

$$\dfrac{\begin{matrix}[A], \left[\begin{matrix}[A]\\ \alpha_2\\ B\end{matrix}\right]\\ \beta_1\\ \neg C\end{matrix} \qquad \begin{matrix}[A], \left[\begin{matrix}[A]\\ \alpha_2\\ B\end{matrix}\right]\\ \beta_2\\ C\end{matrix}}{\neg A}\,\neg I$$

[19] Dummett (1977, 154) seems not to address this worry.

By successive simplifications of this kind, we can ensure that β_1 does not conclude in an application of $\neg$I and so $\neg$B in (†) is not a maximum formula.

5.4 Fregean Absolutism

We have now shown the analytic validity of $\to$I, MPP, RAA (*reductio ad absurdum*, or $\neg$I) and EFQ (or $\neg$E). $\to$I and $\neg$I are justified directly, in defining the meaning of '$\to$' and '$\neg$', while MPP and EFQ are justified indirectly, as admissible rules, by showing that whenever their premises are derivable, so too is their conclusion. Consequently, whatever we can prove with these rules is analytic, by proof-theoretic justification of the first grade, in particular:

$$\cfrac{\cfrac{\overline{A}\ (1)}{B \to A}\ {\to}\mathrm{I}}{A \to (B \to A)}\ {\to}\mathrm{I}(1) \qquad \text{and} \qquad \cfrac{\cfrac{\cfrac{\overline{\neg A}\ (1) \quad \overline{A}\ (2)}{B}\ \neg\mathrm{E}}{A \to B}\ {\to}\mathrm{I}(2)}{\neg A \to (A \to B)}\ {\to}\mathrm{I}(1)$$

But by my lights, (*) $A \to (B \to A)$ and (**) $\neg A \to (A \to B)$, as general forms of inference, are false. As Ackermann put it:

> One would reject the validity of the formula $A \to (B \to A)$, since it permits the inference from A of $B \to A$, and since the truth of A has nothing to do with whether a logical connection holds between B and A.[20]

Indeed, EFQ is also invalid, allowing one apparently to infer anything whatever from a pair of contradictories, for in conjunction with other plausible theses, it can be shown that EFQ leads to the validation of invalid arguments.[21]

Where does the mistake lie? Arguably, it lies in the introduction-rules postulated for '$\to$' and '$\neg$' and the meanings thereby conferred on these connectives. That's what results in the theses (*) and (**), which follow analytically from those rules, as we saw in sections 5.2–5.3.

It is tempting to think that there's nothing in themselves wrong with (*) and (**); what is wrong is to suppose those theses hold of 'if' and 'not'. That is the moral often drawn from the Paradoxes of Material Implication.[22] It is thought that they are indeed "true" (analytically) of '$\to$' and '$\neg$', but they are paradoxical (counter-intuitive, that is, false) as applied to our pre-theoretic conceptions of implication

[20] Ackermann (1956, 113): "So würde man die Allgemeingültigkeit einer Formel $A \to (B \to A)$ ablehnen, da sie den Schluß von A auf $B \to A$ einschließt und da die Richtigkeit von A nichts damit zu tun hat, ob zwischen B und A ein logischer Zusammenhang besteht."

[21] See, for example, Read (1988, ch. 2).

[22] See, for example, Haack (1978, 200).

and negation, since there are false implications with false antecedent and others with true consequent.

This is a mistake. Take ¬I first. As we saw, ¬I justifies EFQ: that is, the claim that each statement is paired with another, its contradictory, which together entail everything. To be sure, A and 'If A then everything is true' together entail that everything is true, and so entail everything. But we can deny A without claiming that if A then everything is true. Not every falsehood is *that* false! A and ¬A cannot both be true, but their conjunction does not necessarily bring the heavens crashing down, as Tennyson's Lady of Shalott feared:

> Out flew the web and floated wide;
> The mirror cracked from side to side;
> 'The curse is come upon me', cried
> The Lady of Shalott.

We can justify denying A simply by showing that A implies a falsehood, without necessarily showing it implies everything. The contradictory of A is the weakest proposition inconsistent with A, and 'If A then everything is true' is too strong.

The concept 'If A then everything is true' is often referred to as Boolean negation.[23] It is a dangerous and unhelpful concept, which threatens to trivialize any theory. One does not need to be a dialetheist to want to reject it. Whether a theory contains both A and ¬A (where '¬' is Boolean negation), for some A, is in general undecidable. But closing under consequence (as one does to form a theory) results in the trivial theory if both such statements are in it. This is unhelpful. A better account of negation is given by De Morgan negation.[24]

Jean van Heijenoort contrasted absolutism with relativism in logic:

> Absolutism, . . . is the doctrine that there is one logic, that this logic is what has become known as classical logic, and, moreover, that such a logic is all-embracing and universal. Relativism is the opposite doctrine. (van Heijenoort 1985, 75)

He attributed absolutism to Kant, Frege, and Russell. Kant seems in strange company as an adherent of classical logic. But in light of this, and the fact that van Heijenoort recognizes the two doctrines as "tendencies", one might adapt the title of absolutism to the belief in one universal logic, whatever it is, which I have elsewhere termed "logical monism"(Read 2006).

We may well tolerate other logics, such as intuitionistic or dialetheic logic, or even classical logic with its commitment to EFQ, but that does not mean we accept them, or believe that they are suitable for the task of determining logical validity.

[23] See, for example, Priest (1990, 203).

[24] See, for example, Read (1988, section 7.6).

They give the wrong answer about certain arguments, either validating invalid arguments (over-generation), or invalidating valid arguments (under-generation). Nonetheless, this raises the question why these arguments are valid or invalid, even though certain logics give a contrary verdict. They are invalid when the premises can be true while the conclusion is false; they are valid when this is impossible. Classical and intuitionistic logic fail as logics of the conditional, since the truth of the consequent does not suffice to make a conditional true, nor does the falsity of the antecedent. There are false conditionals with false antecedent, such as 'If I didn't post the letter I burned it', assuming merely that I posted it; and with true consequent, such as 'If second order logic is undecidable it is incomplete', since other logics, such as first-order logic, are undecidable but still complete. So the compound conditionals, 'If I posted the letter, then if I didn't post it I burned it' and 'If second-order logic is incomplete, then if it's undecidable then it's incomplete' (of the form (**) and (*) respectively), are false, since they have (or could have) true antecedent and false consequent.

A first thought concerning what is wrong with →I (as given above) is that it allows vacuous discharge of the assumption, as in the above proof of (*). Prawitz (1965, 84) showed that we can circumvent this restriction and still prove (*), by use of the rules for '∧' and '∨':

$$\cfrac{\cfrac{\cfrac{\cfrac{\overline{A}\,(1) \qquad \overline{B}\,(2)}{A \wedge B}\wedge I}{A}\wedge E}{B \to A}\to I(2)}{A \to (B \to A)}\to I(1)
\qquad\qquad
\cfrac{\cfrac{\cfrac{\cfrac{\overline{A}\,(1)}{A \vee (B \to A)}\vee I \qquad \overline{A}\,(3) \qquad \cfrac{\overline{B}\,(2) \qquad \overline{B \to A}\,(4)}{A}\to E}{A}\vee E}{B \to A}\to I(2)}{A \to (B \to A)}\to I(1)$$

Of course, these proofs are not in normal form, but, more importantly, they cannot be normalized without permitting vacuous discharge of assumptions in →I. With this restriction, →I does not interact holistically with the rules for '∧' and '∨' to allow normalization. When the maximum formula is removed by the reduction step, the later application of →I to derive B → A is no longer legitimate if vacuous discharge is disallowed. Prawitz (1965, 84 n.2) floats the idea that one could simply restrict proofs to proofs in normal form, but realizes that this would prevent us chaining proofs together in ways we find convenient and natural. For chaining two proofs in normal form together might result in a non-normal derivation, for example:

$$\cfrac{\cfrac{A \qquad \overline{B}}{A \wedge B}}{B \to (A \wedge B)}
\qquad\qquad
\cfrac{\cfrac{\cfrac{B \to (A \wedge B) \qquad \overline{B}}{A \wedge B}}{A}}{B \to A}$$

Chaining these proofs together and eliminating the maximum formula $B \to (A \wedge B)$ results in the non-normalizable derivation of $B \to A$ from A given above.

The moral is that harmony is not enough to guarantee validity. Harmony ensures that the consequences of an assertion are no more and no less than the meaning encapsulated in the introduction-rule warrants. But that meaning may itself be corrupt. An example is the I-rule for • in Read (2000, section 2.8), a formal Liar whose assertion warrants its own denial. One might try to bar such monsters by a restriction on the form of I-rules. The •I-rule breaches Dummett's proposed constraint on I-rules, his "complexity condition":

> The minimal demand we should make on an introduction rule intended to be self-justifying is that its form be such as to guarantee that, in any application of it, the conclusion will be of higher logical complexity [i.e. degree] than any of its premises and than any discharged hypothesis. (Dummett 1991, 258)

As with many proposed solutions to the Liar, out goes the baby with the bath-water: for example, Gentzen's rule R (i.e. $\neg$I) does not satisfy Dummett's condition, yet seems otherwise harmless. Moreover, the complexity condition does not help with our present difficulty, since $\to$I satisfies the condition, yet leads directly to (*).

Prawitz (1985; 2006) and, following him, Schroeder-Heister (2006) try to restrict further the proof-theoretic validity which results from the I-rules by relativizing it to the existence of a justification procedure:[25]

> My approach is now to let the arguments for which validity is defined consist of argument skeletons together with proposed justifications of all the inferences that are non-canonical. (Prawitz 2006, 514)

As we have seen, whether there are such justifications, or reductions, will depend on the form of the E-rules and on the interaction of the various rules for the different connectives (in the end, on the Fundamental Assumption). But if the E-rules are in harmony with the I-rules (and the Fundamental Assumption can be accepted for argument-schemata as in section 5.3) and so are admissible, then they are justified and the appropriate reductions do exist. In that case, the justification cannot be denied, and the consequent derivations are analytically valid, even if they fail to preserve truth. So such a relativization is ineffective.

What has been realized over the past forty years or so (in the theory of relevance logic, in linear logic, and in the theory of sub-structural logics generally) is that one needs to distinguish two different ways of combining premises. They may be used side by side to entail a conclusion, or one may be applied to the other. For example,

[25] Prawitz accepts Dummett's complexity condition on the I-rules: see Prawitz (2006, 515).

$A \wedge B$ follows from A and B in tandem, whereas B follows from applying $A \to B$ to A. In this way, we can develop a theory of '$\to$' and '$\neg$' which does not have such problematic consequences as (*) and (**). What is valid is then analytically valid in virtue of the meanings of '$\to$', '$\neg$', and so on. This is not the place to spell out the details, which require making the assumptions in the rules explicit and the way they are combined to yield the conclusion. There is nothing new in this: the task is to find the right (formulation of the) rules, and hence the real meaning of the logical terms, by checking their consequences and revising accordingly.[26] The moral we can draw, however, is that what logic one is committed to depends on the meaning one gives to the logical particles, encapsulated in the rules for their assertion, and so adopting the wrong logic may result in asserting falsehoods. The latter is something to be avoided.

5.5 Conclusion

A dominant account of validity was given shape by Tarski, saying that an inference is valid if it preserves truth through all substitutions for the non-logical vocabulary. This is a reductionist enterprise, attempting to reduce the real modality in the criterion, 'the premises cannot be true without the conclusion', to the possibility of re-interpretation. It is a hostage to fortune whether the reduction is available (as Etchemendy (1990) observed), and at best turns the obscure question of validity into the more obscure one of the possibility of true premise and false conclusion. In contrast, the idea of proof-theoretic validity is that validity of inference is based on rules of proof and the meanings of the logical constants encapsulated in those rules. Unlike Tarski's account, no division is required between the logical and descriptive vocabulary. In Tarski-validity, the essential idea is truth preservation regardless of the meaning of the descriptive terms, while in proof-theoretic validity it is proof in accordance with the meanings of the logical terms given by the proof-rules. Those inferences simply in virtue of the meanings of the logical terms constitute the formal validities; others are materially valid in virtue of the meanings of (both logical and) non-logical terms.[27]

Thus a proof-theoretically valid inference is analytically valid in virtue of the meanings of the logical constants specified by the rules for their application. Prawitz wrote:

[26] See, for example, Read (1988, ch. 1).

[27] See, for example, Read (1994).

Once we know the condition for correctly asserting a sentence, we also know when to accept an inference and when to accept that a sentence follows logically from a set of premisses.[28]

However, despite Prawitz's use of 'know' here, analytic truth, and analytical validity, does not guarantee truth or validity. Use of an expression with a certain meaning can commit its user to the a priori assertion of falsehoods or to the endorsement of invalid inferences. The classic examples are Tarski's claim that natural languages are inconsistent, by virtue of the meaning they give to the term 'true', and Prior's observation that validity can be trivialized by adoption of his rules for the term 'tonk'. An understanding of 'true' (if Tarski is right) and of 'tonk' commits their users to a justified belief in the correctness of certain statements and of certain inferences. But justified belief is famously not enough to guarantee truth. Hence Tarski refrains from using 'true' in its natural language guise and opts for an account of truth in stratified formal languages; Prior and others refrain from using 'tonk' with Prior's rules.

Prawitz and, following him, Dummett set out to articulate the mechanisms by which the rules for a logical term result in an analytical commitment to the validity of the resulting inference. The core idea is that all aspects of the term's meaning should be in harmony. Meaning can be conferred by I-rules, E-rules, and the proofs composed of such rules. But when the rules are in harmony, they all determine the same meaning. In particular, the whole meaning is then contained in the introduction-rule or rules. The elimination-rule is in harmony and is justified by the meaning so conferred if it is admissible: that is, if its conclusion is provable (without the rule, from some non-logical base) whenever its premises are. This is Dummett's second grade of proof-theoretic justification. Even here, there is a crucial assumption, what Dummett terms the "Fundamental Assumption": namely, that the elimination-rule can be permuted with the other E-rules so that its major premise is an application of the corresponding I-rule. In the familiar systems, this is true, but nonetheless it is a fact which needs to be proved. In general, it is a holistic assumption about the interaction between the rules for the different logical terms, so that local plateaux (or "hillocks") can be reduced to mere peaks.

Moreover, natural deduction proofs consist not only of closed proofs, but also of open proofs: that is, of derivations of conclusions from assumptions, and derivations generalizing on free variables. The Fundamental Assumption is simply

[28] Prawitz (1985, 168). Cf. (Prawitz, 1973, 232): "An argument that is built up . . . of other arguments or argument schemata is thus valid by the very meaning of the logical constants . . . ; it is valid by definition so to speak."

false of open proofs. To apply the Fundamental Assumption in order to show the E-rules admissible, it needs to be shown that the proof can be articulated into a succession of branches, each ending in the application of an E-rule, and then the open assumptions and free variables in each branch can be closed off to obtain a closed proof for which the Fundamental Assumption is available.

What is good about the notion of proof-theoretic validity is that it recognizes that what rules one adopts determines the meaning of the logical terms involved and commits one to accepting certain inferences as valid. What is bad is to infer from this that those inferences really are valid. Proof-theoretic validity serves an epistemological function to reveal how those inferences result from the meaning-determining rules alone. But it cannot serve the metaphysical function of actually making those inferences valid. Validity is truth preservation, and proof must respect that fact.

Nonetheless, this is not to equate validity with preservation of truth through arbitrary replacement of the non-logical vocabulary. That is hostage to fortune through the richness or poverty of the vocabulary and availability of suitable models, and so can also result in the validation of invalid inferences (and failure to validate valid ones). Moreover, it is hollow and unworkable since it equates validity simply with membership of a class of arguments all of which are valid, as Etchemendy (2008) observed. Validity is necessary truth preservation, in itself dependent on the meanings of the constituent propositions. To that extent, Eklund and Scharp are right: be careful what you wish for—or choose to mean—for you may receive it, even though it may not be what you want, or what is true or valid.[29]

References

Ackermann, Wilhelm. 1956. "Begründung einer strengen Implikation." *Journal of Symbolic Logic* 21:113–28.

Azzouni, Jody. 2007. "The inconsistency of natural languages: how we live with it." *Inquiry* 50:590–605.

Boghossian, Paul. 1997. "Analyticity." In B. Hale and C. Wright (eds.), *A Companion to the Philosophy of Language*, 331–68. Oxford: Blackwell.

Dummett, Michael. 1973. *Frege: Philosophy of Language*. London: Duckworth.

Dummett, Michael. 1977. *Elements of Intuitionism*. Oxford: Oxford University Press.

Dummett, Michael. 1978. "The philosophical basis of intuitionistic logic." In *Truth and Other Enigmas*, 215–47. London: Duckworth.

Dummett, Michael. 1991. *Logical Basis of Metaphysics*. London: Duckworth.

[29] This work is supported by Research Grant AH/F018398/1 (Foundations of Logical Consequence) from the Arts and Humanities Research Council, UK.

Dyckhoff, Roy. 1988. "Implementing a simple proof assistant." In J. Derrick and H. Lewis (eds.), *Proceedings of the Workshop on Programming for Logic Teaching, Leeds 1987*, 49–59. University of Leeds.

Eklund, Matti. 2002. "Inconsistent languages." *Philosophy and Phenomenological Research* 64:251–75.

Etchemendy, John. 1990. *The Concept of Logical Consequence.* Cambridge, MA: Harvard University Press.

Etchemendy, John. 2008. "Reflections on consequence." In D. Patterson (ed.), *New Essays on Tarski and Philosophy*, 263–99. Oxford: Oxford University Press.

Francez, Nissim and Dyckhoff, Roy. 2012. "A note on harmony." *Journal of Philosophical Logic* 41:613–28.

Gentzen, Gerhard. 1932. "Untersuchungen über das logische Schliessen." Manuscript 974:271 in Bernays Archive, Eidgenössische Technische Hochschule Zürich.

Gentzen, Gerhard. 1935. "Untersuchungen über das logische Schliessen." *Mathematische Zeitschrift* 39:175–210, 405–31. English translation 'Investigations concerning logical deduction' in Szabo (1969, 68–131).

Haack, Susan. 1978. *Philosophy of Logics.* Cambridge: Cambridge University Press.

Humberstone, Lloyd. 2011. *The Connectives.* Cambridge, MA: MIT Press.

Negri, Sara and von Plato, Jan. 2001. *Structural Proof Theory.* Cambridge: Cambridge University Press.

Patterson, Douglas. 2007. "Inconsistency theories: the importance of semantic ascent." *Inquiry* 50:552–8.

Prawitz, Dag. 1965. *Natural Deduction: A Proof-Theoretical Study.* Stockholm: Almqvist and Wiksell.

Prawitz, Dag. 1973. "Towards the Foundation of a General Proof Theory." In P. Suppes, L. Henkin, A. Joja, and G.C. Moisil (eds.), *Logic, Methodology and Philosophy of Science IV: Proceedings of the 1971 International Congress*, 225–50. Amsterdam: North-Holland.

Prawitz, Dag. 1985. "Remarks on some approaches to the concept of logical consequence." *Synthese* 62:153–71.

Prawitz, Dag. 2006. "Meaning approached via proofs." *Synthese* 148:507–24.

Prawitz, Dag. n.d. "Validity of inferences." Revised version of a paper presented to the 2nd Launer Symposium on the Occasion of the Presentation of the Launer Prize 2006 to Dagfinn Føllesdal.

Priest, Graham. 1990. "Boolean negation and all that." *Journal of Philosophical Logic* 19:201–15.

Prior, Arthur N. 1960. "The runabout inference-ticket." *Analysis* 21:38–9.

Read, Stephen. 1988. *Relevant Logic.* Oxford: Blackwell.

Read, Stephen. 1994. "Formal and material consequence." *Journal of Philosophical Logic* 23:247–65.

Read, Stephen. 2000. "Harmony and autonomy in classical logic." *Journal of Philosophical Logic* 29:123–54.

Read, Stephen. 2006. "Monism: the one true logic." In D. DeVidi and T. Kenyon (eds.), *A Logical Approach to Philosophy: Essays in Honour of Graham Solomon*, 193–209. New York: Springer.

Read, Stephen. 2008. "The truth schema and the Liar." In S. Rahman, T. Tulenheimo, and E. Genot (eds.), *Unity, Truth and the Liar: The Modern Relevance of Medieval Solutions to the Liar Paradox*, 3–17. New York: Springer.

Read, Stephen. 2009. "Plural signification and the Liar paradox." *Philosophical Studies* 145:363–75.

Read, Stephen. 2010. "General-elimination harmony and the meaning of the logical constants." *Journal of Philosophical Logic* 39:557–76.

Read, Stephen. 2015. "General-elimination harmony and higher-level rules." In H. Wansing (ed.), *Dag Prawitz on Proofs and Meaning*, 293–312. New York: Springer.

Scharp, Kevin. 2007. "Replacing truth." *Inquiry* 50:606–21.

Schroeder-Heister, Peter. 1984. "A natural extension of natural deduction", 1–30. *Journal of Symbolic Logic* 49:1284–300.

Schroeder-Heister, Peter. 2006. "Validity concepts in proof-theoretic semantics." *Synthese* 148:525–71.

Schroeder-Heister, Peter. 2012. "Generalized elimination inference, higher-level rules, and the implications-as-rules interpretation of the sequent calculus." In E. Haeusler, L.C. Pereira, and V. de Palva (eds.), *Advances in Natural Deduction*, 1–30. New York: Springer.

Szabo, M.E. (ed.). 1969. *The Collected Papers of Gerhard Gentzen*. Amsterdam and London: North-Holland.

Tarski, Alfred. 1933. "The concept of truth in formalized languages." In *Logic, Semantics, Metamathematics*, 152–278. Translation by J. H. Woodger, 1956. Oxford: Clarendon Press.

van Heijenoort, Jean. 1985. "Absolutism and relativism in logic." In *Selected Essays*, 75–83. Naples: Bibliopolis.

von Plato, Jan. 2008. "Gentzen's proof of normalization for natural deduction." *Bulletin of Symbolic Logic* 14:245–57.

PART III

Properties and Structure of Logical Consequence

6

The Categoricity of Logic

Vann McGee

Apart from such oddities as the stylized cries of barnyard animals, terms of a natural language don't have a natural meaning. They get their meanings from our use. Sometimes "use" is understood narrowly, restricted to the use of language to modify behavior, but it can also be taken more liberally, to include uses that are private and mental as well as those that are public and behavioral. On the broad understanding of "use," once we make the observation that the etymological impact of a pattern of usage may depend on the environment in which it occurs, that meanings result from use seems like a harmless platitude.

What exactly "meanings" are is up for grabs. A natural thesis is that the meanings of the logical connectives are given by determining their contributions to the truth conditions of sentences, and that those contributions are described by the truth tables. Unfortunately, it's a thesis that's hard to sustain, mainly on account of vagueness. Vagueness occurs when usage fails to determine sharp boundaries for a term. If "poor" is such a term, then it is possible for Clare to be in a mildly impecunious state in which there is no fact of the matter whether "poor" applies to her. There are people of whom usage dictates that "poor," as it's used in a particular context, applies, and others of whom it dictates that it doesn't apply, but there are yet others for whom linguistic custom fails to determine whether "poor" applies to them. If usage prescribes that "poor," as it's used in a particular context, can be applied to people in Clare's circumstances, then "Clare is poor" is true in that context. If usage prescribes that "poor" can't be applied, then "Clare is poor" is false in the context. What happens if Clare is on the border, so that usage doesn't determine whether "poor" applies to people like her? The natural thing to say is that, in such a case, "Clare is poor" will be neither true nor false. However, the truth tables, as they've traditionally been written, leave no room for gaps.

Epistemicists, like Timothy Williamson (1994) and Roy Sorenson (1988; 2001), deny that usage leaves gaps. They contend that usage determines, for each context in which the word "poor" can be meaningfully used, a down-to-the-last-penny-and-stick-of-gum partition into people who satisfy "poor," as it's used in that context, and those who satisfy "not poor." If you're an epistemicist, you won't share my reason for wariness of simple compositional semantics. The casual ease with which we are able to employ the word "poor," without first learning advanced accounting techniques or studying the detailed nuances of property law, makes it hard for me to believe that usage established the sharp partition that epistemicists postulate. So does the fluidity with which our usage adapts itself to varying contexts. Whether someone counts as "poor" varies from one context to another, depending mainly on the comparison class, and we are able effortlessly to adapt our use of the word to changing contexts.

Even epistemicists have occasional failures of bivalence[1] to worry about, for example, when there are nondenoting names, but this only means that the application of classical semantics will require the prior verification of appropriate presuppositions. Bivalence will also fail for moral and aesthetic judgments, according to expressivists. But an epistemicist expressivist can give a standard, bivalent semantics for all the sentences that don't contain moral or aesthetic vocabulary, and the sentences that can't be given the standard treatment will be marked by their vocabulary as requiring special handling. Accounting for the logic of moral discourse will remain an urgent problem, but at least it's a problem that can be isolated.

Vagueness cannot be isolated, because vagueness is everywhere, or at least, everywhere outside pure mathematics. Russell (1923) argues persuasively that our efforts at scientific exactitude succeed in making scientific terminology significantly more precise than everyday speech, but they never succeed in eliminating inexactitude entirely. We should be reluctant to go on to conclude, as Russell (1923, 88f.) does, that traditional logic is "not applicable to this terrestrial life, but only to an imagined celestial existence." We should be reluctant precisely because vagueness is ubiquitous, so that if we abjure classical reasoning whenever we are reasoning with vague terms, we renounce all employment of the classical modes of reasoning outside pure mathematics. We should be especially resolute in our resistance if we think the alternative to classical logic is a multivalent

[1] Bivalence is the doctrine that every sentence is either true or false, but not both. In saying this, I am being sloppy in two ways. First, I am restricting "sentence" to the sentences used to make assertions, and second, I am ignoring context.

compositional semantics along the lines of the strong 3-valued logic of Kleene (1952, section 54). Such logics are far too feeble to fulfill our cognitive needs.

Vagueness leaves truth-value gaps, and there are differences in semantic status among the sentences in the gap. Keep in mind that a sentence isn't simply true, false, or neither, but true, false, or neither in a context. Thus on Tuesday morning, Clare might be a borderline case of "poor," but on Tuesday afternoon she might find a winning lottery ticket. If the prize is large enough, it might elevate her into the class of people to whom "poor" definitely does not apply, so that, whereas on Tuesday morning, "Clare is poor" is neither true nor false, on Tuesday night it's false. A smaller prize will still leave her on the border, but she's still noticeably better off on Tuesday night than Tuesday morning. We want to say that "Clare is poor" is truer, or perhaps we should say, more nearly true, on Tuesday morning than Tuesday night. Vagueness produces degrees of truth, intermediate between truth and falsity. The degrees aren't always comparable. If Clare is a borderline case of "poor" and also a borderline case of "tall," we wouldn't expect to be able to make sense of the question "Is Clare as poor as she is tall?" or "Is 'Clare is poor' as true as 'Clare is tall'?" but sometimes the comparisons make sense. They make sense for "Clare is poor" before and after her lucky lottery find.

Vagueness leaves truth-value gaps. The view that truth-value gaps are incompatible with classical logic is still widely held, but it's mistaken, as van Fraassen (1966) makes clear. We can sharpen van Fraassen's point a little bit. Let $\mathfrak{B}$ be an algebra with three binary operations $+$, $\cdot$, and $\rightarrow$, a unary operation $-$, a constant 0, and a partial ordering $\leqslant$. Define a *$\mathfrak{B}$-valuation* to be a function v taking sentences to members of $|\mathfrak{B}|$ meeting the conditions that $v(\varphi \vee \psi) = v(\varphi) + v(\psi)$, $v(\varphi \wedge \psi) = v(\varphi) \cdot v(\psi)$, $v(\varphi \supset \psi) = v(\varphi) \rightarrow v(\psi)$, $v(\sim \varphi) = -v(\varphi)$, and $v(\perp) = 0$. An argument is *$\mathfrak{B}$-valid* iff, for any $\mathfrak{B}$-valuation v and any element b of $|\mathfrak{B}|$, if v assigns a value $\geqslant b$ to each of the premises of the argument, it also assigns a value $\geqslant b$ to the conclusion. The theorem, whose proof is long but not hard, is that natural deduction rules of classical logic preserve $\mathfrak{B}$-validity if and only if $\mathfrak{B}$ satisfies the axioms that make it a Boolean algebra.

The slogan is: Boolean-valued semantics are necessary and sufficient for classical logic. Like most slogans, it oversimplifies. First, it takes for granted a compositional, truth-conditional semantics. Second, it presumes that the standard for good arguments is what Field (2008, section 10.6) calls *strong validity* (the value of a conclusion is at least as great as the greatest lower bound of the values of the premises) rather than *weak validity* (the conclusion is assigned value 1 if the premises are assigned value 1). Strong validity is better, both because it acknowledges the degrees intermediate between simple truth and simple falsity and because it maintains the validity of such mainstays of classical reasoning as conditional proof (If you have derived ψ from $\Gamma \cup \{\varphi\}$, you may derive $(\varphi \supset \psi)$ from Γ) and classical *reductio ad absurdum* (If you have derived φ from $\Gamma \cup \{\sim \varphi\}$, you may derive it from Γ alone).

In its fidelity to the norms of classical reasoning, 2-valued semantics isn't any better than any other Boolean valued-semantics. Nonetheless, 2-valued semantics has a huge advantage: you can see where the norms come from. The truth values of compound sentences are determined by the truth values of simple sentences as described by the truth tables, and the valid arguments are the ones that are invariably truth preserving. Once we introduce the additional truth values, such straightforward explanations are no longer available. The disjunction, "Either Clare is poor, or she is not poor but very nearly so," is true, but neither of its disjuncts is true. The disjunction, "Either Clare is poor or not poor," is not only true but logically valid. But what makes it true, when neither disjunct is true?

There aren't any brute semantic facts, semantic features of the world that are unsupported by non-semantic features. In general, if a sentence is true (in a context), it is made true by facts about usage that give the sentence the truth conditions it has (in the context), and facts about what the world is like that ensure that the truth conditions are met. For logical truths, the facts about the world and the considerations of context fall out of the picture. Usage somehow makes the logically valid sentences true, but for "Clare is poor or not poor," the usual pathway by which a disjunction is made true—a disjunction is made truth by making one or both disjuncts true—is blocked off.

I would like to propose reversing the usual order of explanation of logical consequence and logical validity, which starts from truth conditions and identifies the logically valid inferences as the ones that are reliably truth preserving. Instead, following a program loosely attributed to Gerhard Gentzen (1935), let us start with deductive rules and use them to generate truth conditions. We introduce logical symbols into a formal language from which such terms were previously absent by stipulating that their use is to be governed by the rules of inference.

Introducing the new rules permits us to assert sentences that weren't part of the original language. Assertion is governed by the overarching principle that one is only permitted to assert sentences that are true, so if adopting the rules permits us to assert a sentence, the sentence asserted has to be true. But it can't have been true before we adopted the rules, because before we adopted the rules, sentences containing the new symbols weren't meaningful at all. So adopting the rules has made the derivable sentences true. The sentences derivable from the empty set are true by stipulation, in the same way that a definition is true by stipulation. The rules "implicitly define" the logical connectives.

The program starts with the platitude that the meanings of expressions are given by their use, and combines it with the observation that the preeminent use of logical words is their role in deductive inferences, to get to Gentzen's key idea that the meanings of the logical connectives are given by the rules of inference. In the end, I'll want to identify the meaning of a word with its contribution to the truth conditions of sentences. I don't propose this seriously as an analysis of "meaning," but only as identifying the aspects of meaning that will be relevant to our discussion here.

The sleight-of-hand, whereby we raise a question about natural languages and respond with an answer about formal languages, should not go unnoted. The purpose of the formal languages is to provide a simplified and idealized model of ordinary reasoning, but the match-up between the behavior of "$\vee$," "$\wedge$," and "$\sim$" and that of "or," "and," and "not" is by no means perfect. ("$\supset$" and "if..., then" are, in my opinion, hopelessly mismatched, but that's a story for another day.) Neither in natural languages nor in formal languages are the connectives introduced by the adoption of explicit rules. When authors invent artificial connectives, they typically introduce them by providing natural-language translations. The real story I want to tell isn't an account of how the connectives were introduced but a story about their role in the mature language. The story is that the meanings of the connectives are determined by the inferences one is disposed to make. The fictional history, recounting how the rules have been brought into the language by explicit stipulation, is only intended to make the real story vivid. Origin myths are ahistorical, but they can be useful nonetheless.

According to the mythical history, we give the connectives their meanings by stipulating how they are to be used. There have to be some constraints on our stipulative prerogative. This is demonstrated dramatically by Arthur Prior's (1960) "runabout inference ticket," which introduces the connective "TONK" by two rules:

From $\{\varphi\}$, you may deduce (φ TONK ψ).

From $\{(\varphi \text{ TONK } \psi)\}$, you may deduce ψ.

He notes that, given the standard structural principles that permit you to form chains of inferences, these rules permit you to derive anything from anything. The introduction of connectives by rules of inference was supposed to play a role analogous to explicit definitions, but explicit definitions never make this type of mischief. The reason is that explicit definitions are conservative: any conclusion that doesn't contain the defined term that you can derive from premises that don't contain the defined term with the aid of the definition, you can also derive without it. Simply replace the defined term by the defining expression throughout the derivation. (This shows that an explicit definition provides a conservative extension "in the syntactical sense" of Craig and Vaught (1958). Explicit definitions are also conservative "in the semantic sense," which requires that every model of the original language can be expanded to a model of the extended language. Within classical logic, semantically conservative extensions are always syntactically conservative, on account of the completeness theorem, but not conversely.)

Nuel Belnap (1962) proposed conservativeness as a constraint on introducing new connectives by stipulated rules. If, using the rules, you can derive a conclusion not containing the new connectives from premises that don't contain them, then it must be possible to reach the same conclusion using only the inferential capacities you had before the new rules' introduction. Belnap doesn't specify exactly how the old and new rules are to be combined, but the simplest proposal is this: Define a *consequence relation* to be a relation between sets of sentences and sentences that meets the conditions that all the elements of a set are consequences of it, and that, if all the members of Δ are consequences of Γ and φ is a consequence of Δ, then φ is a consequence of Γ. Start with a base consequence relation on a language $\mathcal{L}$, and form the smallest consequence relation that contains the base relation and is closed under the new rules. Then the enlarged consequence relation shouldn't introduce any new pairings of sets of sentences of $\mathcal{L}$ with sentences of $\mathcal{L}$ that weren't part of the base relation.

It's easy to convince ourselves that we can conservatively add the rules of the classical sentential calculus to any consequence relation in which the classical connectives are not already present. Suppose that χ isn't derivable from Ω by the original rules. Form a classical valuation ν by stipulating that a sentence of the original language is assigned the value "true" by ν if and only if it's derivable from Ω by the original rules, that a negation is assigned "true" by ν if and only if the negatum is assigned "false" by ν, that a disjunction is assigned "true" by ν if and only if one or both disjuncts are assigned "true" by ν, and so on. Define $\vdash_\nu$ to be material implication according to ν, so that a sentence φ of the extended language counts as a consequence of Γ iff either the valuation assigns the value "true" to φ or it assigns "false" to some of the members of Γ. $\vdash_\nu$ is a consequence relation that includes the base consequence relation for the original language and is closed under the rules of the classical sentential calculus, so it contains the smallest such relation, and it excludes the pair $\langle \Omega, \chi \rangle$.

Belnap proposed a further condition, intended to legitimate talk about "the" logical operation introduced by a system of rules, by guaranteeing that the rules pin down a unique meaning for the new connective. The idea is that, if the rules fail to single out a unique meaning, so that there are several candidates for the semantic value of the new connective (whatever the "semantic value" of a connective might turn out to be), it should be possible to disambiguate, by introducing different connectives for different candidates. In the same way that we can disambiguate "bank" by introducing two words "riverbank" and "moneybank," we can distinguish two candidates for what "$\supset$" means by introducing two symbols "$\supset_1$" and "$\supset_2$." Belnap's uniqueness condition tells us that, if, contrariwise, our rules nail down the meaning of "$\supset$" precisely, then there won't be two distinct candidates, and our efforts at disambiguation will falter. If we introduce two connectives, "$\supset_1$" and "$\supset_2$," that play the inferential role the rules prescribe for "$\supset$," they can't designate different candidates because there's only one candidate. If φ_1 and φ_2 are respectively obtained from the ordinary formula φ by substituting "$\supset_1$" and "$\supset_2$" for "$\supset$," then φ_1 and φ_2 will be logically equivalent, and it will be possible to derive φ_2 from $\{\varphi_1\}$ and to derive φ_1 from $\{\varphi_2\}$. The uniqueness condition is less urgent than conservativeness, since we can live comfortably enough with a little vagueness, but we can't be content with a regime that allows us to make substantive judgments true by stipulation. Even so, other things being equal, less vagueness is to be preferred to more, and an optimal system of rules will satisfy both uniqueness and conservativeness. Belnap's rule for "PLONK"—From $\{\varphi\}$, you may derive (φ PLONK ψ)—surely hasn't done enough to tell us what "PLONK" means, since it is compatible with the rule that (φ PLONK ψ) should mean what one ordinarily means by $\varphi \vee \psi$, by $\varphi \vee \sim \psi$, or by φ.

The classical rules for sentential calculus satisfy uniqueness as well as conservativeness. This is shown by an induction on the complexity of formulas. To see how it's done, it will be enough to look at one illustrative case, the derivation of $(\psi \supset_2 \theta)$ from $\{(\psi \supset_1 \theta)\}$. From $\{(\psi \supset_1 \theta), \psi)\}$, we can derive θ by *modus ponens* for "$\supset_1$." Consequently, using conditional proof for "$\supset_2$," we can derive $(\psi \supset_2 \theta)$ from $\{(\psi \supset_1 \theta)\}$.

The proof is easy. Too easy. The only rules it uses are conditional proof and *modus ponens* (You may derive ψ from $\{\varphi, (\varphi \supset \psi)\}$. But these rules are permitted by both intuitionist and classical logic, leading us to the conclusion that the conditional as used by intuitionists is logically equivalent to the conditional as used classically. But that's absurd. The two conditionals can't be logically equivalent, because Peirce's Law, which let's you derive φ from $\{((\varphi \supset \psi) \supset \varphi)\}$, is valid classically but not intuitionistically.

There is a similar situation with negation. *Ex contradictione quodlibet* (From $\{\varphi, \sim \varphi\}$ you may derive anything you like) and the intuitionistic version of *reductio ad absurdum* (If from $\Gamma \cup \{\varphi\}$ you have derived $\sim \varphi$, you may derive $\sim \varphi$ from Γ alone) suffice for Belnap's uniqueness condition, and they are valid in both intuitionistic and classical logic, but only the latter gives us double negation elimination (From $\{\sim\sim \varphi\}$ you may derive φ) and classical *reductio ad absurdum.*

The uniqueness proof, which is due to J.H. Harris (1982), proves too much, and once we have it, we see that the conservativeness proof proves too little. It was supposed to assure us that the introduction of the classical connectives wouldn't allow us to make any inferences not involving the classical connectives that we couldn't make before. But we have no such assurance. The Harris construction enables us to derive the classical conditional $((\varphi \supset_C \psi) \supset_C \varphi)$ from the intuitionistic conditional $\{((\varphi \supset_I \psi) \supset_I \varphi)\}$. Because Peirce's Law is legitimate classically, we can derive φ from $\{((\varphi \supset_C \psi) \supset_C \varphi)\}$. Transitivity permits us to go on to derive φ from $\{((\varphi \supset_I \psi) \supset_I \varphi)\}$, in spite of the intuitionistic invalidity of Peirce's Law.

Imagine yourself an intuitionist. As such, you'll have misgivings about the conservativeness proof I gave earlier, regarding it as establishing, at best, the contrapositive of conservativeness. However, Gentzen (1935), along the way to proving bigger things, gives a constructive argument for conservativeness that you will find persuasive. Taking the conservativeness argument at its advertised value, you'll be entitled to introduce new connectives governed by the classical rules of inference, in addition to the familiar connectives governed by the intuitionistic rules. Your understanding of what "or," "not," and "if" mean won't have changed—you'll still understand them intuitionistically—but you'll have additional connectives available for use purely as aids to calculation. But once you have the new connectives, you'll be able to use the Harris argument to establish Peirce's Law, double negation elimination, and classical *reductio* not only for the new connectives, but for the familiar ones. This is an unwelcome result. Your intuitionism has been overturned by a mere stipulation. The conservativeness argument has failed to do its job.

To understand the discrepancy between what the conservativeness proof promises and what it delivers, we need to look more carefully at what is required for a consequence relation $\vdash^+$ on a language $\mathcal{L}^+$ to extend a consequence relation $\vdash$ on a smaller language $\mathcal{L}$. Natural deduction systems have two kinds of rules of inference: *direct* rules, which Prawitz (1965) calls *proper* rules, which stipulate that certain conclusions are derivable from certain premises; and *conditional* or *improper* rules, which tell you that, if you've derived some things, you may derive some further things. Examples of direct rules are *modus ponens*, Peirce's law, *ex contraditione quodlibet*, double negation elimination, and the structural rule that tells you that a set of premises entails each of its elements.

Conditional rules include conditional proof, both versions of *reductio ad absurdum*, and the structural rule that tells you that, if you have derived each member of Δ from Γ and you have derived φ from Δ, you may derive φ from Γ. Given a consequence relation $\vdash$ on a language $\mathcal{L}$, a simple extension of $\vdash$ to a consequence relation on a larger language $\vdash^+$ is a consequence relation on $\mathcal{L}^+$ that includes $\vdash$. A rule-closed extension will be a consequence relation on $\vdash^+$ that includes $\vdash$ and, moreover, is closed under the conditional rules of $\vdash$. If the base consequence relation $\vdash$ includes an intuitionistic conditional "$\supset_I$" governed by conditional proof and *modus ponens*, all that the conditional proof rule will require of a simple extension is that, if φ, ψ, and the members of Γ are sentences of $\mathcal{L}$ with $\Gamma \cup \{\varphi\} \vdash \psi$, we'll have $\Gamma \vdash^+ (\varphi \supset_I \psi)$. A rule-closed extension will require that, if φ and ψ are sentences in $\mathcal{L}$ and Γ is a set of sentences of $\mathcal{L}^+$ with $\Gamma \cup \{\varphi\} \vdash^+ \psi$, we have $\Gamma \vdash^+ (\varphi \supset_I \psi)$.[2]

The conservativeness proof showed that intuitionist logic, like every other logical system that doesn't already contain the classical connectives, has a conservative simple extension that includes classical logic. The proof that intuitionistic logic doesn't have a conservative extension that includes classical logic, since if there were such an extension, there would be an intuitionistic proof of Peirce's Law, showed the nonexistence of a conservative rule-closed extension. The argument required us to use conditional proof for "$\supset_I$" in cases in which Γ contains sentences outside the intuitionistic language.

For talk about rule-based extensions even to make sense, we need to think of a logical system as a system of rules, and not merely as a collection of ordered pairs from $\wp(\mathcal{L}) \times \mathcal{L}$. We need to think of the rules as principles of correct reasoning, and not merely as a device for describing the outcomes of correct reasoning.[3] Rule-closed extensions will be what are germane to the present chapter, where we are developing the thesis that the rules fix the meanings of the connectives. I don't intend my neglect of simple extensions as a disparagement of the philosophical merits of an instrumentalist approach.

Thinking of a logical system as a system of rules, rather than as a pairing of sentences, raises the stakes when we're deciding to adopt a new rule. A feature of human languages is that they are forever changing. We enlarge the language by enabling it to express concepts that weren't expressible before. We do this when we develop a new scientific theory with concomitant vocabulary, and also when we bring a new puppy home from the pound. There is nothing objectionable about having rules that are specific to a particular language. There is, however, an objection to regarding such language-specific rules as *logical* principles. If logic is

[2] One could go further and require the extended language to be closed under the syntactic formation rules of the base language, so that, whenever φ and ψ are sentences of $\mathcal{L}^+$, $(\varphi \supset_I \psi)$ will be a sentence of $\mathcal{L}^+$. We won't be concerned with these syntactic extensions here.

[3] Joshua Schechter (2011) gives a system of algebraic rules for generating the classical validities that doesn't yield uniqueness in Belnap's sense.

to fulfill its Aristotelean ambition of providing fully general, universally applicable principles of correct reasoning, the methods it provides can't be dependent on the quirks of the language we happen to be speaking at the moment. Logical principles need to be resilient enough that they will continue to serve us faithfully even as we advance beyond our current vocabulary.

The distinction I want to draw is familiar from the axiomatization of theories by schemata. Acceptance of the induction axiom schema, "$((N0 \wedge (\forall x)((Nx \supset (Rx \supset Rs(x)))) \supset (\forall x)(Nx \supset Rx))$," as part of Peano Arithmetic (PA) requires at least this much: you should be willing to accept the universal closure of every open sentence obtained from the schema by substituting an arithmetical formula for "R." So understood, PA has nonstandard models, in which the familiar natural numbers are isomorphic to a proper initial segment of the things the model designates "N." While the nonstandard model satisfies all the instances of the induction axiom schema, it doesn't satisfy the principle of mathematical induction. The standard initial segment, which contains 0 and is closed under successor but doesn't exhaust "N," is a counterexample. To capture the principle of mathematical induction, you need to understand the induction axiom schema in an open-ended way, so that it will continue to be upheld even if we enrich the language by adding new predicates that aren't part of the language of arithmetic. The nonstandard model fails to satisfy open-ended induction, since it will falter when we introduce a predicate true of the standard initial segment. Open-ended induction is, as Kreisel (1967) noted, tantamount to second-order induction, which we can express by prefixing the second-order quantifier "$\forall R$" to the schema.

An axiom schema is a rule that permits you to assert sentences of a certain syntactic form, and it can be understood either in a restrictive, language-immanent way or in an expansive, language-transcendent version. We can apply the same distinction to other rules, among them the rules of natural deduction. For the rules to count as genuinely logical principles they must have language-transcendent applicability.

Following up on the general idea that the meaning of a word is fixed by speakers' practices in using the word, our thesis is that the meanings of logical words are given by speakers' inferential practices. More specifically, the meanings of the sentential-calculus connectives (in our idealized language) are given by the natural deduction rules. The meanings of the connectives are so robust that we'll continue to know what the connectives mean even as the other parts of the language change, and our commitment to the rules is so resolute that it will be maintained throughout the changes. In proposing this, I don't want to go to the extreme of saying that any inferential system whose methods of reasoning with the classical connectives don't include the rules is deficient or defective. Normal systems of modal logic include the rule of necessitation (From φ, you may infer $\Box\varphi$). If they also included conditional proof, you could assume φ, derive $\Box\varphi$ by necessitation, then discharge the assumption to derive the invalid conclusion $(\varphi \supset \Box\varphi)$. The fact that modal logicians aren't willing to draw the unwanted conclusion isn't a sign of faintheartedness in their commitment to classical sentential calculus. The rules are intended to preserve modal validity. The natural deduction rules preserve the property that, if you are entitled to accept the premises of the argument, you are entitled to accept the conclusion, on the basis of the premises alone. They do not, however, preserve modal validity.

The rules of deduction, I want to claim, fix the meanings of the connectives, and they are installed in the language by stipulation, the way explicit definitions are made available by stipulation.[4] For this contention to make sense, we need to be assured that the new rules are harmless. While they instruct us in the use of the new symbol, they don't have any untoward implications for sentences not involving the new symbol. If we intend the new rules as fundamental principles of reasoning, not specific to any particular language, we need assurance that the rules will remain harmless even as the language grows. We'll want to be sure that introduction of the new symbol isn't impeding the advancement of science by blocking the future development of useful theoretical tools that otherwise would have been available. This is a strenuous requirement, and the condition that the rule has to yield a conservative extension (using either the simple or rule-closed sense of "extension") doesn't guarantee it. To see this, let the base implication $\vdash$ be the trivial consequence relation on a language $\mathcal{L}$, according to which $\Gamma \vdash \varphi$ iff $\varphi \in \Gamma$. Introduce a new connective "$\not{c}$" by the following rules:

If you have derived φ from $\{\psi\}$, you may derive $\not{c}\psi$ from $\{\varphi\}$.

You may derive ψ from $\not{c}\psi$.

Since the trivial consequence relation doesn't have any nonstructural rules, over $\vdash$ the two notions of extension coincide. To see that the rules for "$\not{c}$" are conservative over $\vdash$, define the *base* of a sentence to be the result of erasing all the "$\not{c}$"s, so that, $\mathrm{base}(\varphi) = \varphi$ for φ in the base language, and $\mathrm{base}(\not{c}\psi) = \mathrm{base}(\psi)$, and define a consequence relation "$\vdash^{+}$" by stipulating that $\Gamma \vdash^{+} \varphi$ iff $\mathrm{base}(\varphi)$ is

[4] I am being noncommittal about whether we are talking about a public language or an individual's idiolect. With a public language, we have the possibility of outliers, speakers who, despite their intention to conform to community logical standards, use the connectives eccentrically. It seems to me that the best explanation of why such people fail to engage in normal inferential practices is that they don't have a perfect grasp of the meanings of the connectives. Williamson (2007, ch. 4) (who very kindly takes me as the exemplar of an outlier, an honor I greatly appreciate) disagrees, for reasons I won't go into here.

an element of the image under *base* of Γ. Then $\vdash^+$ is a conservative extension of $\vdash$, and it is closed under the "$\not\subset$" rules. If $\{\psi\} \vdash^+ \varphi$, then $\text{base}(\varphi) = \text{base}(\psi)$, and so $\{\varphi\} \vdash^+ \not\subset\psi$.

The "$\not\subset$" rules are conservative, but they are by no means harmless, for they prevent our developing the language in other ways that might have proved advantageous. Before "$\not\subset$," we were free to introduce the symbol for disjunction, obeying (among other rules) "$\vee$"-introduction, which permits the derivation of $(\varphi \vee \psi)$ from $\{\varphi\}$, and also from $\{\psi\}$. After "$\not\subset$," "$\vee$"- introduction leads to tonkish catastrophe. You may derive $(\varphi \vee \psi)$ from $\{\varphi\}$, and you can also derive it from $\{\psi\}$, which means that you can derive $\not\subset\psi$ from $\{(\varphi \vee \psi)\}$. You can derive ψ from $\{\not\subset\psi\}$. Linking the chain together yields a derivation of ψ from $\{\varphi\}$. Looking at it the other way, "$\vee$"-introduction is conservative over $\vdash$, but it ties our hands, leaving us unable to introduce "$\not\subset$," in the same way that introducing "$\supset_C$" over $\vdash$ renders us unable to introduce "$\supset_I$."

Looking back at the paradigm case of implicit definition, the introduction of a new predicate by means of axioms, we find that there a conservative extension doesn't tie our hands. It doesn't prevent us from freely expanding the language in other ways. Suppose we start with a theory Γ on a language $\mathcal{L}$, and we introduce a new predicate "R" by adopting a theory $\Delta(R)$ in such a way that $\Gamma \cup \Delta(R)$ is a conservative extension (in either the syntactical or the semantic sense) of Γ. If we go on to extend Γ to a stronger theory Ω, whose axioms may include new vocabulary (although they don't contain "R"), $\Omega \cup \Delta(R)$ will be conservative (in the same sense) over Ω.[5] The "$\vee$" and "$\not\subset$" example shows that we don't have the same kind of stability of conservativeness when we're introducing new logical operators, which means that, in devising new sentential connectives, we can't afford to be nearly so freewheeling.

Conservativeness, with respect to either notion of extension, is a language-immanent condition, so it is perhaps not surprising that it doesn't suffice to vindicate language-transcendent rules of inference. Arnold Koslow (1992) has a way of thinking about the logical connectives that will be helpful in trying to develop a language-transcendent perspective. Conditional proof and *modus ponens*, he says, position $(\varphi \supset \psi)$ as the weakest sentence that, together with φ,

[5] For conservativeness in the semantic sense, the proof is trivial. For syntactical conservativeness, suppose that $\Gamma \cup \Delta(R)$ is syntactically conservative over Γ, let Ω be an extension of Γ not containing "R," and let φ be a consequence of $\Omega \cup \Delta(R)$ that doesn't contain "R." We can find sentences $\omega_1, \omega_2, \ldots, \omega_m$ in Ω so that the conditional $((\omega_1 \wedge \omega_2 \wedge \ldots \wedge \omega_m) \supset \varphi)$ is a consequence of $\Delta(R)$. By the Craig Interpolation Theorem, there is a sentence θ of $\mathcal{L}$ such that θ is a consequence of $\Delta(R)$ and $((\omega_1 \wedge \omega_2 \wedge \ldots \wedge \omega_m) \supset \varphi)$ is a consequence of $\{\theta\}$. By conservativeness, θ is a consequence of Γ, and so φ is a consequence of Ω. This result continues to hold if we are introducing many new predicates, and not just "R."

implies ψ. More precisely, what the rules do is to assign the sentence a role that can only be played by the sentence and other sentences logically equivalent to it. Similarly, intuitionistic *reductio* and *ex contradictione quodlibet* fix $\sim \varphi$ as the weakest (up to logical equivalence) sentence that, together with φ, implies every sentence; equivalently, $\sim \varphi$ is the weakest (up to logical equivalence) sentence that, together with φ, implies falsum, "$\perp$," which implies every sentence. "$\vee$"-introduction and proof by cases establish $(\varphi \vee \psi)$ as the strongest (up to equivalence) sentence that is implied by both $\{\varphi\}$ and $\{\psi\}$.

In classical logicians' estimation, the classical conditional is strictly weaker than the intuitionistic conditional. For $(\varphi \supset_I \psi)$ to be true, $(\varphi \supset_C \psi)$ has to be, not merely true, but provable. $\{\varphi, (\varphi \supset_I \psi)\}$ and $\{\varphi, (\varphi \supset_C \psi)\}$ both entail ψ, but $(\varphi \supset_C \psi)$ is a strictly weaker sentence that, together which φ, entails ψ, so "$\supset_C$" better fits our intentions in using the conditional. Conditional proof is valid intuitionistically because "$(\varphi \supset_I \psi)$" is the weakest sentence in the intuitionistic language that, together with φ, entails ψ, but that's just because of the expressive limitations of the intuitionistic language. Enrich the language by introducing the classical connectives, and you'll be able to construct a still weaker sentence that, together with φ, entails ψ, so $(\varphi \supset_I \psi)$ will relinquish its status. Conditional proof is valid intuitionistically, but only in a language-immanent way. Similarly with "$\sim_I$." Intuitionistic *reductio* is valid intuitionistically only because of the expressive limitations of the intuitionists' language.

The intuitionist sees things differently, judging that many of the expressions that the classical logicians call "sentences" have the right syntactic form, but not the proper epistemic backing, to count as meaningful sentences. The classicist hasn't provided the verification conditions that would be required to make her sentences genuinely meaningful. Once illegitimate sentences are cast out of the language, the purported violations of intuitionistic conditional proof disappear.

This standoff shows just how difficult it will be to settle the controversy between intuitionists and classicists by rational argument. We would like to compare the merits of the proposals put forward by each side, but such a comparison presumes that partisans of both sides are speaking meaningfully, a presumption that stacks the deck in the classicists' favor. Even if the dispute were easy, however, it wouldn't be our task to judge it here. Here our concern is a question internal to classicism: can the classical logician give an account according to which the rules of inference affix meanings to the sentential connectives in such a way as to render the classically endorsed inferences legitimate?

We intend to exercise the power that we get from embracing the rules open-endedly, so that they'll continue to hold even after we expand the language by adding new sentences. Our exemplar is the open-ended acceptance of an axiom schema. We know what the effect of the unimpeded acceptance of an axiom schema is. It amounts to the acceptance of the schema's second-order universal closure. We can say this because we know what the relevant extensions of the language are—we add to the language by introducing new predicates—and we have a model theory—a generalized semantic theory—that tells us what predicates can do. At present, we are expanding the language by adding new sentences, so we need an understanding of what sentences can do.

A purely syntactic approach, which counts as a new sentence for the purpose of the rules any expression that behaves like a sentence syntactically, has a ruinous outcome. Take the Annihilator to be an expression that is grammatically a perfectly ordinary atomic sentence, whose use is governed by the extraordinary rule that, from any sentence that contains the Annihilator as a part, you may derive any sentence you like. Admitting the Annihilator into a language with an open-ended disjunction-introduction rule will enable you to derive anything from anything.

Introducing the Annihilator by itself, over a trivial consequence relation, gives you a conservative extension. There being no way to derive a sentence with the Annihilator as a part, its awesome destructive power is never triggered. The disjunction-introduction rule is likewise innocuous, but put the two together and you produce a tonkish collapse. The difficulty here echoes a familiar objection to consistency as the criterion of mathematical existence. It is possible to make two existence claims, each consistent but jointly inconsistent.

We have an understanding of the semantic role of predicates, and this lets us know what the effect would be of adding new predicates to the language of arithmetic. The new predicates would enable us to construct formulas that had as their extensions sets of numbers that weren't previously definable. To get an analogous notion of a language-transcendent rule of inference, we'll need to have an understanding of the semantic role of sentences, so that we can say what the possibilities are for newly minted sentences to fill the role.

The theory we want is ready to hand. Indeed, one of the core tenets of analytic philosophy is the doctrine that sentences express propositions. Propositions express the contents of thoughts and sentences, but they are themselves independent of mind and language. They are necessary entities that have their truth conditions essentially. The modern theory of propositions was developed by Frege (1892), who was particularly interested in getting the truth conditions for mental attitude attributions and indirect speech reports, and it was embraced enthusiastically. There have been a few dissenters, most notably Quine (1960), but propositionalism has been a dominant philosophical thesis of the twentieth century, and thus far of the twenty-first.

The propositionalist consensus is broad, but it isn't deep. People agree in saying that there are propositions, but they have different things in mind when they say it.

They have a variety of conceptions of propositions, which they employ for a variety of purposes, some of them incompatible. To try to discern from the many voices the one true theory of propositions is a fool's errand, and I'm not that big a fool. Here I want to utilize one particular conception of proposition that is especially useful for present purposes, without venturing any judgment about whether other conceptions might have served as well or better. The conception, whose most prominent advocate is Robert Stalnaker (1984), is a very simple one: a proposition is a collection of possible worlds. Stalnaker spends a lot of time trying to figure out what possible worlds are, but for our purposes, it will scarcely matter what answer he gets. Here we are only talking at the level of the sentential calculus, and what we say won't be sensitive to differences among the various notions of possible world, or to disagreement over whether possible-world talk is to be understood as literal or figurative. Also, we'll take it for granted that the possible worlds aren't too numerous to form a set. This is a highly dubious assumption, but to investigate it seriously would require a comprehensive understanding of possible-world metaphysics, and it would also require an inquiry into what resources are available for the mathematical description of totalities too big to constitute a set. Such investigations lie well beyond our present reach.

The identification of propositions with sets of possible worlds is well suited to the treatment of propositions as the primary bearers of truth values. It is not so well suited to the treatment of propositions as the contents of knowledge and beliefs, since it implies that, whenever two belief states have necessarily equivalent content, they have identical content. In particular, all necessarily true beliefs have the same content, so that, if somebody believes that $2 + 2 = 4$, she thereby believes the Poincaré conjecture. It's hard to explain how mathematicians can continue to draw their paychecks, since they are only telling us things we all already knew.

The reason why the Stalnaker-style propositions are well suited to our purposes is that they give us a straightforward semantics for possible languages. The possible sentences of possible languages do the same thing that the actual sentences of actual languages do: they express propositions, and questions about the semantic properties of the sentences of possible expansions of our language can be rephrased as questions about propositions. Truth and falsity are primarily attributes of propositions, and they are ascribed to sentences only derivatively: a sentence counts as true because the proposition it expresses is true. Actually, we should be talking about a sentence being true in a language in a context, but contextual variation won't play a role here. The implication relation among sentences is likewise derivative. A proposition p is implied by a set S of propositions if and only if the intersection of S is included in p, so that p is true in every world in which all the members of S are true. Sentential implication is derivative:

a sentence φ is implied by a set of sentences Γ if and only if the proposition expressed by φ, which (following Horwich 1990) we designate "$\langle\varphi\rangle$," is implied by $\{\langle\gamma\rangle : \gamma \in \Gamma\}$.

"Clare is poor" expresses a different proposition on Tuesday evening than on Tuesday morning, and an explicit notation should reflect this, writing "$\langle\varphi\rangle_C$," with a parameter for context, rather than "$\langle\varphi\rangle$." However, dragging the subscripts around makes the notation terribly messy, and contextual variation won't play a significant role in our discussion here, so I hope you'll forgive me if I write simply "$\langle\varphi\rangle$."

We want to reduce sentential calculus to propositional calculus, so that direct rules have the property that the proposition expressed by the conclusion of the argument is implied by the propositions expressed by the premises, and conditional rules preserve the property. For "$\supset$" this requires that we have the propositional version of *modus ponens*:

$$\{\langle\varphi\rangle, \langle(\varphi \supset \psi)\rangle\} \vdash \langle\psi\rangle$$

(where "$\vdash$" represents implication) and the propositional version of conditional proof:

If $\{\langle\gamma\rangle : \gamma \in \Gamma\} \cup \{\langle\varphi\rangle\} \vdash \langle\psi\rangle$, then $\{\langle\gamma\rangle : \gamma \in \Gamma\} \vdash \langle(\varphi \supset \psi)\rangle$.

These conditions assure us that, if we attempt to define an binary operation $\supset_{\text{PROP}}$ on the propositions expressed by sentences by stipulating:

$$\langle\varphi\rangle \supset_{\text{PROP}} \langle\psi\rangle =_{\text{Def}} \langle(\varphi \supset \psi)\rangle,$$

our stipulation will be well defined. These conditions constrain what proposition $(\varphi \supset \psi)$ might express, but they do not uniquely determine it, so they do not pin down a definite value for $\supset_{\text{PROP}}$. For that we need to go beyond the language-immanent rules.

To get a language-transcendent version, we need to extend $\supset_{\text{PROP}}$ into an operator acting on arbitrary propositions, and we need to extend the rules accordingly. The generalized version of modus ponens is this:

$$\{p, (p \supset_{\text{PROP}} q)\} \vdash q.$$

For conditional proof, we have:

Whenever $S \cup \{p\} \vdash q$, we have $S \vdash (p \supset_{\text{PROP}} q)$.

How we get from the sentential rules to the propositional rules is by observing that the sentential rules will continue to be upheld if we expand the language by adding new sentences that express the members of $S \cup \{p, q\}$.

There is one and only one operation on propositions that satisfies the propositional rules. $(p \supset_{\text{PROP}} q)$ has to be the union of all propositions of r with $\{p, r\} \vdash q$. The rules, understood in an open-ended way, uniquely pin down the semantic value of the conditional.

Let r be the set of worlds that are in $((p \supset_{\text{PROP}} q) \supset_{\text{PROP}} p)$ but outside p. Since $p \cap r$ is empty, p and r together entail every proposition. In particular, $\{p, r\} \vdash q$. By conditional proof, $\{r\} \vdash (p \supset_{\text{PROP}} q)$. Since $\{r\}$ implies $((p \supset_{\text{PROP}} q) \supset_{\text{PROP}} p)$, it follows by *modus ponens* that $\{r\}$ implies p: that is, that $r \subseteq p$. Since we already noted that $p \cap r = \emptyset$, it follows that $r = \emptyset$, and so $\{((p \supset_{\text{PROP}} q) \supset_{\text{PROP}} p)\}$ implies p, and $\{((\varphi \supset \psi) \supset \varphi)\}$ implies φ.

Does this argument show that intuitionists are mistaken in rejecting Peirce's Law? Far from it. The ideas that the content of a sentence is given by a mind-independent proposition and that worlds can be thrown together to make possible contents of sentences without any consideration of how one might know when the sentences are true in the worlds are repellant to intuitionists. Metaphysics and epistemology aside, intuitionists will regard the argument just given as committing a simple logical fallacy. Using "$\wedge_{\text{PROP}}$" and "$\sim_{\text{PROP}}$" in a natural way—see below—we can define r as $(((p \supset_{\text{PROP}} q) \supset_{\text{PROP}} p) \wedge_{\text{PROP}} \sim_{\text{PROP}} p)$ and get an argument that there is no world in which the premise of Peirce's law holds and the conclusion fails. But it's an intuitionistic error to proceed to the further judgment that the conclusion holds in every world in which the premise holds.

What we have here is an unabashedly realist picture that presumes from the outset a realist understanding of how language works, and takes for granted the availability, within the metatheory, of classical logic. Assuming the classically sanctioned inferences are valid, can we give an account of how speaker usage attaches meanings to the connectives so as to make them valid? If you aren't willing to grant the assumption, the question will have no interest for you. It's like the question why the bear has a short tail (Williams 1941). If you think that bears sometimes have long tails, or you think short-tailed mammals are unremarkable, you'll find the question pointless.

The propositional analogue of a binary sentential connective is a function taking pairs of propositions to propositions. Propositionalism gives us the space of candidates, and the rules of inference determine which of the candidates is the propositional analogue of "$\supset$." We have language-specific rules for $\supset_{\text{PROP}}$:

$\{\langle\varphi\rangle, \langle\varphi\rangle \supset_{\text{PROP}} \langle\psi\rangle\} \vdash \langle\psi\rangle$.

If $\{\langle\gamma\rangle : \gamma \in \Gamma\} \cup \{\langle\varphi\rangle\} \vdash \langle\psi\rangle$, then $\{\langle\gamma\rangle : \gamma \in \Gamma\} \vdash \langle\varphi\rangle \supset_{\text{PROP}} \langle\psi\rangle$.

These rules narrow the range of candidates, but they don't narrow it down to one. The intuitionist and classical conditionals both remain viable. However, if

"$\supset$" is taken to represent the intuitionistic conditional, conditional proof won't be maintained when the classical conditional is added to the language, whereas conditional proof for the classical conditional does survive the introduction of the intuitionistic connectives. So the open-ended application of the rules shows the classical conditional to be the better candidate. When two "$\supset$"-candidates fight it out *mano a mano*, the weaker candidate always wins. The fully general form of the rules—

$\{p, p \supset_{PROP} q\} \vdash q.$

If $S \cup \{p\} \vdash q$, then $S \vdash (p \supset_{PROP} q)$.

—singles out a unique candidate.

The propositionalist analysis gives an understanding of conditional proof that is open-endedly open-ended. The validity of the rule is maintained not only when we add new sentences but also when we add new rules. The original version told us that, whenever you have derived ψ from φ, together with background assumptions from Γ, you can discharge the assumption to derive $(\varphi \supset \psi)$ from Γ. As long as we were working within a fixed, stable system of rules, the meaning of the phrase "whenever you have derived ψ from φ" was clear. But once we start talking about the introduction of new connectives governed by new rules, uncertainty arises over whether the new rules ought to be permitted within conditional proofs. Our perplexity is dissolved by the treatment of implication in terms of propositions, which provides a uniform, language-independent standard of good argument: the good arguments are the ones that preserve, at every step, the property that the conclusion of the argument is implied by its premises. The general version of conditional proof tells us that, if we assume φ and derive ψ by a good argument, we may discharge the assumption to conclude $(\varphi \supset \psi)$. This condition ensures that conditional proofs are good arguments.

We can give propositionalist analyses of connectives other than "$\supset$." The language-independent form of the inference rule for "$\bot$" tells us that $\{\bot_{PROP}\}$ entails every proposition: that is, that $\bot_{PROP} = \emptyset$. The language-independent form of intuitionistic *reductio* (formulated in terms of "$\bot$") tells us that, if $S \cup \{p\}$ entails $\bot_{PROP}$, then S entails $\sim_{PROP}$ p. Let r be the set of worlds in which $\sim_{PROP}\sim_{PROP}$ p is true but p isn't true. Since r and p are incompatible, we have $\{r, p\} \vdash \bot_{PROP}$, and hence $\{r\} \vdash \sim_{PROP}$ p. Since the propositional version of *ex contradictione quodlibet* ($\{p, \sim_{PROP} p\} \vdash \bot_{PROP}$) tells us that $\sim_{PROP}\sim_{PROP}$ p and $\sim_{PROP}$ p are incompatible, $\sim_{PROP}\sim_{PROP}$ p and r are incompatible. But $\{r\}$ entails $\sim_{PROP}\sim_{PROP}$ p, so r must be impossible. Thus there aren't any worlds in which $\sim_{PROP}\sim_{PROP}$ p is true and p isn't true, and $\{\sim_{PROP}\sim_{PROP} p\}$ entails p. Once we have double negation elimination, classical *reductio ad absurdum* (which tells us that S entails p whenever $S \cup \{\sim_{PROP} p\}$ entails $\bot_{PROP}$) readily follows.

Let s be the set of worlds in which neither p nor $\sim_{\text{PROP}}$ p is true. Since there are no worlds in which s and p are both true, we have $\{s, p\} \vdash \perp_{\text{PROP}}$, and hence, by intuitionistic *reductio ad absurdum*, $\{s\} \vdash \sim_{\text{PROP}} p$. Since there are no worlds in which s and $\sim_{\text{PROP}}$ p are both true, we have $\{s, \sim_{\text{PROP}} p\} \vdash \perp_{\text{PROP}}$. Hence $\{s\} \vdash \perp_{\text{PROP}}$, so that there are no worlds in which neither p nor $\sim_{\text{PROP}}$ p is true. Since $\{p, \sim_{\text{PROP}} p\} \vdash \perp_{\text{PROP}}$, there are no worlds in which p and $\sim_{\text{PROP}}$ p are both true. Consequently $\sim_{\text{PROP}}$ p is true in a world if and only if p is not, and we have the principle of bivalence: within any world, every proposition is either true or false, but not both. Once we've gotten this far, it's easy to go on to extract the expected truth conditions for the other propositional connectives from propositional versions of the standard inference rules: $(p \vee_{\text{PROP}} q)$ is true in a world iff one or both of p and q are true in that world. $(p \wedge_{\text{PROP}} q)$ is true in a world iff both p and q are. $(p \supset_{\text{PROP}} q)$ is true in a world iff p isn't true or q is true in the world. $(p \equiv_{\text{PROP}} q)$ is true in a world iff both or neither of p and q is true in the world.

Although I won't undertake it here, the story we've been going through for the sentential connectives can be extended to take account of the quantifiers. One way of doing this is by the open-ended application of the rule for Hilbert's (1927) ϵ-operator; see McGee and McLaughlin (forthcoming).

We have a sufficient condition under which it is permitted to introduce a new sentential connective by open-ended rules. There needs to be an operation on propositions that satisfies the propositional generalization of the rules. Preferably, there will be only one such operation, for then we can say that the rules pin down that meaning of the connective uniquely. Whether there is such an operation is often a purely mathematical question. A positive answer for $\perp_{\text{PROP}}$ is guaranteed by the null set axiom. The union axiom gets us the unique determination of $\supset_{\text{PROP}}$ and $\sim_{\text{PROP}}$, whereas the axioms that ensure the existence of intersections are needed for $\wedge_{\text{PROP}}$ and $\vee_{\text{PROP}}$. On the other hand, because the subset relation isn't symmetrical, the implication relation isn't symmetrical, which means that there isn't any operator $\not\subset_{\text{PROP}}$ that satisfies the propositional analogues of the rules for "$\not\subset$":

If $\{q\} \vdash p$, then $\{p\} \vdash \not\subset_{\text{PROP}} q$.

$\{\not\subset_{\text{PROP}} q\} \vdash q$.

Also, since there is no proposition such that it and its complement both imply $\perp_{\text{PROP}}$, the Annihilator doesn't express a proposition.

The outcome we have reached is not what we were looking for. Our starting point was the repudiation of traditional truth-functional semantics on the grounds that we need to get rid of bivalence in order to make room for vagueness, but now we've wound up with bivalent, truth-functional semantics after all. This is not such a misfortune as it first appears. The outcome we get is that every proposition is either true or false, whereas the outcome we want to avoid is that every sentence is either true or false. Offhand, there are two ways that a sentence might be vague. It might be that the sentence expresses a vague proposition, but it might also be that propositions aren't vague but there is some slippage between sentences and the propositions they express. It now appears that an account of the semantics of vagueness needs to look to the second possibility.

That there are no vague propositions shouldn't come as a surprise. Propositions, as Stalnaker explains them, are just sets, and his understanding of set theory is in no way eccentric. There isn't any room in the universe of sets for vague sets in addition to ordinary sets. It will be helpful to look at a different case of a purportedly vague set that's somewhat simpler because everything is finite, the set of poor people. The reason why this case draws our attention is this: "Poor" is an unmistakably vague adjective, and so $\{x : x \text{ is a poor person}\}$, assuming it exists, is a vague set, if there are any vague sets. Is there such a set as $\{x : x \text{ is a poor person}\}$? The most direct way to prove its existence is to start with $\{x : x \text{ is a person}\}$ and apply the separation principle, but one might be doubtful of the proof, suspecting that vague predicates aren't allowed for substitution into the separation axiom schema, "$(\forall \text{ set } x)(\exists \text{ set } y)(\forall z)(z \in y \equiv (z \in x \wedge Rz))$." Indeed, the suspicion was already raised by Zermelo (1908), who required that "Rz" denote a propositional function for which it is "definite" for every member of x whether the function obtains. We can evade this worry, following the example of Dummett (1975), who used a similar maneuver to circumvent analogous worries about the applicability to vague predicates of mathematical induction. We take advantage of the fact that we can put a bound on how many people there are. To derive the conclusion "$(\exists \text{ set } x)(\forall y)(y \in x \equiv y \text{ is a poor person})$" from the auxiliary premise that there are fewer than a trillion persons (cashed out in terms of quantification theory), we require only the postulates of so-called adjunctive set theory (Visser 2009), axioms that are so weak that no one who is willing to talk about sets at all is going to deny them:

$$(\exists \text{ set } x)(\forall y) \sim y \in x.$$

$$(\forall \text{ set } x)(\forall y)(\exists \text{ set } z)(\forall w)(w \in z \equiv (w \in x \vee w = y)).$$

Each set of persons can be, in principle, named precisely, presumably, by listing its elements (the presumption being that each person can be named precisely), so if each set that can be specified precisely is a precise set, the axiom of extensionality forces us to the conclusion that the vague set $\{x : x \text{ is a poor person}\}$ is identical to some precise set. We avoid this absurdity by denying that there are two kinds of set, vague and precise.[6] Sets are neither vague nor precise, but they can be vaguely or precisely specified. Sets of worlds aren't different, in this regard, from other sets, and propositions are sets of worlds. There aren't vague or precise propositions; instead, there are propositions that are vaguely or precisely specified.

We get the outcome that there are no vague propositions because we have chosen to use "proposition" to refer to sets of possible worlds. We wouldn't get the same outcome if we took propositions to be the referents of "that" clauses. If we take the direct object of the true sentence, "Carlos said that Clare is poor" to refer to a Stalnakerian proposition, we'll get the erroneous outcome that Carlos's judgment partitions the totality of possible worlds into two neat pieces. "That" clauses refer, if they refer at all, to the kind of thing that can be the contents of judgments and speech acts, and the contents of judgments and speech acts can surely be vague.

In saying that propositions, as we are using the notion here, are never vague, I don't mean to suggest that they are precise. As Russell (1923) insisted, vagueness and precision are attributes of representations. I suppose that just about anything can be used to represent just about anything by adopting appropriate conventions, but without some special stage setting, it's hard to see how a set, whether of worlds or of anything else, can count as a representation; a set is just a set. Fregean senses, which is where the whole propositionalist craze got started, can be vague or precise. Stalnakerean propositions cannot.

There are many good candidates for what set "$\{x : x \text{ is a poor person}\}$" might refer to. Some of the candidates will position the actual world as an element of the proposition expressed by "Clare is poor," and others will leave it outside. Our notation "$\langle\varphi\rangle$" only succeeds in naming a well-defined proposition if the sentence φ is precise. There are many good choices for what "$\langle$Clare is poor$\rangle$" might refer to, and no best choice. Giving up the pretense that "$\langle\varphi\rangle$" is unambiguous, let us define an interpretation of our language to be a function assigning a proposition to each sentence. An acceptable interpretation will be one that respects the meanings of the sentences. The plan is to understand the semantic role of a sentence by examining the propositions it is assigned by the various acceptable interpretations. (We should really think of an interpretation as function assigning a proposition to a pair consisting of a sentence and a context, but I'm continuing to neglect contextual variation.)

Respecting the meanings of the sentences is a one-way street. If a sentence is true in a world, then the proposition assigned to the sentence by an acceptable

[6] The argument here is a adaptation of Evans (1978).

assignment has to be true in the world, and if a sentence is false in a world, the proposition assigned to it by an acceptable assignment will be false in the world. The converses of these conditionals do not obtain, since a sentence might be neither true nor false in a world, but propositions are bivalent. The basic strategy is to get the proposition associated with the sentence φ by an acceptable model by partitioning the worlds in which φ is neither true nor false. The divvying-up process is not entirely arbitrary, being confined by what Fine (1975) calls "penumbral constraints." If an acceptable interpretation puts the actual world, where Clare finds the winning ticket on Tuesday afternoon, into the proposition associated with "Clare is poor Tuesday evening," it will also put it into the propositions associated with "Clare is poor Tuesday morning." If we follow Clare's ancestral tree back far enough, we'll find forebears that are borderline cases of "human being"; let's say Oogaa is such an ancestor. Then the fact that humanity is an essential attribute of those who have it tells us that, if the proposition assigned to "Oogaa is human" by an acceptable interpretation includes some worlds in which Oogaa exists, it includes all such worlds.

The central theoretical task of people trying to give an account of vagueness that doesn't relinquish classical logic has been to understand what makes an interpretation acceptable. The principal architects of the program were David Lewis (1970), Hans Kamp (1975), and especially Kit Fine. They made use of van Fraassen's technique of "supervaluations," which enables us to employ the methods of classical model theory in non-bivalent settings. The program is governed by the grand principle that interpretations are made acceptable by speaker usage.

Let $|\mathfrak{B}|$ be the set of all sets of acceptable interpretations. Make $|\mathfrak{B}|$ into the domain of a Boolean algebra by stipulating that $+$ is union, $-$ is complementation, $\leqslant$ is inclusion, and so on. It is easy to verify that, for each world w, the function that takes a sentence φ to the set of acceptable interpretations in which the proposition assigned to φ is true in w is a $\mathfrak{B}$-valuation. It follows that, whether we are inquiring about the actual world or about some counterfactual situation, the natural-deduction rules are good, $\mathfrak{B}$-valid ways to reason.

The rules of inference, understood in such a way that they apply to arbitrary Stalnakerean propositions, determine, for each classical connective $\bigstar$, a unique operation $\bigstar_{\text{PROP}}$ on propositions such that, if $\bigstar_{\text{PROP}}$ is taken to be the semantic value of $\bigstar$, then the rules of inference will be implication preserving. Our open-ended disposition to apply the rules establishes $\bigstar_{\text{PROP}}$ as what we mean by $\bigstar$. Moreover, the fact that the rules are implication preserving assures us that, if the premises of a classically valid argument are true under an acceptable interpretation, its conclusion will be true under the interpretation as well. Consequently, if the premises are true under every acceptable interpretation, which means that they are, as the supervaluationists assess the situation, true,[7] the conclusion will be true as well.

Before we adopted the rules, "inferences" involving the logical connectives were meaningless jumbles of symbols. Adopting the rules gave the connectives meaning, by giving each of them a semantic role. These semantic roles affirm the legitimacy of the rules.

One can't help feeling overburdened by the heavy assumptions I needed to get here: not only a highly controversial theory of propositions, but the uninhibited application of classical logic in the metatheory. If you're someone who finds classical logic distasteful, nothing I've said here will improve its flavor. What we have here is a classical realist story, told from a classical realist perspective, something classical logicians can whisper reassuringly in one another's ears, but of no interest to outsiders.

Embarrassed by my overweight assumptions, I feel the need to point out, a little defensively, what I haven't assumed. I haven't assumed anything about the connectives other than that they obey the rules. I've made free metatheoretical use of classical logic, but I've not presupposed any connection between "if..., then" and "⊃" or between "not" and "~." I've adopted the supervaluationist strategy, but I haven't followed the supervaluationist custom of helping oneself to the classical compositional definition of truth under an interpretation. Instead, the laws of classical model theory, insofar as they are concerned with sentential calculus, were derived from the thesis that the use of the connectives is governed by the natural-deduction rules. The principles of compositional semantics—a disjunction is true under an interpretation iff one or both disjuncts are true under the interpretation, and so on—were obtained, not by theft, but by honest, if not particularly arduous, toil.[8]

References

Belnap, Nuel. 1962. "Tonk, plonk and plink." *Analysis* 22:130–4.

Craig, William and Vaught, Robert L. 1958. "Finite axiomatizability using additional predicates." *Journal of Symbolic Logic* 23:289–308.

Dummett, Michael. 1975. "Wang's Paradox." *Synthese* 30:301–24. Reprinted in Dummett (1978, 248–68) and in Keefe and Smith (1997, 99–118).

[7] Supervaluationists identify truth with having Boolean value 1 in the Boolean algebra of sets of acceptable interpretations.

[8] A version of this chapter was presented at the Arché Conference on Logical Consequence at the University of St Andrews in June 2010, where I gratefully received useful comments and advice, especially from the commentator, Graham Priest. I also got helpful comments when I presented it at MIT. I would like to thank Brian McLaughlin for his help.

Dummett, Michael. 1978. *Truth and Other Enigmas*. London: Duckworth.
Evans, Gareth. 1978. "Can there be vague objects?" *Analysis* 38:208. Reprinted in Keefe and Smith (1997, 317).
Field, Hartry. 2008. *Saving Truth from Paradox*. Oxford: Oxford University Press.
Fine, Kit. 1975. "Vagueness, truth and logic." *Synthese* 30:265–300. Reprinted in Keefe and Smith (1997, 119–50).
Frege, Gottlob. 1892. "Über Sinn und Bedeutung." *Zeitschrift für Philosophie und philosophische Kritik* 100:25–50. English translation by Peter Geach in *The Frege Reader*, Michael Beaney (ed.), Oxford: Blackwell, 1997, 151–71.
Gentzen, Gerhard. 1935. "Untersuchungen über das logische Schliessen." *Mathematische Zeitschrift* 39:175–210, 405–31. English translation 'Investigations concerning logical deduction' in Szabo (1969, 68–131).
Harris, J.H. 1982. "What's so logical about the logical axioms?" *Studia Logica* 41:159–71.
Hilbert, David. 1927. "Die Grundlagen der Mathematik." *Abhandlungen aus dem mathematischen Seminar der Hamburgischen Universität* 6:65–85. English translation by Stefan Bauer-Mengelberg and Dagfinn Føllesdal in van Heijenoort (1967, 464–79).
Horwich, Paul. 1990. *Truth*. Oxford: Oxford University Press.
Kamp, Hans. 1975. "Two theories about adjectives." In E.L. Keenan (ed.), *Formal Semantics of Natural Language*, 123–55. Cambridge: Cambridge University Press.
Keefe, R. and Smith, P. (eds.). 1997. *Vagueness: A Reader*. Cambridge, MA: MIT Press.
Kleene, Stephen Cole. 1952. *Introduction to Metamathematics*. New York: Van Nostrand.
Koslow, Arnold. 1992. *A Structuralist Theory of Logic*. Cambridge: Cambridge University Press.
Kreisel, Georg. 1967. "Informal rigour and completeness proofs." In Jaakko Hintikka (ed.), *Problems in the Philosophy of Mathematics*, 138–57. Amsterdam: North-Holland Publishing Co.
Lewis, David K. 1970. "General semantics." *Synthese* 22:18–67. Reprinted in Lewis (1983, 189–229).
Lewis, David K. 1983. *Philosophical Papers*, volume 1. Oxford: Oxford University Press.
McGee, V. and McLaughlin, B. forthcoming. *Terrestrial Logic*. Oxford: Oxford University Press.
Prawitz, Dag. 1965. *Natural Deduction: A Proof-Theoretical Study*. Stockholm: Almqvist and Wiksell.
Prior, Arthur N. 1960. "The runabout inference-ticket." *Analysis* 21:38–39.
Quine, W.V. 1960. *Word and Object*. Cambridge, MA: MIT Press.
Russell, Bertrand. 1923. "Vagueness." *Australasian Journal of Philosophy* 1:84–92. Reprinted in *Vagueness: A Reader*, R. Keefe and P. Smith (eds.), Cambridge, MA: MIT Press, 1997, 61–8. Page references are to the original.
Schechter, Joshua. 2011. "Juxtaposition: a new way to combine logics." *Review of Symbolic Logic* 4:550–606.
Sorensen, Roy. 1988. *Blindspots*. Oxford: Clarendon.
Sorensen, Roy. 2001. *Vagueness and Contradiction*. Oxford: Oxford University Press.
Stalnaker, Robert. 1984. *Inquiry*. Cambridge, MA: MIT Press.
Szabo, M.E. (ed.). 1969. *The Collected Papers of Gerhard Gentzen*. Amsterdam and London: North-Holland.

van Fraassen, Bas C. 1966. "Singular terms, truth value gaps, and free logic." *Journal of Philosophy* 63:464–95.

van Heijenoort, Jean. 1967. *From Frege to Gödel*. Cambridge, MA: Harvard University Press.

Visser, Albert. 2009. "Cardinal arithmetic in the style of Baron von Münchhausen." *Review of Symbolic Logic* 2:570–89.

Williams, Louise. 1941. *Why the Bear Has a Short Tail*. New York: Wonder Books.

Williamson, Timothy. 1994. *Vagueness*. London: Routledge.

Williamson, Timothy. 2007. *The Philosophy of Philosophy*. Oxford: Blackwell.

Zermelo, Ernst. 1908. "Untersuchungen über die Grundlagen der Mengenlehre I." *Mathematische Annalen* 59:261–81. English translation by Stefan Bauer-Mengelberg in van Heijenoort (1967, 199–215).

7

The Meaning of Logical Terms

Stewart Shapiro

7.1 Introduction

Recently, I have been arguing for a sort of pluralism or relativism concerning logic (Shapiro 2011; 2014).[1] The idea is that there are different logics for different mathematical structures or, to put it otherwise, there is nothing illegitimate about structures that invoke non-classical logics and are, in a strong sense, inconsistent with classical logic. The purpose of this chapter is to explore the consequences (if that is the right word) for this view concerning the meanings of logical terminology, at least as the terms are deployed in typical mathematical contexts. It is a twist on the old question of whether there is a substantial disagreement between, say, classicists and intuitionists, or whether they are merely talking past each other, as they attach different meanings to the crucial logical terminology.

To foreshadow some of the conclusions, I should note that I am skeptical that we can sharply assign unique and sharp meanings to each sentence, even in the relatively pristine case of live mathematics. Arguably, in the reconstructions of mathematics in formal logical languages, we just *stipulate* the meanings of the logical terms. A logician will just assert that such and such a connective has such and such a truth table and/or that it obeys such and such inference rules. Then our question focuses on the relationship between the formal languages and the mathematics it reconstructs. How do the stipulated rules or truth tables match up

[1] The present chapter is based on material in Chapters 4 and 5 of Shapiro (2014). For the most part, we are not concerned here with the connections to the literature on contextualism and relativism. Let us define *pluralism* about a given subject, such as truth, logic, ethics, or etiquette, to be the view that different accounts of the subject are equally correct, equally good, equally legitimate, or perhaps even true. So a pluralist about logic is someone who countenances different, mutually incompatible logics. The opponent, a logical *monist*, holds that there is but One True Logic. We won't engage issues of logical relativism here.

with the proper use of words like "or", "not", and "for all" as they occur in this or that piece of live mathematics (pure or applied)?

Of course, we regularly speak of "meaning", and even "the meaning" of an expression in a natural language, or a natural language of mathematics. I do not mean to suggest that such talk is incoherent, or flawed; nor am I out to ban it. But I suspect that this talk is context sensitive and interest relative, in part because it is vague. So I am skeptical of the very question of whether the classicist and intuitionist, for example, talk past each other—if we insist that this is a sharp and objective matter, to be determined once and for all.

This, then, is the present agenda. I'll begin with a brief sketch of my underlying pluralism. Then we will turn to questions of meaning.

7.2 Relativity to Structure

Since at least the end of the nineteenth century, there has been a trend in mathematics that any consistent axiomatization characterizes a structure, one at least potentially worthy of mathematical study. A key element in the development was the publication of David Hilbert's *Grundlagen der Geometrie* (1899). In that book, Hilbert provided (relative) consistency proofs for his axiomatization, as well as a number of independence proofs, showing that various combinations of axioms are consistent. In a brief, but much-studied correspondence, Gottlob Frege claimed that there is no need to worry about the consistency of the axioms of geometry, since the axioms are all true (presumably of space).[2] Hilbert replied:

> As long as I have been thinking, writing and lecturing on these things, I have been saying the exact reverse: if the arbitrarily given axioms do not contradict each other with all their consequences, then they are true and the things defined by them exist. This is for me the criterion of truth and existence.

It seems clear, at least by now, that this Hilbertian approach applies, at least approximately, to large chunks of mathematics, if not the bulk of it or even all of it. Consistency, or some mathematical explication thereof, like satisfiability in set theory, is the only formal criterion for legitimacy—for existence, if you will. One might dismiss a proposed area of mathematical study as uninteresting, or inelegant, or unfruitful, but if it is consistent, or satisfiable, then there is no further metaphysical, formal, or mathematical hoop the proposed theory must jump through before being legitimate mathematics.

[2] The correspondence is published in Frege (1976) and translated in Frege (1980). The passage to follow is in a letter from Hilbert to Frege, dated December 29, 1899.

The crucial, if obvious, observation is that consistency is tied to logic. A given theory may be consistent with respect to one logic, and inconsistent with respect to another. In general, the weaker the logic, the more theories that are consistent, and thus legitimate for the Hilbertian.

There are a number of interesting and, I think, fruitful theories that invoke intuitionistic logic, and are rendered inconsistent if excluded middle is added. I'll briefly present one such here, smooth infinitesimal analysis, a sub-theory of its richer cousin, Kock–Lawvere's synthetic differential geometry (see, for example, Bell 1998). This is a fascinating theory of infinitesimals, but very different from the standard Robinson-style non-standard analysis (which makes heavy use of classical logic). Smooth infinitesimal analysis is also very different from intuitionistic analysis, both in the mathematics and in the philosophical underpinnings.

In the spirit of the Hilbertian perspective, Bell presents the theory axiomatically, albeit informally. Begin with axioms for a field, and let Δ be the collection of "nilsquares", numbers n such that $n^2 = 0$. Of course, in both classical and intuitionistic analysis, it is easy to show that 0 is the only nilsquare:[3] if $n^2 = 0$, then $n = 0$. But not here. Among the new axioms to be added, the most interesting is the principle of micro-affineness, that every function is linear on the nilsquares. Its interesting consequence is this:

> Let f be a function and x a number. Then there is a *unique* number d such that for any nilsquare α, $f(x + \alpha) = fx + d\alpha$.

This number d is the derivative of f at x. As Bell (1998) puts it, the nilsquares constitute an infinitesimal region that can have an orientation, but is too short to be bent. It follows from the principle of micro-affineness that every function is differentiable everywhere on its domain, and that the derivative is itself differentiable, etc. The slogan is that all functions are smooth.

If follows from the principle of micro-affineness that 0 is not the only nilsquare:

$$\neg(\forall\alpha)(\alpha^2 = 0 \rightarrow \alpha = 0).$$

Otherwise, the value d would not be unique, for any function. Recall, however, that in any field, every element distinct from zero has a multiplicative inverse. It is

[3] In any field where the underlying logic is classical, zero is the only nilsquare. In intuitionistic analysis, one can identify each real number with an equivalence class of Cauchy sequences of rational numbers. One does not need excluded middle to show that if r is a Cauchy sequence and r^2 converges to zero, then r converges to zero. Two other intuitionistic theories of note are intuitionistic analysis and intuitionistic arithmetic, sometimes called "Heyting arithmetic", augmented with an object language verison of Church's thesis. Like smooth infinitesimal analysis, both of these theories become inconsistent if excluded middle is added.

easy to see that no nilsquare can have a multiplicative inverse. So no nilsquare is distinct from zero. In other words, there are no nilsquares other than 0:

$(\forall\alpha)(\alpha^2 = 0 \rightarrow \neg\neg(\alpha = 0))$, which is just
$(\forall\alpha)(\alpha^2 = 0 \rightarrow \neg(\alpha \neq 0))$.

To repeat, zero is not the only nilsquare, and there are no nilsquares other than zero. Of course, this would lead to a contradiction if we also had $(\forall x)(x = 0 \vee x \neq 0)$, and so smooth infinitesimal analysis is inconsistent with classical logic. Indeed, $\neg(\forall x)(x = 0 \vee x \neq 0)$ is a theorem of the theory.

Smooth infinitesimal analysis is an elegant theory of infinitesimals, showing that at least some of the prejudice against them can be traced to the use of classical logic—Robinson's non-standard analysis notwithstanding. Bell shows how smooth infinitesimal analysis captures a number of intuitions about continuity, many of which are violated in the classical theory of the reals (and also in non-standard analysis). Some of these intuitions have been articulated, and maintained throughout the history of philosophy and science, but have been dropped in the main contemporary account of continuity, due to Cantor and Dedekind. To take one simple example, a number of historical mathematicians and philosophers held that a continuous substance, such as a line segment, cannot be divided cleanly into two parts, with nothing created or left over. Continua have a sort of cohesiveness, or viscosity. *This* intuition is maintained in smooth infinitesimal analysis (and also in intuitionistic analysis), but not, of course, in classical analysis, which views a continuous substance as a set of points, which can be divided at will, cleanly, anywhere.

Smooth infinitesimal analysis is an interesting field with the look and feel of mathematics. It has attracted the attention of mainstream mathematicians, people whose credentials cannot be questioned. One would think that those folks would recognize their subject when they see it. The theory also seems to be useful in articulating and developing at least some conceptions of the continuum. So one would think smooth infinitesimal analysis should count as mathematics, despite its reliance on intuitionistic logic (see also Hellman 2006).

In holding that smooth infinitesimal analysis (and synthetic differential geometry), along with other, similar theories, are legitimate branches of mathematics, I am thus conceding that the law of excluded middle, and thus classical logic, is not universally valid. That is, classical logic is not correct in *all* discourses, about all subject matters, etc. I agree with the intuitionist about that much.

One reaction to this is to maintain a logical monism, but to insist that intuitionistic logic, or something weaker, is the One True Logic. The viability of this would depend on there being no interesting, viable theories that invoke a logic different from those two. Admittedly, I know of no examples that are as compelling (at least to me) as the ones that invoke intuitionistic logic. For example, I do not know of any interesting mathematical theories that are consistent with a quantum logic, but become inconsistent if the distributive principle is added. Nevertheless, it does not seem wise to legislate for future generations, telling them what logic they must use, at least not without a compelling argument that only such and such a logic gives rise to legitimate structures. One hard lesson we have learned from history is that it is dangerous to try to provide a priori, armchair arguments concerning what the future of science and mathematics must be.

I also do not know of any theories that are consistent with a relevance logic but become inconsistent if disjunctive syllogism or *ex falso quodlibet* are added. In fact, with many relevance logics, there can be no such theories. In many cases, if a sentence Φ follows from set Γ of sentences in classical logic then, in the logic in question, there is disjunction of contradictions C such that $\Phi \vee C$ follows from Γ in the logic in question. So if Γ is classically inconsistent, then in the logic in question it entails a disjunction of contradictions in the indicated relevance logic. When it comes to contradictions, classical and intuitionistic logic are actually conservative over the relevance logics favored by Neil Tennant (e.g. 1997). That is, if a contradiction is derivable in a classical (resp. intuitionistic) theory Γ, then a contradiction can be derived in Tennant's classical relevant logic CR (resp. intuitionistic relevant logic IR, now called "core logic") from a sub-theory of Γ.

However, these logics suggest a variation on the Hilbertian theme. If a set Γ of sentences entails a contradiction in classical, or intuitionistic, logic, then for every sentence Ψ, Γ entails Ψ. In other words, in classical and intuitionistic logic, any inconsistent theory is trivial. A logic is called paraconsistent if it does not sanction the ill-named *ex falso quodlibet*. Typical relevance logics are *paraconsistent*, but there are paraconsistent logics that fail the strictures of relevance. The main observation here is that with paraconsistent logics, there are inconsistent, but non-trivial theories.

So if we are to countenance paraconsistent logics, then perhaps we should change the Hilbertian slogan from "consistency implies existence" to something like "non-triviality implies existence". To transpose the themes, on this view, non-triviality is the only formal criterion for mathematical legitimacy. One might dismiss a proposed area of mathematical study as uninteresting, or inelegant, or unfruitful, but if it is non-trivial, then there is no further metaphysical, formal, or mathematical hoop the proposed theory must jump through.

To carry this a small step further, a trivial theory can be dismissed on the pragmatic ground that it is uninteresting and unfruitful. So the liberal Hilbertian, who countenances paraconsistent logics, might hold that there are no criteria for mathematical legitimacy. There is no metaphysical, formal, or mathematical hoop that a proposed theory must jump through. There are only pragmatic criteria of interest and usefulness.

So are there any interesting and/or fruitful inconsistent mathematical theories, invoking paraconsistent logics of course? There is indeed an industry of developing and studying such theories.[4] It is claimed that such theories may even have applications, perhaps in computer science and psychology. I will not comment here on the viability of this project, nor on how interesting and fruitful the systems may be, nor on their supposed applications. I do wonder, however, what sort of argument one might give to dismiss them out of hand, in advance of seeing what sort of fruit they may bear.

7.3 Logical Terms

We can now turn to words like "or", "not", and "for all", at least as those are used in mathematics, to connect entire formulas. Questions about meaning come to the fore here, and cannot be simply ignored. Is the classical "or", for example, the same as, or different from, its intuitionistic counterpart? Is the fundamental theorem of calculus the same proposition in classical and in smooth infinitesimal analysis? In the next section, I argue, tentatively, that the re-identification of logical terms, and thus propositions, is itself a context-sensitive, interest-relative matter. There simply is no hard and fast answer to be had, independent of conversational goals, as to whether the connectives and quantifiers are the same. But let us first see what the issues and options are.

7.3.1 Meanings are different?

Consider, first, the simplest of the logical connectives, conjunction. There is a natural and widely held assumption that the usual introduction and elimination rules and/or the simple truth table are somehow essential to this connective, at least as it appears in mathematics when connecting formulas. That is, if a connective does not obey the &-introduction and &-elimination rules, or the usual truth table that goes with them, then it is not conjunction.[5]

[4] See, for example, da Costa (1974), Mortensen (1995; 2010), Priest (2006), Brady (2006), Berto (2007), and the papers in Batens et al. (2000). Weber (2010) is a nice overview of the enterprise, and the state of the art at that time.

[5] This has been challenged. In most dynamic semantics, conjunctions are not commutative, as predicted by the truth table. Among logicians, Stephen Read (1988) has developed a binary connective, called "fusion", which is an "intensional conjunction". The usual conjunction-introduction and conjunction-elimination rules do not hold for it, in full generality (see also Mares 2004, 166–71).

One *might* say the same thing about the other connectives and quantifiers. Accordingly, if *ex falso quodlibet* is not valid for a given one-place connective, then that connective is not the negation we all know and use (in mathematical contexts anyway). And if excluded middle is not valid, then either we do not have disjunction or we do not have negation—at least not the disjunction/negation we are (supposedly) familiar with.

Of course, this is a standard line to take. It is often claimed that the intuitionist and the classicist, for example, talk past each other—that the dispute between them is merely verbal. The common, but not universal, view is that excluded middle is valid for the classical connectives, invalid for their intuitionistic counterparts, and that is that.

Michael Dummett (1991, 17) puts the matter in these terms, but argues that there is a serious issue nonetheless:

> The intuitionists held, and continue to hold, that certain methods of reasoning actually employed by classical mathematicians in proving theorems are invalid: the premisses do not justify the conclusion. The immediate effect of a challenge to fundamental accustomed modes of reasoning is perplexity: on what basis can we argue the matter, if we are not in agreement about what constitutes a valid argument? In any case how can such a basic principle of rational thought be rationally put in doubt?
>
> The affront to which the challenge gives rise is quickly allayed by a resolve to take no notice. The challenger must mean something different by the logical constants; so he is not really challenging the laws that we have always accepted and may therefore continue to accept. This attempt to brush the challenge aside works no better when the issue concerns logic than in any other case. Perhaps a polytheist cannot mean the same by 'God' as a monotheist; but there is disagreement between them all the same. Each denies that the other has got hold of a coherent meaning; and that is just the charge made by the intuitionist against the classical mathematician.
>
> He acknowledges that he attaches meanings to mathematical terms different from those the classical mathematician ascribes to them; but he maintains that the classical meanings are incoherent and arise out of a misconception on the part of the classical mathematician about how mathematical language functions.

I am not sure Dummett is right in his example concerning religion. As a sort of quasi-Quinean, I don't feel a big need to decide if a typical monotheist and a typical polytheist (who both speak English) mean the same thing by "god", when it is used as a predicate, or whether they accuse each other of conceptual incoherence. It is hard to make sense of that question independent of conversational context.[6] Some monotheists do hold that unity is somehow essential to divine beings, say

Read argues that fusion captures at least some of the proper use of the ordinary English word "and". I do not know if either dynamic semantics or fusion is to extend to mathematics.

[6] Thanks to Michael Miller for suggesting this line of thought.

following Maimonides (e.g. *Guide for the Perplexed*, Part I, Chapter 51). For them, it follows from the nature of the concept of "god" that there cannot be more than one of them (and possibly also that there cannot be fewer than one of them). Accordingly, the polytheist is conceptually confused. From this perspective, the proposition that there is more than one god is like the proposition that there is a square circle, a triangle with four angles, or a married bachelor. As Dummett puts it, this monotheist "denies that the [polytheist] has got hold of a coherent meaning", or at least the latter is confused about what follows from the meaning of the terms he uses.

Other monotheists, however, do not hold that unity is essential to divinity. For those thinkers, the dispute with the polytheist is substantial, not verbal. They both have the same concept of "god", and disagree over how many of them there are, each holding a conceptually coherent view.

Notice that, when the issue is put this way, it looks like our first monotheists (those who take unity to be essential to divinity) and our second monotheists (those who don't) would also be talking past each other, if they got together to compare notes. In particular, our first monotheist would accuse the second of not having a coherent meaning, and the second might accuse the first of conceptual confusion as well (thinking that something follows from a concept that, in fact, does not). Presumably, this would not prevent our two monotheists from having a mutually satisfiable discussion of internal theological matters, say over whether God can take human form.

Of course, Dummett is not a pluralist concerning logic, nor does he adopt the foregoing Hilbertian theme concerning mathematics. For Dummett, there is but One True Logic and it isn't classical. The classicist is accused of conceptual confusion: there just is no coherent meaning for her to have.

Dummett is not alone in arguing that there is some sort of conceptual incoherence in the classical connectives and quantifiers. The most common target is the classical, so-called Boolean negation. From vastly different perspectives, Priest (1990) and Field (2008) both claim that there simply is *no* coherent connective that satisfies the law of excluded middle and the inference of *ex falso quodlibet*. The reason, in short, is that such a connective conflicts with a naive truth-predicate. From the relevantist camp, Anderson et al. (1992, 490) display typical wit:

> There are some things one sometimes wishes had never been invented or discovered: e.g., nuclear energy, irrational numbers, plastic, cigarettes, mouthwash, and Boolean negation. The reader may not yet have heard of this last-named threat, and it is our present purpose to inform and caution him regarding it.

See also Restall (1999, section 5).

Of course, the foregoing Hilbertian orientation takes it that classical and intuitionistic mathematics are both legitimate, probably along with the various relevant, paraconsistent, and paracomplete logics, especially those that find interesting application. This presupposes that at least the classical and intuitionistic theories, and thus their underlying languages and logics, are conceptually coherent.

To even ask the question about meaning-shift here, we need some sort of analytic-synthetic distinction. We need some idea of what goes into the meaning of a logical term. Nevertheless, W.V.O. Quine also understands the matter of alternate logics in terms of the meanings of the logical terminology. He agrees with Dummett that the classicist and the non-classicist talk past each other, since they give different meanings to the logical connectives. But Quine does not have one of them accusing the other of simply having no coherent meaning. For Quine, the meanings they assign to the logical terms are just different. Concerning the debate over paraconsistent logics, he wrote:

> My view of this dialogue is that neither party knows what he is talking about. They think they are talking about negation, '~', 'not'; but surely the notation ceased to be recognizable as negation when they took to regarding some conjunctions in the form 'p.~ p' as true, and stopped regarding such sentences as implying all others. Here, evidently, is the deviant logician's predicament: when he tries to deny the doctrine he only changes the subject. (Quine 1986, 81)

Notice that Quine does not object (at least not here) to the deviant logician changing the subject. She just has to acknowledge that that is what she is doing—changing the subject. Quine admits that, at least in principle, there may be reasons to change the logic—to change the subject.

This passage is, I think, confusing. It has different readings, but as far as I can see, none advances the dialectic substantially. One reading, suggested by the mention of the word "not", is that Quine is making a claim about the established meaning of an English word, perhaps what is sometimes called the term's "conventional meaning". On this reading, Quine is saying that when the word "not" is used to operate on sentences, it, in fact, obeys the classical inference rule of *ex falso quodlibet*.

If this is the intended reading of the passage, then Quine simply begs the question against at least some advocates of paraconsistent systems. They argue that the inference of *ex falso quodlibet* is not valid for the negation of natural language (nor the natural language of mathematics). Priest (1990) claims that if the negation of natural language—the word "not" in some uses—is coherent, then it is not classical (since, as above, Priest holds that there is no coherent Boolean negation).

If Quine is, in fact, making a claim here concerning the meaning of an English word, it would seem that we should invoke the techniques of empirical linguists or lexicographers to adjudicate the matter. It is not all that clear, a priori, that the word 'not' has a single use. Many linguists think it does not (see Horn 1989).

In any case, this first reading of the passage makes for a strange point for Quine, of all philosophers, to be making (see also Arnold and Shapiro 2007). Quine insists that there is no difference in kind between matters of meaning and matters of fact. Suppose, for the sake of argument, that Quine is correct that the utterances of competent English speakers, or competent English-speaking mathematicians, do conform to the principles of non-contradiction and *ex falso quodlibet*. Even so, what business does Quine have for saying that this pattern of use and correction is due to the meaning of the word "not" and not due to, say, some widely shared beliefs among the speakers? Maybe they believe that, say, *ex falso quodlibet* is materially valid, that its instances *happen* to be truth preserving? Why is Quine here entitled to insist on a distinction between changing one's beliefs about a given subject and changing the very subject, which is, in this case, to change the meaning of the logical words we use?

In the program for linguistics described in Chapter 2 of *Word and Object* (1960), Quine seems to be saying that, in effect, truth-functional connectives are exempt from the indeterminacy of translation. We must translate others as accepting our logic. Quine finds it incoherent to think that there might be a "pre-logical" people who do not agree with us about the rules of logic:

> To take the extreme case, let us suppose that certain natives are said to accept as true certain sentences translatable in the form 'p and not p'. Now this claim is absurd under our semantic criteria. And, not to be dogmatic about them, what criteria might one prefer? Wanton translation can make natives sound as queer as one pleases. Better translation imposes our logic upon them, and would beg the question of prelogicality, if there were a question to beg. (Quine 1960, 58)

As a sort of test case, take the "natives" to be living in Australia, and have some of them assert contradictions in the course of developing inconsistent mathematics: Mortensen, Priest, Brady, Weber (see note 4 above)—although we should not call those authors "pre-logical". Or suppose that other natives (in Holland or in London, Ontario) are pursuing intuitionistic analysis or smooth infinitesimal analysis, and so they refuse to accept every instance of excluded middle.[7]

[7] In *some* cases, perhaps, it may be possible to "impose our logic upon them" when translating, say by invoking a modal interpretation. However, that would miss or at least distort much of what they are trying to say or do. See Chapter 3 of Shapiro (2014) for details.

A second reading of the above passage from Quine (1986, 81) is suggested by the mention of the connective "~". Perhaps Quine is making a claim about mainstream logical systems. If so, his claim is surely correct, but not very interesting or informative—or relevant to the matter at hand. Most logical systems are classical, and were so then as well. On this second reading of the passage, it is hard to see what the big deal is, and why Quine would bother stating something so obvious. There are, of course, legitimate mathematical theories whose underlying logic is correctly described—well modeled—by the classical texts. From the present, pluralist perspective, there are also legitimate mathematical theories whose underlying logic is correctly described by the intuitionistic texts and not by the classical texts. Our real question is whether the various logical particles across these various theories share a meaning. Do the words "not" and "or", when used in, say, classical analysis have the same meaning as the same words when used in intuitionistic analysis and smooth infinitesimal analysis?

It will prove instructive briefly to broach a third interpretation of the above passage from Quine (1986, 81).[8] We might understand the passage as articulating what *Quine* means (or intends to mean) by the word "not" and by the negation sign, "~". He is saying that, as he intends to use those terms, they conform to the truth table of classical, logical systems. His negation is Boolean. This, of course, is a stipulation, and so is not subject to refutation (assuming that the proposed usage is coherent). A substantial claim behind the stipulation is that classical logic is appropriate for certain theoretical purposes, possibly having to do with regimentation, mutual translation, or the like. If those theoretical purposes would be frustrated by abandoning the law of non-contradiction or the rule of *ex falso quodlibet*, then, given that Quine has those theoretical purposes, he is right to insist on those principles—and to reject the deviant logician's proposal out of hand. On this reading of the passage, the deviant logician cannot actually be charged with *changing* the subject—changing the meaning of those terms—since it has not been established what that meaning is (or was). Nevertheless, the deviant logician is charged with operating with terms whose meaning is foreign to those being stipulated.

John Burgess (2004) formulates a notion of analyticity, supposedly available to Quineans and other skeptics about meaning. Burgess suggests that a principle should be regarded as "basic" or as "part of the meaning or concept attached to a term", when, "in case of disagreement over the [principle], it would be helpful for the minority or perhaps even both sides to stop using the term, or at least to attach some distinguishing modifier to it" (p. 54). Burgess has disputes over the law

[8] Thanks to Michael Miller here.

of excluded middle explicitly in mind. The idea seems to be that it makes sense, even for a Quinean, to call a sentence "analytic", if to deny the sentence would call so much into question that we cannot pursue ordinary communication—thus the suggestion that new terms be coined in such a case (see also Chalmers 2011). Our third reading of the above passage is that Quine is claiming that non-contradiction and *ex falso quodlibet* are "analytic" (for him), in that sense. Given his theoretical purposes, to challenge those principles would be to call too much into question, and to frustrate ordinary communication. So other terms, with different meanings, should be used instead for whatever theoretical purposes the deviant logician may have. The deviant logician should "change the subject", by using a different set of connectives.

Another version of the view that meaning shifts between the theories is found in Rudolf Carnap's principle of tolerance, as articulated in *The Logical Syntax of Language*:

> *In logic, there are no morals.* Everyone is at liberty to build his own logic, i.e. his own form of language, as he wishes. All that is required of him is that, if he wishes to discuss it, he must state his methods clearly, and give syntactical rules instead of philosophical arguments. (Carnap 1934, section 17)

See also Carnap (1950). For Carnap, different logics are tied to different "linguistic frameworks". The meaning of a given term, in a given linguistic framework, is given by explicit syntactic rules particular to the framework. Different languages, associated with different logics, give different syntactic rules to the connectives and quantifiers. Hence, the meanings of said connectives and quantifiers differ.

Greg Restall (2002, 430) puts the Carnapian perspective well:

> The important constraint, if it is indeed a constraint, is that the language builder explain what principles are being used, in a clear, syntactic manner, as befits the choice of a language. If you wish to build a language containing a symbol, written '¬', that satisfies the rules of the classical account of negation, that is your prerogative. If, instead, you wish to build a language containing a symbol, also written '¬', that satisfies the rules of a relevant account of negation, that is fine, too. For Carnap, the choice of which language is to be preferred is pragmatic, to be answered by asking: What is the language for? Is your aim to give a comprehensive theory, able to prove as much as you can in a simple way? Classical negation might be for you. Is your aim to avoid contradiction and limit it if it arises? Then relevant negation might be for you.

To port this idea to the present framework, if one is out to capture the basics of classical real analysis, then classical negation (etc.) is appropriate. If, instead, the aim is to capture the basics of smooth infinitesimal analysis, then intuitionistic negation (etc.) is needed. The crucial Carnapian conclusion is that these are different *negations* (see also Hellman 1989).

7.3.2 *Meanings Are the Same (Even Though the Logics Are Different)?*

This Dummett–Quine–Carnap perspective is at odds with the pluralism of Jc Beall and Restall (2006). They take it that there are different logics, equally correct, *for one and the same language*. As Restall (2002, 432) put it:

> If accepting different logics commits one to accepting different languages for those logics, then my pluralism is primarily one of languages (which come with their logics in tow) instead of logics. To put it graphically, as a pluralist, I wish to say that
>
> $$A, \neg A \vdash_C B, \text{but } A, \neg A \nvdash_R B$$
>
> A and ¬A together, classically entail B, but A and ¬A together do not relevantly entail B. On the other hand, Carnap wishes to say that
>
> $$A, \neg_C A \vdash B, \text{but } A, \neg_R A \nvdash B$$
>
> A together with its classical negation entails B, but A together with its relevant negation need not entail B.

As Restall notes, this last assumes that the sentences A and B have the same content in both systems. That matter will loom large in the next section.

Restall (2002) gives several reasons why one might adopt the Dummett–Quine–Carnap perspective and he argues that none of these "leads to insurmountable problems" for his opposing perspective that there are different logics for one and the same language, with a single batch of logical terms.

Note that Restall does not claim to have a compelling argument for his opposing view, only that his opponents—advocates of the Dummett–Quine–Carnap perspective—have not put up "insurmountable problems" for him. This aspect of Restall's conclusion is, I suggest, support for (or is at least consistent with) the view developed in the next section, that the question of meaning-shift is itself context sensitive. There just is no fact of the matter, independent of conversational goals, whether the meaning is the same or different, in the different languages/theories.

Clearly, the issue here turns on what, exactly, meaning is and, in particular, on when it is that two particles have the same meaning. When it comes to formal languages, there are two main models for meaning. Some argue that meaning is given by inferential role, typically in the form of introduction and elimination rules. Others propose that meaning is given by truth conditions.

We'll begin with truth conditions. Priest (2006, Chapter 12) argues against the pluralism of Beall and Restall (2006), insisting that since the different logics attribute different truth conditions to the various connectives and quantifiers, they also attribute different meanings, thus siding with Dummett, Quine, and Carnap above. Priest also argues that the different logics, with their differing truth conditions, be understood as different *theories of* the meaning of the corresponding ordinary language connectives and quantifiers. At most one such theory is correct, assuming that the terms in ordinary (mathematical) language have a single meaning.

Against Priest's perspective, Restall (1999, 2002) proposes that we can develop minimal truth conditions for each connective (see also Beall and Restall 2006, 50, 78–9), truth conditions that do not settle the logic. His proposed clause for negation is the one given in the semantics for some relevance logics, involving a primitive compatibility relation among interpretations. The details do not matter here. Restall's point is that we can get either classical logic or a relevance logic, using the same clauses for the connectives and quantifiers, by varying the collection of interpretations. In particular, we get classical logic if we use Tarskian models and we get relevance logic if we use situations (together with a compatibility relation). The idea is that in both cases, the *meaning* of the negation sign is the same—given by the clause. What varies is something else: namely, the background class of interpretations. Of course, this all depends on what it is that goes into meaning, a matter that is fraught with controversy.[9]

Restall (2014) turns to proof-theoretic accounts of meaning. There, he proposes that the appropriate framework is a sequent calculus. There, too, we can formulate a single clause for each connective—the usual sequent rules. Again, the details do not matter here. The point is that we get different logics by varying structural rules. Classical logic results if we allow multiple premises and multiple conclusions, and intuitionistic logic results if we allow multiple premises but only single conclusions.

Ole Hjortland (2013) also provides ways to think of different logics as applying to the same formal language (or batch of languages), or, as he puts it, we get pluralism within a single *theory* of logic. For a sort of toy example, begin with the so-called strong Kleene 3-valued logic, sometimes called K3. It has three truth values, T, F, and I, given by truth tables. Only the first truth value T is designated; the third value I stands for indeterminacy, neither true nor false. If we use the very same truth tables, but think of both T and I as designated, then we get Graham Priest's (2006) logic of paradox LP. In that case, we think of I as *both*

[9] We are not told how to get intuitionistic logic into Restall's framework, although one can speculate as to how it would go. We might think of stages in Kripke structures as interpretations, and formulate a compatibility relation, among stages, in such a way that the clause for negation is, word for word, the same as the clause in the relevant framework. Alternatively, one might formulate a different compatibility relation on stages in Kripke structures. Maybe it is the compatibility relation that can vary, keeping meaning fixed. In any case, it is not clear how plausible the package is, as in interpretation of intuitionism. The devil would lie in the details.

true *and* false. So we have one set of truth tables, and two logics. The idea is that we think of *meaning* as given by the common truth tables, and we can define *validity*, uniformly, as the necessary preservation of a designated truth value. We get different logics by specifying different sets of designated values. And we get classical logic by restricting ourselves to interpretations that eschew the middle value I. Intuitionistic logic is, apparently, out of bounds, or at least out of these bounds, since it cannot be formulated with truth tables.

Hjortland (2013) goes on to develop a more sophisticated sort of pluralism, from a proof-theoretic perspective. It is a generalization of the sequent calculus, where we allow more than just premises and conclusions in sequents. The main idea is that we can formulate different logics all operating within a single formal language and a single proof-theoretic framework. The upshot is similar to that in the Restall articles. We vary the logic, within a single semantical/logical framework, by varying something that is, supposedly, unrelated to meaning, the class of interpretations, the designated values, the structural rules, etc.

Notice that all of these proposals focus exclusively on formal languages. Each takes some aspect of a formal system—sequent rules, truth tables, truth conditions—and uses that to formulate the "meaning" of the logical terminology. As noted above, however, when it comes to formal languages, matters of meaning are given by stipulation. The author of the formal system gives rules of inference and/or model-theoretic semantics. It does not matter much which elements in the package are designated as "meaning" or "meaning constitutive". The interest of the system lies elsewhere.

In contrast, present concern is with natural languages or, better, the natural languages of mathematics. Part of the issue, then, is to figure out how the meaning of natural language terms, even when restricted to use in mathematical deductions, relate to the central items of various formal logical systems. Is this an empirical question, to be settled by the techniques of contemporary linguistics?

There is nothing but controversy concerning what counts as the meaning of even the logical terminology as it occurs in live mathematics. One might look for some analogue of, say, sequent calculus introduction and elimination rules in the pursuit of ordinary mathematical reasoning, and designate that as "meaning". Or with some analogue of truth conditions, say in a sort of Montague-style framework. But, surely, there is no consensus to be found, and matters will not be nearly as clean as they are when we just focus on formal languages.

Notice also that, even with the focus on formal languages, none of the aforementioned proposals by Restall and Hjortland captures *all* of the logics of present interest. Restall (2014) recovers classical and intuitionistic (and dual intuitionistic) logics, but none of the paraconsistent or quantum species; Restall (1999) gets classical logic, one of the relevant logics, and, just maybe, intuitionistic logic; Hjortland captures several logics, depending on how the "theory" is formulated. Of course, it is an option for a pluralist to adopt one of these frameworks, and just rule out any logic that it does not capture. From the present, Hilbertian perspective, there is, or at least might be, more to the pluralism and folk relativism than the few samples captured in this or that framework.

7.4 Context-Sensitive Resolution

I'd like to float a thesis that the matter of meaning-shift is itself context sensitive, and, moreover, interest-relative. Whether we say that the logical terms have the same meaning, or different meanings, in the different structures or theories, depends on what is salient in a conversation comparing the structures or theories. For some purposes—in some conversational situations—it makes sense to say that the classical connectives and quantifiers have different meanings than their counterparts in intuitionistic, paraconsistent, quantum, etc. systems. In other conversational contexts, it makes sense to say that the meaning of the logical terminology is the same in the different systems.

I claim, first, that there are situations in which it is at least natural to speak of the logical terminology as having different meanings in the different mathematical theories, at least when the logics are different. I claim, second, that there are other conversational situations in which it is more natural to speak of the logical terminology as having the same meaning, for the very same pair of theories. Of course, even if I am right about that, it remains to be seen whether these prima facie "natural" thoughts can be sustained in a full-blown account of meaning—a task that is, I am afraid, beyond the scope of this work (not to mention beyond the scope of anything I am competent to pronounce on). The rest of the chapter is devoted to developing the idea, and locating the thesis in some wider, philosophical perspectives.

We start with the first claim. It seems to me natural to speak of meaning-shift when basic logical differences between the systems are explicitly in focus. Consider a student, Dudley, who has a normal training in classical mathematics. Suppose he is shown the basis of smooth infinitesimal analysis.[10] He is told that, in that theory, zero is not the only nilsquare or, in other words, he is told that it is not the case that every nilsquare is identical to zero:

$$\neg(\forall\alpha)(\alpha^2 = 0 \rightarrow \alpha = 0).$$

[10] The thought experiment that follows is based on reactions I have gotten when explaining smooth infinitesimal analysis to some audiences. The prototype for Dudley's guide is, of course, John Bell. The role of Dudley was played by more than one audience member.

Dudley responds, "Ah, so there is another nilsquare, one that is different from zero". Dudley's guide corrects him on this. She says, "I'm afraid you are mistaken there. In fact, no nilsquare is distinct from zero:

$$(\forall \alpha)(\alpha^2 = 0 \rightarrow \neg(\alpha \neq 0))."$$

Dudley's first reaction might be that he is confronting a new form of madness, that everyone who works on this so-called branch of mathematics is nuts. On reflection, however, Dudley may adopt a more charitable orientation, based on the solid reputation of his guide in the mathematical community, her substantial publications in ordinary, classical mathematics, and the fact that she seems quite adept in all non-mathematical linguistic matters. So, Dudley charitably concludes, the guide must not mean the same as he does with words like "only", "not", and perhaps "identical to". As *he* understands these terms, it simply *follows from* "zero is not the only nilsquare" that "there is a nilsquare different from zero". Dudley maintains that anyone who thinks otherwise must not know what they are talking about.

Dudley may be inspired here by a passage from Quine's *Word and Object*, invoked above:

> Let us suppose that certain natives are said to accept as true certain sentences translatable in the form 'p and not p'. Now this claim is absurd under our semantic criteria. And, not to be dogmatic about them, what criteria might one prefer? Wanton translation can make natives sound as queer as one pleases. Better translation imposes our logic upon them.
>
> (Quine 1960, 58)

The "natives" here, those who pursue smooth infinitesimal analysis, do not assert (what look like) contradictions, but they do violate other equally dear tenets of classical logic. To Dudley, the "homophonic" translation, where, say, "not-for all" means "not-for al", makes them sound weird to him, almost as weird as Quine's imagined natives sound to Quine.

In a more charitable moment, Dudley may conclude instead that his guide must be using the logical words differently from how he uses them. He might adopt the aforementioned proposal from Burgess (2004) and tag his guide's logical words to indicate that they have a radically different meaning from his own. If Dudley is sufficiently interested, he might work to understand the meaning his guide seems to attach to these so-called "logical particles", so that he can digest the theory properly. It would be an open question whether this can be accomplished by translating the guide's language into his own—say by following guidelines for radical translation or radical interpretation, or, more formally, via a modal interpretation. Alternatively, the two languages/theories may somehow be incommensurable, and Dudley will have to "go native" in order to get the hang of the interesting new language/theory.

To be sure, this conclusion concerning meaning-shift can be, and will be, resisted. For example, a monistic intuitionist (like Heyting or Dummett) might claim that Dudley is simply mistaken in holding that the principle of excluded middle (or something equivalent) is constitutive of the meaning of "not" and "or". For such an intuitionist, excluded middle is, at most, a deeply held belief, perhaps based on some metaphysical or epistemological view that Dudley, along with just about every classical mathematician, tacitly holds, even unbeknownst to themselves (see Tennant 1996). My only claim here, so far, is that in contexts like this one, where crucial differences in rather basic inference patterns are salient, it is natural to speak in terms of a change in meaning between the various mathematical theories. That would just be to take some straightforward remarks at face value.

That is our first scenario. In other conversational contexts, it is more natural to take the logical terminology in the different theories to have the *same* meaning. That happens if one is discussing what the theories (appear to) have in common, or if one is highlighting some less basic, and more directly mathematical differences between the theories—differences that show up logically downstream, so to speak. For example, when sketching smooth infinitesimal analysis above, we began with the axioms for a field. Notice that those axioms contain logical terminology—the very terms whose meaning is being called into question. One of the axioms, for example, is that every number has an additive inverse. Readers have to understand what is meant by "every". So, it seems, "understanding what was meant", which, I presume, at least some readers of Bell (1998) did, need not include interpreting the logical terms classically. To say the least, it would be weird to protest later: "Hey. You said to start with the axioms for a field. But in your theory, *those* axioms do not hold, as the words you use do not have the same meaning in your theory as they do in the axioms of (classical) field theory."

The Fundamental Theorem of calculus is that if $A(x)$ is a function giving the area under the curve given by a continuous function f, between 0 and x, then the derivative of A is f. Bell (1998, 30) remarks that the fundamental theorem holds in smooth infinitesimal analysis. In fact, it has a lovely proof, turning on the axiom of micro-affineness. This seems like the correct thing to say, as the corresponding theorems—one in classical analysis and one in smooth infinitesimal analysis—at least appear to have the same content, at least for these communicative purposes. Both theorems say what seems to be the same thing about the areas under continuous curves. It at least *seems* correct to say that the fundamental theorem—the very same proposition—is provable in both classical analysis and smooth infinitesimal analysis, and, moreover *it*—this proposition—has a lovely proof in the latter.

Along similar lines, Bell (1998, 105) provides an insightful and illuminating treatment of the intermediate value theorem, in the context of smooth infinitesimal analysis. Although it—the intermediate value theorem—holds for quadratic functions, it fails in general. Bell provides a nice explanation for this, involving projection functions. For this talk to be understood at face value we have to assume that it is coherent to speak of the classical intermediate value theorem and its smooth counterpart as having the same content, so we can state where it holds and where it does not.

Restall (2002, 432 n. 14) makes a similar point, focusing on how we typically compare classical and intuitionistic theories:

> Instead of taking the intuitionistic *language* to differ from a classical language, we can identify them. The difference between Heyting arithmetic . . . and classical Peano arithmetic is then what it appears to be on the surface. Some theorems of Peano arithmetic are not theorems of Heyting arithmetic, but all theorems of Heyting arithmetic are also theorems of Peano arithmetic. If we were to take Carnap's doctrine [that meaning shifts between the theories] literally, this could not be the case. For Carnap, all theorems of Heyting arithmetic featuring "intuitionistic negation" would not be expressible in the language of classical arithmetic, which features no such connective . . . It is common in metamathematical practice to consider the relative strengths of theories couched in different logics.

To compare the relative strengths of two theories, at least in the most straightforward way, they should share some content, or at least one should be easily translatable into the other.

As with the first claim of this section, the present conclusion concerning shared content can be resisted, and, of course, it would be resisted by someone who agrees with Dummett, Quine, or Carnap. For example, one might protest that *the* Fundamental Theorem does not hold in smooth infinitesimal analysis, since the statement of the Fundamental theorem that we all know contains logical terminology—as does just about any non-trivial statement in any theory—and this logical terminology does not mean the same thing in the two theories. In fact, some philosophers have insisted that the elements of smooth infinitesimal analysis do not constitute a field, just because the logic is not classical. They concede that it is an "intuitionistic field". This is of-a-piece with the aforementioned suggestion from John Burgess (2004), where the logical terminology gets tagged somehow.

Our resisting thinker will concede that there are some striking analogies between certain propositions from the different theories. Both fundamental theorems, for example, say something about the areas under continuous functions. Well, even this assumes that "area" and "continuous function" have the same meaning in the two theories. Unless those terms are primitive, their definitions will contain logical terms, which, according to the resistor, have different meanings in the two theories. A more careful statement, from the one who insists on meaning shift, would be that the respective theorems make analogous claims about structurally analogous things, classical-areas and smooth-areas, classical-continuous-functions and smooth-functions, and the like. Restall (2002) claims, perhaps for this reason, that this interpretive move is "strained". Sometimes it is. However, in scenarios more like our first, the Burgess-inspired interpretive move is not strained, but completely natural.

Notice that we might say that two cars cost "the same" even if their price differs by a few dollars, and we might say that two people have the "same height", even if they are a few millimeters apart. This loose talk is felicitous in a conversation if the differences between the items do not matter in that context, say when we are talking about what someone with a given income can purchase or whether someone is a better candidate for playing center on a given basketball team. But, strictly speaking, we all know that the two cars do not cost the same and the two people are not of the same height. Someone sympathetic to the Dummett–Quine–Carnap perspective might similarly allow a loose use of the locution "same meaning", in certain conversational contexts. Just as we do not need to take loose talk of "same price" and "same height" seriously, we also might not take the loose talk of "same meaning" seriously—or so someone might argue. Alternately, one might claim that expressions like "same meaning" are ambiguous (or polysemous), sometimes used strictly and sometimes used loosely, with both uses legitimate. Then, one might add, the "strict" reading is the primary one for semantics.[11]

My only point, for now, is that in some contexts, it *is* natural to speak of the logical terminology as meaning the same in the two languages and, in fact, it is common to speak that way sometimes. I'd *like* to take that talk at face value, just as I would in the other cases, where it is more natural to speak of a meaning-shift. I now proceed to (try to) make that more plausible.

7.5 Vagueness

As noted above, the very question of whether various terms have the same meaning or different meanings presupposes something like an analytic-synthetic distinction. This, of course, has come under attack lately, at least since Quine (1951). I noted that John Burgess (2004) formulates a notion of analyticity, supposedly

[11] Thanks to Herman Cappelen for this suggestion.

available to Quineans and other skeptics about meaning. The crucial passage is this, focusing on the law of excluded middle:

> My proposal is that the law should be regarded as "basic", as "part of the meaning or concept attached to the term", when in case of disagreement over the law, it would be helpful for the minority or perhaps even both sides to stop using the term, or at least to attach some distinguishing modifier to it. Such basic statements would then count as analytic . . . This proposal makes the notion of analyticity vague, a matter of degree, and relative to interests and purposes: just as vague, just as much a matter of degree, and just as relative to interests and purposes, as "helpful" . . . [T]he notion, if vague, and a matter of degree, and relative, is also pragmatic. (Burgess 2004, 54)

The scenarios above illustrate the context-sensitivity of Burgess's notion. As noted, in the case of our student Dudley and his guide, it would indeed be "helpful" for the two of them "to attach some distinguishing modifier to" the logical terms. They would speak of "intuitionistic" and "classical" connectives and quantifiers. That would surely facilitate Dudley in following the developments in the new theory, and to keep him from making erroneous logical moves—moves he has come to regard as automatic.

In the second scenario, however, we are comparing our two theories at a more advanced level, noting, for example, that the fundamental theorem of the calculus holds in both, and has an especially nice proof in smooth infinitesimal analysis, and also noting that the intermediate value theorem fails in the latter. In those cases, I suggest, it would be decidedly *unhelpful* to always attach distinguishing modifiers to the logical terminology. *That* would be a distraction, since it would get in the way of making the intended re-identifications and properly comparing the theories. For each classical theorem, we want to be able to easily locate its smooth counterpart. Presumably, the interlocutors to the latter conversations have fully digested the basic logical differences, and are focusing their attention elsewhere.

Burgess says that his notion of analyticity is not only context-sensitive, but also vague. To frame the present theme from a larger perspective, I also suggest that the notion of "same meaning" is vague. In particular, I submit that the following relation is a vague one:

> lexical item a, as used in language/theory T1, means the same as, or is synonymous with, lexical item b, as used in language/theory T2.

To be more specific, I suggest that if a is a logical particle in the language of a mathematical theory T1, and b is its counterpart in a different theory T2, with a different logic, then we have a borderline case of the above relation.

To be sure, it is only helpful to invoke vagueness here if one has a sufficiently comprehensive account of vagueness, and even then only if that account ends up supporting the present thesis concerning meaning-shift. Here, however, I can be helpful, and do not have to plea that these matters are beyond the scope of this work and my own competence, or at least my own interests.

It is widely agreed that at least typical vague terms, such as gradable adjectives, exhibit context-sensitivity. The truth value of a statement in the form, "a is tall" or "b is rich", will vary depending on the comparison class, paradigm cases, and the like. The same person can be filthy rich when compared with philosophy professors and downright poor when compared with college athletic coaches. The same person can be bald when compared to Jerry Garcia and not bald when compared with Dwight Eisenhower.

This much, I think, is uncontroversial. It is also widely, but perhaps not universally, held that vagueness remains even after a comparison class, paradigm cases, and the like, are fixed. There are borderline cases of, say, rich philosophy professors, and those borderline cases are not resolved by articulating a further, more restrictive comparison class.

There is a range of views that postulate a different sort of variability in the proper use of vague terms, one that kicks in even when things like comparison class and paradigm cases are held fixed. Delia Graff Fara (2000), for example, argues that the extensions of vague terms vary according to the interests of speakers at a given moment. As interests change from moment to moment, so do the extensions of vague predicates, even if the comparison class remains the same. Scott Soames (1998) holds that vague terms are indexical expressions whose contents, and thus, extensions vary from context of utterance to context of utterance, again even when we hold comparison class and the like fixed.

Some accounts of vagueness have it that competent speakers of the language can go either way in the borderline regions of a predicate or relation, at least sometimes, without compromising their competence in the use of these words (Raffman 1994; 1996; Shapiro 2006). To focus in on my own account, suppose that a certain man, Harry, is a borderline bald. Then, as I see it, a speaker can usually call Harry "bald" and, if this is accepted in a conversation, then in that context, Harry is bald. In other conversations, a different speaker—or even the same speaker—can call Harry not bald and, if that is accepted, then in those contexts, Harry is not bald. This is despite that fact that the nothing has changed regarding the number and arrangement of Harry's hair, and the comparison class is held fixed.[12]

[12] Raffman postulates a range of acceptable extensions for each vague predicate. Competent speakers can choose from among those. Shapiro (2006) develops the view in terms of Lewis's (1979) notion of conversational score.

The views in question here—those of Shapiro, Raffman, Graff, and Soames—differ on whether it is the *content* of a vague term or just its *extension* that shifts from context to context. Those are the very issues that give rise to various forms of indexical contextualism, non-indexical contextualism, and assessment-sensitive relativism. But we can put those differences aside here, as all four of us agree that extensions shift from context to context. That supports the speculative conclusion here.

Recall the present, tentative thesis that the following relation is vague:

> lexical item a, as used in language/theory T1, means the same as, or is synonymous with, lexical item b, as used in language/theory T2.

This thesis, together with at least some of the broadly "contextualist" views on vagueness (especially that of Shapiro 2006), and, of course, the present logical pluralism, would "predict" that there are conversational situations in which it is natural to speak of the logical particles as having the same meaning across the theories of interest, and such views would likewise "predict" that there are conversational situations in which it is natural to speak of a meaning-shift, following Dummett, Quine, and Carnap. As we saw in the previous section, such predictions are borne out. There *are* such situations.

The accounts of vagueness in question have it that what is said in the "natural" conversational moves presented in the previous section are in fact correct. There are conversational contexts in which the meaning of the logical particles is the same and there are contexts in which the meaning is different. It is not a loose manner of speaking, which is, strictly speaking, false. The "natural" pronouncements all come out true—each in its respective context. In particular, in the first scenario, involving Dudley and his guide, the logical terminology has different meanings. In the second scenario, the meanings are the same.

I do not claim, here, that a broadly contextualist account of vagueness, along the lines of Shapiro (2006), is the only viable option. The point here is that this account of vagueness dovetails nicely with the present thesis that the question of the meaning-shift of logical particles is itself context-sensitive. Indeed, given the vagueness of the above expression concerning sameness of meaning, the account of vagueness predicts the present context-sensitivity.

7.6 Open-Texture, Analyticity, and All That

The speculative proposal of this chapter also dovetails with Friedrich Waismann's (1945) philosophy of language. Let P be a predicate from natural language.

According to Waismann, P exhibits *open-texture* if there are possible objects p such that nothing concerning the established use of P, or the non-linguistic facts, determines that P holds of p, nor does anything determine that P fails to hold of p. In effect, the sentence, or proposition, Pp is left open by the use of the language and the non-linguistic facts. Nothing languages users have said or done to date—whether by way of the ordinary use of the term in communication or in an attempt to stipulate its meaning—fixes how the terms should be applied to the new cases.

Waismann introduced the notion of open-texture as part of an attack on a crude phenomenalism, a view that all contingent concepts can be understood in terms of verification conditions. However, he notes that the interest and applicability of the notion of open-texture extends well beyond that. Here is a thought experiment he uses to illustrate the idea:

> Suppose I have to verify a statement such as 'There is a cat next door'; suppose I go over to the next room, open the door, look into it and actually see a cat. Is this enough to prove my statement? ... What ... should I say when that creature later on grew to a gigantic size? Or if it showed some queer behavior usually not to be found with cats, say, if, under certain conditions it could be revived from death whereas normal cats could not? Shall I, in such a case, say that a new species has come into being? Or that it was a cat with extraordinary properties? ... The fact that in many cases there is no such thing as a conclusive verification is connected to the fact that most of our empirical concepts are not delimited in all possible directions. (Waismann 1945, 121–2)

The last observation in the passage is the key: our empirical concepts are not delimited in all possible directions.

I suspect that, at least nowadays, many philosophers would concede that many, perhaps most, everyday empirical terms are subject to open-texture. The exceptions, it will be argued, include the so-called natural-kind terms. Those supposedly pick out properties which are sharp, in the sense that they have fixed extensions in all metaphysically possible worlds. The progress of science (or metaphysics) tells us, or will tell us, whether various hitherto unconsidered cases fall under the kind in question, and this future study will correct any mistakes we now make with the terms (assuming that the future study gets it right). Waismann would reject views like this. In fact, he illustrates the thesis with what is usually taken to be a natural kind term, gold:[13]

[13] Waismann adds: "*Vagueness* should be distinguished from *open-texture*. A word which is actually used in a fluctuating way (such as 'heap' or 'pink') is said to be vague; a term like 'gold', though its actual use may not be vague, is non-exhaustive or of an open-texture in that we can never fill up all the possible gaps though which a doubt may seep in. Open texture, then, is something like *possibility of vagueness*. Vagueness can be remedied by giving more accurate rules, open-texture cannot." This difference is set aside here, as I am suggesting that "means the same" is both vague and subject to open-texture.

The notion of gold seems to be defined with absolute precision, say by the spectrum of gold with its characteristic lines. Now what would you say if a substance was discovered that looked like gold, satisfied all the chemical tests for gold, whilst it emitted a new sort of radiation? 'But such things do not happen.' Quite so; but they *might* happen, and that is enough to show that we can never exclude altogether the possibility of some unforseen situation arising in which we shall have to modify our definition. Try as we may, no concept is limited in such a way that there is no room for any doubt. We introduce a concept and limit it in *some* directions; for instance we define gold in contrast to some other metals such as alloys. This suffices for our present needs, and we do not probe any farther. We tend to *overlook* the fact that there are always other directions in which the concept has not been defined.... we could easily imagine conditions which would necessitate new limitations. In short, it is not possible to define a concept like gold with absolute precision; i.e., in such a way that every nook and cranny is blocked against entry of doubt. That is what is meant by the open-texture of a concept. (Waismann 1945, 122–3)

The present contention is that the notion of open-texture, or something closely related to it, applies to the locution "has the same meaning". As noted in the previous section, this invokes one of the hottest topics in late twentieth-century philosophy, the notion of analyticity, or truth-in-virtue-of-meaning. The phrase "open-texture" does not appear in Waismann's treatment of the analytic-synthetic distinction in a lengthy article published serially in *Analysis* (1949; 1950; 1951a; 1951b; 1952; 1953), but the notion—or something clearly related to it—plays a central role there. The underlying theme is extended and developed there, in interesting directions.

Waismann observes that language is an evolving phenomenon. As new situations are encountered, and as new scientific theories develop, the extensions of predicates *change*. As new, unexpected cases are encountered, the predicate in question is extended to cover them, one way or the other. When things like this happen, there is often no need to decide—and no point in deciding—whether the application of a given predicate, or its negation, to a novel case represents a change in its meaning or a discovery concerning the term's old meaning. The contrary thesis is what Mark Wilson (2006) derides as the "classical picture", the view that the concepts we deploy are precisely "delimited in all possible situations"—that there is one and only one correct way to go on consistent with the meaning of any given term. For Waismann, as for Wilson, this "classical picture" badly misrepresents the nature of language:

Simply... to refer to "the" ordinary use [of a term] is naive... [The] whole picture is in a state of flux. One must indeed be blind not to see that there is something unsettled about language; that it is a living and growing thing, adapting itself to new sorts of situations, groping for new sorts of expression, forever changing. (Waismann 1951a, 122–3)

Toward the end of the series, Waismann writes: "What lies at the root of this is something of great significance, the fact, namely, that language is never complete for the expression of all ideas, on the contrary, that it has an essential *openness*" (Waismann 1953, 81–2).

The dynamic nature of language goes well beyond the application of empirical predicates (or their negations) to hitherto unconsidered, and perhaps undreamt of, cases. In the analyticity series, Waismann notes that sometimes there are outright *changes* in the application of various words and expressions. Major advances in science sometimes—indeed usually—demand *revisions* in the accepted use of common terms: "breaking away from the norm is sometimes the *only way* of making oneself understood" (1953, 84).

Waismann (1952) illustrates the point in some detail with the evolution of the word "simultaneous". Is there any *semantic* fallout of the theory of relativity? Are we to say that a brand new word, with a new meaning, was coined (even though it has the same spelling as an old word), or should we say instead that we have discovered some new and interesting features of an old word? Did Einstein discover a hidden and previously unnoticed context-sensitivity in the established meaning of the established word "simultaneous" (or its German equivalent), even though he showed no special interest in language as such? Or did Einstein introduce a brand new theoretical term, to replace an old term whose use had scientifically false presuppositions?

According to Waismann, there is often no need, and no reason, to decide what counts as a change in meaning and what counts as the extension of an old meaning to new cases—going on as before, as Wittgenstein might put it. Waismann said, in an earlier installment in the series: "there are no *precise rules* governing the use of words like 'time', 'pain', etc., and that consequently to speak of the 'meaning' of a word, and to ask whether it has, or has not changed in meaning, is to operate with too blurred an expression" (Waismann 1951a, 53). The key term here, I think, is his word "precise", in "precise rules". The "too blurred an expression" we are trying to "operate with" is something like "means that same as", the very relation under scrutiny here.

I'd like to extend what Waismann claims about "time" and "simultaneous" to the logical terminology in the various mathematical theories, at least in the absence of some theoretical framework to make sharp distinctions. To ask, independent of conversational context, or explicit theoretical definitions, whether words like "there exists", "or", and "not" have the same meaning in, say, classical analysis and smooth infinitesimal analysis is to operate with too blurred an expression: namely, "has the same meaning as". The expression is "too blurred" for us to expect a sharp fact of the matter, one that is correct in all contexts. In sum, the view that Wilson dubs the "classical picture" of concepts is false even for logical terms.

Waismann himself comes to a related conclusion, when he argues that the very notion of analyticity is subject to what he earlier had called "open-texture":

> I have defined 'analytic' in terms of 'logical truth', and further in terms of certain 'operators' used in transforming a given sentence into a truth of logic. The accuracy of this definition will thus essentially depend on the clarity and precision of the terms used in the definition. If these were precisely bounded concepts, the same would hold of 'analytic'; if, on the other hand, they should turn out to be even ever so slightly ambiguous, blurred, or indeterminate, this would affect the concept of analytic with exactly the same degree of inaccuracy...
>
> I shall try to show that both concepts ["operator" and "logical truth"] are more or less blurred, and that in consequence of this the conception of analytic, too, cannot be precisely defined... [I]t is significant that we do not only 'find out' that a given statement *is* analytic; we more often precisify the use of language, chart the logical force of an expression, by *declaring* such-and-such a statement to be analytic... It is precisely because, in the case of 'analytic', the boundary is *left open* somewhat that, in a special instance, we may, or may not, recognize a statement as analytic. (Waismann 1950, 25)

The publication of Waismann's series coincides with Quine's (e.g. 1951) celebrated attack on the notion of analyticity. Waismann agrees (or would agree) with Quine that the notion of analyticity cannot play the role assigned to it in logical positivism. In Waismann's case, this is probably for the simple reason that analyticity is too blurred and context-dependent. Quine and Waismann also agree that analyticity is not a source of a priori, incorrigible knowledge. They also seem to agree, at least in broad terms, on the dynamic nature of language, on how words, and their meanings, evolve over time, sometimes in response to developments in empirical science.

Unlike Quine, however, Waismann allows that the notion of analyticity—of being true in virtue of meaning—*does* have an important role to play in how we understand linguistic activity, and in how we interpret each other. Waismann is not out to disparage the general notion.

Of course, the literature on the notion of analyticity, and related notions, is extensive and there is no point in rehearsing it here. It will prove instructive, however, to conclude with some aspects of one of the more spirited defenses of the traditional notion of analyticity, that of H.P. Grice and P.F. Strawson (1956). Some of the observations and insights they bring actually support the dynamic nature of language and, indeed, the vagueness, the open-texture, and even the context-sensitivity of relations like "means the same as" between linguistic entities. Indirectly, this supports the present theme concerning the meaning of logical terms in various mathematical theories.

Recall that Quine expresses "indecision" over whether the statement "Everything green is extended" should count as analytic, and claims that his hesitation is not due to "an incomplete grasp of the 'meanings' of 'green' and 'extended' ". The problem, Quine says, is with the notion of analyticity itself (Quine 1951, section 4). Grice and Strawson concede the hesitation over this case, but give another diagnosis of it:

> The indecision over "analytic" . . . arises . . . from a further indecision: viz., that which we feel when confronted with such questions as "Should we count a *point* of green light as *extended* or not?" As is frequent in such cases, the hesitation arises from the fact that the boundaries of words are not determined by usage in all possible directions. (Grice and Strawson 1956, 153)

Note that this last just is the main insight behind Waismann's notion of open-texture, almost word for word. The boundaries of our words are *not determined in all possible directions*, just as Waismann pointed out. For present purposes, the task at hand is to determine the fallout of this for the dynamic nature of language generally, and for the notions of analyticity and meaning. And, in the present case, the fallout concerning the locution "same meaning" and the question of whether the logical terminology has the same meaning in various mathematical theories.

Famously, Grice and Strawson appeal "to the fact that those who use the terms 'analytic' and 'synthetic' do to a very considerable extent agree in the applications they make of them", and conclude that those terms "have a more or less established philosophical use; and this seems to suggest that it is absurd, even senseless, to say that there is no such distinction" (1956, 143). Waismann, if not Quine, agrees, but goes on to say something about how the notion actually functions in the dynamic evolution of language, and what role it plays in how speakers interpret each other.

Grice and Strawson invite us to consider a pair of English expressions, "means the same as" and "does not mean the same as", the very locutions under scrutiny here, in our study of the uses of logical particles in different mathematical theories:

> Now since [Quine] cannot claim this time that the pair of expressions in question is the special property of philosophers , the strategy . . . of countering the presumption in favor of their marking a genuine distinction is not available here (or is at least enormously less plausible). Yet the denial that the distinction . . . really exists is extremely paradoxical. It involves saying, for example, that anyone who seriously remarks that "bachelor" means the same as "unmarried man" but that "creature with kidneys" does not mean the same as "creature with a heart" . . . *either* is not in fact drawing attention to any distinction at all between the relations between the members of each pair of expressions *or* is making a philosophical mistake about the nature of the distinction between them . . . [W]e frequently talk of the presence or absence of synonymy between kinds of expressions—e.g., conjunctions, particles of many kinds, whole sentences—where there does not appear to be

any obvious substitute for the ordinary notion of synonymy . . . Is all such talk meaningless? (Grice and Strawson 1956, 145–6, emphasis in original).

Of course, neither Quine nor Waismann claims that the use of locutions like "has the same meaning", both in ordinary discourse and in linguistics, is itself meaningless. Grice and Strawson argue that Quine falls into what they call a "philosopher's paradox": "Instead of examining the actual use that we make of the notion *meaning the same*, the philosopher measures it by some perhaps inappropriate standard (in this case some standard of clarifiability)" (1956, 147). Waismann, for his part, *does* examine the actual use of phrases like "analytic", "synonymous", and, indeed, "means the same as". He finds them to be vague, dynamic, and subject to open-texture. This, again, is what I wish to claim (speculatively and tentatively) about the logical particles in various mathematical theories.

Grice and Strawson provide a famous thought experiment, involving two conversations. In one of them, a speaker (X) makes the following claim:

(1) My neighbor's three-year-old child understands Russell's theory of types.

In the second conversation, another speaker (Y) says

(1') My neighbor's three-year-old child is an adult.

Grice and Strawson point out, plausibly, that with X we would know what is being said, even if we find it extremely unlikely or perhaps psychologically impossible for it to be true. In contrast, "we shall be inclined to say that we just don't understand what Y is saying". Indeed, "whatever kind of creature is ultimately produced for our inspection, it will not lead us to say that what Y said was literally true". If "like Pascal, we thought it prudent to prepare against long chances, we should in the first case know what to prepare for; in the second, we should have no idea". Perhaps a more charitable conclusion, still in line with Grice and Strawson's agenda, would be that Y must not be using the word "adult" the same way we do. Grice and Strawson conclude:

> The distinction in which we ultimately come to rest is that between not believing something and not understanding something; or between incredulity yielding to conviction, and incomprehension yielding to comprehension. (1956, 151)

Of course, the distinction Grice and Strawson point to is a real one. There is an important difference between not believing someone, because what she claims is wildly implausible, and not understanding her (or, perhaps better, thinking that she is using words differently than we do). The present thesis, however, is that the distinction in question—between not believing and not understanding—depends heavily on background beliefs and also on what is salient in a given conversation, and so it is context-dependent. It is also more of a difference in degree rather than a difference in kind. The difference is a vague one.

Actually, it is not hard to produce a scenario in which it is plausible—given the right background beliefs—to say that we do understand what is meant by saying that a three-year-old child is, literally, an adult. Suppose the character (Y)'s neighbor's child has an extreme case of progeria, a rage genetic disorder that results in rapid aging in children. Perhaps there are no cases that are quite that severe, but it does not stretch things too much to imagine them.

Maybe the retort will be that, even in that case, the most unfortunate child is not really an adult. Perhaps. I won't quibble over the meanings of ordinary English words. The point here is that, with a reasonable explanation, we would know what was meant. We might argue over what it is to be "literally" correct.

Other cases, supporting Grice and Strawson's thesis, are perhaps easier to come by. I won't try to produce a scenario in which we might say that we understand a claim that a three-year-old child is, literally, a ten-year-old child.

Situations like this are, I submit, more the exception than the rule. To adapt one of Waismann's examples, we can surely imagine someone rightfully expressing an inability to understand Einstein's claim that the same two events can be simultaneous for one observer and not for another. It is my experience that some folks still have trouble understanding this. Didn't Einstein know what "simultaneous" means? Or consider a statement, made some time ago, that atoms have parts. From a certain perspective, one could be forgiven for thinking that whoever said that does not know what she is talking about. Doesn't she know the meaning of the word "atom"?

For a politically loaded example, anyone sufficiently old can surely think of a time when a statement like "Jane and Jill, my two female cousins, are married to each other" would have produced the same sort of incredulity as one would get from Grice and Strawson's character Y saying that a neighbor's three-year-old child is an adult. One would, literally, not know what to make of the statement—would not know what was being said. More charitably, one might think that the speaker was using words with a different meaning than we do.

An opponent, perhaps a defender of the classical picture of concepts, might retort that these are cases of a change in meaning. The staunchest defender of the traditional notion of analyticity never claimed that words never get new meanings. So, someone might argue, the words "simultaneous", "moves", "atom", and "married" do not mean what they used to mean, thanks to Einstein, Galileo, et al. Under the old meanings, the statements in question were indeed of the sort to produce incredulity, and not because the claims are unlikely. They fall on the "incomprehensible" side of Grice and Strawson's dichotomy. Under the new meanings, however, the statements are understandable and perhaps even true.[14]

As above, in at least some cases, Waismann rejects this diagnosis: "there are no *precise rules* governing the use of words like 'time', 'pain', etc., and that consequently to speak of the 'meaning' of a word, and to ask whether it has, or has not changed in meaning, is to operate with too blurred an expression" (Waismann 1951a, 53). Again, the "too blurred" expression is "has the same meaning as".

This is what I'd like to say about the matter of meaning-shift of the logical terminology between mathematical theories. Recall the thought experiment where our character Dudley expressed incredulity at some of the pronouncements of smooth infinitesimal analysis—because the basic rules concerning the logical connectives are different from what he is used to. When the guide said that "zero is not the only nilsquare" and "there is no nilsquare distinct from zero", Dudley was in the same boat as Grice and Strawson toward their character Y saying that a three-year-old child is an adult. Dudley did not know what was being said. But we also saw other conversational contexts in which it is natural to say that the meaning of the logical particles is the same in the different theories, simply because other aspects of the theories are in focus, and the interlocutors are sufficiently familiar with the two theories.

In the foregoing cases, involving "simultaneous", "moves", "atom", and "married", I had the skeptics focusing on some rather fundamental differences between past and present usage—or at least differences that they took to be fundamental. So it is natural for them to insist that meaning has changed. But, when one takes a broader view and looks at the theoretical and practical role of these terms, it is more natural to say that they have not changed. We speak naturally of when our predecessors were wrong, and not that they used words with different meanings than we do today.

The present contention, inspired by Waismann and Quine (and Lewis Carroll), is not that either reaction is somehow mistaken. The complaint is lodged against a

[14] In the case of "married", of course, the question of whether the meaning is the same or different is politically and socially significant. Some conservatives argue that the word does and always did (and always will) stand for a relation between people of different genders, while some progressives claim that gender difference is in no way part of the meaning, and never was. There was surely a time in the past when some claimed that the relation of marriage can only hold between members of the same race—perhaps also as a matter of meaning. A Waismannian or Quinean would be skeptical of all of these claims. The upshot is that matters of meaning can be important for many purposes, even if there is no antecedent fact of the matter what the meaning is. As one of my heroes in philosophy of language, Lewis Carroll put it, "The question is, said Humpty Dumpty, which is to be master—that's all."

non-context-sensitive account of "same meaning" that someone in the grip of the classical picture might insist on.[15]

Toward the end of their article, Grice and Strawson respond to Quine's argument that adopting his brand of confirmational holism undermines at least some attempts to understand the disputed notions of analyticity and synonymy. Grice and Strawson concede, at least for the sake of argument, "that experience can confirm or disconfirm an individual statement, only given certain assumptions about the truth of other statements". They respond that this

> requires only a slight modification of the definition of statement-synonymy in terms of confirmation and disconfirmation. All we have to say now is that two statements are synonymous if and only if any experiences which *on certain assumptions about the truth-values of other statements* confirm or disconfirm one of the pair, also, *on the same assumptions*, confirm or disconfirm the other to the same degree. (1956, 156)

It thus seems that Grice and Strawson are defining an absolute, context-independent relation between statements, by using a bound variable ranging over background assumptions. The idea is that two statements are synonymous *simpliciter* just in case, for any batch of background assumptions whatsoever, the same experiences will confirm or disconfirm each statement to the same degree, given those background assumptions (see also Chalmers 2011).

One can make a similar move with most context-sensitive terms. A person is "super-rich" if he is rich according to any comparison class whatsoever; someone is "super-ready" if they are ready to do anything; an object is "super-left" if it is on the left side of every perspective. Of course, some of these terms are useless.

A Quinean or Waismannian might respond to Grice and Strawson by noting that there are precious few pairs of statements that meet their strict definition of statement-synonymy. Given the right background beliefs, almost any two distinct statements can be pulled apart—even stock examples like "Benedict is a bachelor" and "Benedict is an unmarried man".

A better move for Grice and Strawson would be to define a relative notion of synonymy. That is, two statements are synonymous relative to certain statements S if and only if any experiences which on the assumption that every member of S is true, confirm or disconfirm one of the pair, also confirm or disconfirm the other to the same degree. If so, they would be moving in the direction of Waismann, if not Quine (and Burgess 2004).

At the very end of the article, Grice and Strawson make an interesting concession, possibly also only for the sake of argument:

[15] Thanks for Michael Miller for suggesting this way of putting the point.

The adherent of the analytic–synthetic distinction may go further and admit that there may be cases (particularly perhaps in the field of science) where it would be pointless to press the question of whether a change in the attributed truth-value of a statement represented a conceptual revision or not, and correspondingly pointless to press the analytic–synthetic distinction... [T]he existence, if they do exist, of statements about which it is pointless to press the question of whether they are analytic or synthetic, does not entail the nonexistence of statements which are clearly classifiable in one or the other of these ways. (1956, 158)

The present contention, inspired by Waismann, is that paying attention to such cases, even in the field of logic, will shed light on the dynamic nature of the analytic–synthetic distinction, and the vagueness and open-texture of the relation "means the same as". This, I think, supports the speculative conclusion of this paper, that the very question of whether the logical particles have the same or different meanings in the different mathematical theories, is context sensitive.

References

Anderson, Alan Ross, Belnap, Nuel D., and Dunn, J. Michael. 1992. *Entailment: The Logic of Relevance and Necessity*, volume II. Princeton, NJ: Princeton University Press.

Arnold, Jack and Shapiro, Stewart. 2007. "Where in the (world wide) web of belief is the law of non-contradiction." *Noûs* 41:276–97.

Batens, D., Mortensen, C., and Van Bendegem, Jean-Paul (eds.). 2000. *Frontiers of Paraconsistent Logic*. Dordrecht: Kluwer.

Beall, Jc and Restall, Greg. 2006. *Logical Pluralism*. Oxford: Oxford University Press.

Bell, John. 1998. *A Primer of Infinitesimal Analysis*. Cambridge: Cambridge University Press.

Berto, Francesco. 2007. *How to Sell a Contradiction*. London: College Publications.

Brady, Ross. 2006. *Universal Logic*. Stanford, CA: CSLI Publications.

Burgess, J. 2004. "Quine, analyticity and philosophy of mathematics." *Philosophical Quarterly* 54:38–55.

Carnap, Rudolf. 1934. *Logische Syntax der Sprache*. Vienna: J. Springer. Reprinted as *The Logical Syntax of Language*, trans. Amethe Smeaton, Paterson, NJ: Littlefield, Adams and Co., 1937.

Carnap, Rudolf. 1950. "Empiricism, semantics, and ontology." *Revue Internationale de Philosophie* 4:20–40.

Chalmers, David J. 2011. "Revisability and conceptual change." *Journal of Philosophy* 108:387–415.

da Costa, Newton C.A. 1974. "On the theory of inconsistent formal systems." *Notre Dame Journal of Formal Logic* 15:497–510.

Dummett, Michael. 1991. *The Logical Basis of Metaphysics*. Cambridge, MA: Harvard University Press.

Fara, Delia Graff. 2000. "Shifting sands: an interest-relative theory of vagueness." *Philosophical Topics* 28:45–81. Originally published under the name "Delia Graff".

Field, Hartry. 2008. *Saving Truth from Paradox*. Oxford: Oxford University Press.
Frege, Gottlob. 1976. *Wissenschaftlicher Briefwechsel*. Hamburg: Meiner. Edited by G. Gabriel, H. Hermes, F. Kambartel, C. Thiel, and A. Veraart.
Frege, Gottlob. 1980. *Philosophical and Mathematical Correspondence*. Oxford: Blackwell.
Grice, H.P. and Strawson, P.F. 1956. "In defense of a dogma." *Philosophical Review* 141–58.
Hellman, Geoffrey. 1989. "Never say never! On the communication problem between intuitionism and classicism." *Philosophical Topics* 17:47–67.
Hellman, Geoffrey. 2006. "Mathematical pluralism: the case of smooth infinitesimal analysis." *Journal of Philosophical Logic* 35:621–51.
Hilbert, David. 1899. *Grundlagen der Geometrie*. Leipzig: Teubner; *Foundations of Geometry*, trans. E. Townsend, La Salee, IL: Open Court, 1959.
Hjortland, Ole Thomassen. 2013. "Logical pluralism, meaning-variance, and verbal disputes." *Australasian Journal of Philosophy* 91:355–73.
Horn, Laurence R. 1989. *A Natural History of Negation*. Chicago: University of Chicago Press.
Lewis, David K. 1979. "Scorekeeping in a language game." *Journal of Philosophical Logic* 8:339–59.
Mares, Edwin. 2004. *Relevant Logic: A Philosophical Interpretation*. Cambridge: Cambridge University Press.
Mortensen, Chris. 1995. *Inconsistent Mathematics*. Dordrecht: Kluwer.
Mortensen, Chris. 2010. *Inconsistent Geometry*. Studies in Logic 27. London: College Publications.
Priest, Graham. 1990. "Boolean negation and all that." *Journal of Philosophical Logic* 19:201–15.
Priest, Graham. 2006. *In Contradiction*. Oxford: Oxford University Press, 2nd edition.
Quine, W.V. 1951. "Two dogmas of empiricism." *Philosophical Review* 60:20–43.
Quine, W.V. 1960. *Word and Object*. Cambridge, MA: MIT Press.
Quine, W.V. 1986. *Philosophy of Logic*. Cambridge, MA: Harvard University Press, 2nd edition.
Raffman, Diana. 1994. "Vagueness without paradox." *Philosophical Review* 103:41–74.
Raffman, Diana. 1996. "Vagueness and context-relativity." *Philosophical Studies* 81:175–92.
Read, Stephen. 1988. *Relevant Logic*. Oxford: Blackwell.
Restall, Greg. 1999. "Negation in relevant logics: how I stopped worrying and learned to love the Routley star." In D. Gabbay and H. Wansing (eds.), *What is Negation?* 53–76. Dordrecht: Kluwer.
Restall, Greg. 2002. "Carnap's tolerance, meaning, and logical pluralism." *Journal of Philosophy* 99:426–43.
Restall, Greg. 2014. "Pluralism and proofs." *Erkenntnis* 79:279–91.
Shapiro, Stewart. 2006. *Vagueness in Context*. Oxford: Oxford University Press.
Shapiro, Stewart. 2011. "Varieties of pluralism and relativism for logic." In S. D. Hales (ed.), *A Companion to Relativism*, 526–55. Oxford: Wiley-Blackwell.
Shapiro, Stewart. 2014. *Varieties of Logic*. Oxford: Oxford University Press.
Soames, Scott. 1998. *Understanding Truth*. Oxford: Oxford University Press.
Tennant, Neil. 1996. "The law of excluded middle is synthetic a priori, if valid." *Philosophical Topics* 24:205–29.

Tennant, Neil. 1997. *The Taming of the True*. Oxford: Oxford University Press.
Waismann, Friedrich. 1945. "Verifiability." *Proceedings of the Aristotelian Society* 19:119–50. Reprinted in *Logic and Language*, A. Flew (ed.), Oxford: Basil Blackwell, 1968, 117–44.
Waismann, Friedrich. 1949. "Analytic-synthetic I." *Analysis* 10:25–40.
Waismann, Friedrich. 1950. "Analytic-synthetic II." *Analysis* 11:25–38.
Waismann, Friedrich. 1951a. "Analytic-synthetic III." *Analysis* 11:49–61.
Waismann, Friedrich. 1951b. "Analytic-synthetic IV." *Analysis* 11:115–24.
Waismann, Friedrich. 1952. "Analytic-synthetic V." *Analysis* 13:1–14.
Waismann, Friedrich. 1953. "Analytic-synthetic VI." *Analysis* 13:73–89.
Weber, Zach. 2010. "Inconsistent mathematics." *Internet Encyclopedia of Philosophy* ISSN 2161-0002, <http://www.iep.utm.edu/math-inc/>.
Wilson, Mark. 2006. *Wandering Significance*. Oxford: Oxford University Press.

8

Breaking the Chains

Following-from and Transitivity

Elia Zardini

8.1 Introduction and Overview

I start with some rather dogmatic statements, simply in order to fix a specific enough framework against which to investigate the topic of this chapter. The reader who does not share some or all of the doctrines thereby expressed is invited to modify the rest of the discussion in this chapter in accordance with her favourite views on logical consequence (I'd sympathize myself in some cases with such a reader).

Sometimes, some things logically *follow from* some things. The former are a logical *consequence* of the latter and, conversely, the latter logically *entail* the former. Something logically[1] following "from nothing" is a *logical truth* (see Zardini [forthcoming a] for some reason to doubt the general correctness of this vulgate about logical truth). I will assume that the things that can stand in such relations are *sentences* (see Zardini [2013a] for a defence of this assumption).

While there may be ways of following-from other than the logical one, even restricting to the logical way of following-from this may be determined by different features of the sentences in question. Assuming a *semantic* individuation of sentences ("two" sentences are the same only if they mean exactly the same things), the relevant features may be:

- The *semantic value* of some expressions or others (together with syntactic structure and the identities of all occurring expressions), as in 'John is unmarried' 's following from 'John is a bachelor';

[1] For readability's sake, I will henceforth mostly drop in such locutions qualifications like 'logically' and its like.

- The *semantic value* of some expressions belonging to a *privileged class* of "*logical constants*" (again together with syntactic structure and the identities of all occurring expressions), as in 'Snow is white' 's following from 'Snow is white and grass is green';
- The *semantic structure* of the sentences, determined by the semantic categories of the expressions occurring in them and their modes of composition (again together with syntactic structure and the identities of all occurring expressions), as in 'New York is a city' 's following from 'New York is a great city' (see Evans [1976]; Sainsbury [2001, 359–64] for some discussion);
- The sheer *identities* of the sentences, as in 'Snow is white' 's following from 'Snow is white' (see Varzi [2002, 213–14]; Moruzzi and Zardini [2007, 180–2] for critical discussion).

The last three cases are usually considered to be cases of "*formal* consequence". In the following, we focus attention on them, even though, on my view and as I have just set up things, the interesting divide between logical following-from and the rest does not coincide with the divide between formal logical following-from and the rest, the latter simply arising from a division of the theoretically basic notion of logical following-from into subnotions which are individuated by the specific features of the sentences which, in each particular case, make it so that the relation of logical following-from obtains (see García-Carpintero [1993] for a similar view). Unfortunately, I won't have anything more to say here about the analysis of either the basic notion of logical following-from or the derivative notion of formal logical following-from (see Moruzzi and Zardini [2007, 161–74] for a critical survey of the main approaches to the analysis of logical following-from and formal logical following-from).

In consequence, sentences are often "*put together*". Their mode of being put together is signalled in English by 'and' in the locution ''Q_0' *and* 'Q_1' *and* 'Q_2'... follow from 'P_0' *and* 'P_1' *and* 'P_2'...'[2] and cannot be assumed to be less structured than the mode of being put together enjoyed by the coordinates of a *sequence*. This is so because many logics (i.e. those that are *non-contractive*) give divergent answers to the questions:

(i) Whether ψ follows from φ;[3]
(ii) Whether ψ follows from φ and φ

[2] Most of the time, I will be assuming a *multiple-conclusion* framework: just like no/one/many premises can occupy the first place of the relation of entailment, so can no/one/many conclusions occupy the second place of that relation (see the informal explanation to follow in this section of what it means to reason in a multiple-conclusion framework).

[3] Throughout, 'φ', 'ψ', 'χ', and 'υ' (possibly with numerical subscripts) are used as metalinguistic variables ranging over the set of formulae of the relevant language.

(I myself have developed and defended one such logic for solving the *semantic paradoxes*: see Zardini [2011]; [2012a]; [2013f]; [2014b]; [forthcoming b]). But φ and φ form the same plurality, set, fusion, aggregate, compound, etc. as φ. Some logics (i.e. those that are *non-commutative*) even give divergent answers to the questions:

(i′) Whether χ follows from φ and ψ;
(ii′) Whether χ follows from ψ and φ.

(I myself have developed and defended one such logic for representing *cross-contextual consequence*: see Zardini [2013d]). But φ and ψ form the same multi-set as ψ and φ (see Restall [2000] and Paoli [2002] for useful overviews of non-contractive and non-commutative logics). We will thus adopt the convention of representing with sequences such a fine-grained mode of putting together sentences (and will in turn adopt the convention of representing sequences with functions whose domain is some suitable initial segment of the ordinals and whose range is a subset of the formulae of the relevant language).

An *argument* is a structure representing some or no sentences (the *conclusions* of the argument) as following from some or no sentences (the *premises* of the argument). In English, an argument is usually expressed with a discourse of the form 'P_0; (*and*) P_1; (*and*) P_2 ... Therefore, Q_0; (*or*) Q_1; (*or*) Q_2 ...'. An argument is *valid* iff the conclusions in effect follow from the premises. An *inference* is an act of drawing outputs (mostly, conclusions or whole arguments) from inputs (mostly, premises or whole arguments). (I should stress for future reference that I am not equating inferring with drawing outputs conforming to a certain collection of syntactic rules.) A *derivation* is an abstract codification of an inference conforming to a certain collection of syntactic rules.

What does it mean to draw (accept, reject, doubt, etc.) conclusions when they are not *one* conclusion, but *many* (as it might be the case in our general multiple-conclusion framework)? As a first approximation, we can say that, while premises have to be treated (accepted, rejected, doubted, etc.) "*conjunctively*", conclusions have to be treated (accepted, rejected, doubted, etc.) "*disjunctively*". Focussing on acceptance, it is important to note that accepting "disjunctively" a sequence should not be interpreted as accepting every (or even some) of its coordinates (i.e. sentences in its range). For example, in most multiple-conclusion logics 'There are 1,963 houses in St Andrews', 'It is not the case that there are 1,963 houses in St Andrews' follows from 'Either there are 1,963 houses in St Andrews or it is not the case that there are 1,963 houses in St Andrews', and in most of these logics one may well rationally accept the latter while having no idea of how many houses there are in St Andrews. Of course, if one is in such a situation, it is not only false that one is committed to accepting *both* 'There are 1,963 houses in St Andrews' *and* 'It is not the case that there are 1,963 houses in St Andrews'; it is also false that *either* one is committed to accepting 'There are 1,963 houses in St Andrews' *or* one is committed to accepting 'It is not the case that there are 1,963 houses in St Andrews'. One is only committed to accepting, as it were, "disjunctively" 'There are 1,963 houses in St Andrews', 'It is not the case that there are 1,963 houses in St Andrews', which can very roughly be characterized as a commitment to accepting *that* either 'There are 1,963 houses in St Andrews' is true or 'It is not the case that there are 1,963 houses in St Andrews' is true.[4] Our multiple-conclusion framework also allows for the cases of *no* premises and *no* conclusions. For reasons we don't need to go into here, and focussing on acceptance and rejection, we can assume that one always accepts and never rejects no premises and that one always rejects and never accepts no conclusions (and, in the terminology to be introduced in section 8.4.3, that one does so for non-inferential reasons).

A major part of the philosophical investigation of consequence consists in an attempt at elucidating its *nature*—what consequence consists in. Yet, consequence is also a relation, and as such one can sensibly ask what its *formal properties* are.[5] Arguably, Tarski's most notorious contribution to the philosophical investigation of consequence is constituted by his theory of what consequence consists in: truth preservation in every model (see Tarski 1936). An at least equally important contribution to such investigation is however represented by his earlier studies concerning an abstract theory of consequence relations, aimed at determining the formal behaviour of any such relation. In Tarski [1930, 64–5], he mentions

[4] Another reason for not equating accepting a sequence with accepting all (or some) of its coordinates emerges already in the *conjunctive* case (where the idea is presumably that of accepting *all* of its coordinates) if we pay heed to non-contractive and non-commutative logics. For accepting φ, φ would then be the same as accepting φ, and accepting φ, ψ would be the same as accepting ψ, φ. Notice that some philosophical applications of such logics may in fact not draw these distinctions for *every* attitude: for example, on my non-contractive view of the semantic paradoxes accepting "conjunctively" φ, φ may be the same as accepting φ—it is for attitudes like supposing that bearing that attitude "conjunctively" towards φ, φ is not the same as bearing it towards φ.

[5] Compare with resemblance: a major task for resemblance theories is to determine what resemblance between two individuals consists in (sharing of universals, matching in tropes, primitive similarity, etc.); yet, the study of the formal properties of resemblance (seriality, reflexivity, symmetry, etc.) can fruitfully be pursued even in the absence of an answer to the question about its ultimate nature.

four properties a consequence relation worthy of this name must have: reflexivity, monotonicity, transitivity, and compactness.[6,7]

I think all of these properties are at least questionable (see Moruzzi and Zardini 2007, 180–7). But here I want to focus on *transitivity*, trying to make sense of a position according to which consequence is not transitive. Although there are many different transitivity properties of various strength (see Zardini [forthcoming a]), for our purposes we can simply work with the following principle:[8]

(T) If, for every $\varphi \in \mathrm{ran}(\Theta)$, $\Gamma \vdash \Delta, \varphi$, and $\Lambda, \Theta \vdash \Xi$, then $\Lambda, \Gamma \vdash \Delta, \Xi$.[9,10]

Having proposed elsewhere a solution to the sorites paradox which consists in placing some principled restrictions on transitivity (see the family of *tolerant logics* developed in Zardini [2008a, 93–173]; [2008b]; [2009]; [2013e]; [2014a]; [2013b]; [2013h]; [forthcoming a]), I do have a heavy investment in the issue. However, here I will not argue directly for any position according to which consequence is not transitive. Rather, I will simply try to make it adequately intelligible and assess what impact its correctness would have on our understanding of consequence. The discussion will typically be conducted at a high level of generality, concerning non-transitive logics as such; however, I cannot deny that, heuristically, I have been led to the conclusions defended here by taking tolerant logics as my main paradigm, and this will no doubt in some cases reflect in the fact that the assumptions made and arguments developed will fit tolerant logics better than other non-transitive logics.[11]

[6] Important as they may be, it is worth stressing that they would still grossly underdetermine the nature and even the extension of consequence. Consider, for example, that the relation which holds between certain premises and conclusions iff the premises entail the conclusions and are non-empty will satisfy all the properties mentioned in the text if consequence does.

[7] Contraction and commutativity are not discussed as they are simply built into Tarski's framework, since he puts premises together in sets.

[8] Throughout, 'Γ', 'Δ', 'Θ', 'Λ', 'Ξ', 'Π', and 'Σ' (possibly with subscripts) are used as metalinguistic variables ranging over the set of sequences of formulae of the relevant language; Γ, Δ is the obvious composition of Γ with Δ; $\langle . \rangle$ is the empty sequence. If no ambiguity threatens, I will identify $\langle \varphi \rangle$ with φ.

[9] Throughout, $\mathrm{ran}(\Gamma)$ is the range of Γ; '$\vdash$' (possibly with subscripts) denotes the relevant consequence relation.

[10] In some respects, (T) does not embody a *pure idea* of transitivity, since it also implies versions of monotonicity and contraction. More guarded (and convoluted) transitivity principles which do not have suchlike implications are available. However, I will mostly stick to (T) for its relative simplicity, which will help me to focus on the issues I want to focus on here.

[11] And also fit some tolerant logics (for example, the tolerant logics in favour of whose adoption I myself have argued in the works just referenced in the text, as well as many other relatively weak tolerant logics) better than others (for example, the very strong tolerant logics that have subsequently been studied in more detail by Cobreros et al. [2012]).

The rest of this chapter is organized as follows. To fix ideas, section 8.2 puts on the table a range of philosophically interesting non-transitive consequence relations, introducing briefly their rationale. Section 8.3 discusses and disposes of two very influential objections of principle to the use of non-transitive consequence relations. Section 8.4 delves into some fine details of the metaphysical and normative structures generated by non-transitivity. Section 8.5 concludes by placing the foregoing investigations within the wider context of the relationships between consequence and rationality.

8.2 Non-transitive Consequence Relations

8.2.1 Relevance

In the following, it will be helpful to have in mind concrete examples of non-transitive consequence relations—the different informal pictures underlying the various examples will help us to see (some of) the different ways in which one can so understand consequence as to make sense of failures of transitivity.[12] Of course, again, the aim here is not to argue for the adoption of any such logic, but only to shed light on their rational motivation.

Consider the two following arguments:

CONTRADICTION Intuitively, 'Graham Priest is wrong' does not follow from 'The Strengthened-Liar sentence is true and the Strengthened-Liar sentence is not true'. Graham Priest's error should not be entailed by the correctness of one of his most famous doctrines. Yet:

(i) Both 'The Strengthened-Liar sentence is true' and 'The Strengthened-Liar sentence is not true' do seem to follow from 'The Strengthened-Liar sentence is true and the Strengthened-Liar sentence is not true' (by simplification);
(ii) 'The Strengthened-Liar sentence is true or Graham Priest is wrong' does seem to follow from 'The Strengthened-Liar sentence is true' (by addition);
(iii) 'Graham Priest is wrong' does seem to follow from 'The Strengthened-Liar sentence is true or Graham Priest is wrong' and 'The Strengthened-Liar sentence is not true' (by disjunctive syllogism).

[12] The reader should keep in mind that, throughout, what is meant by 'failures of transitivity' and its like is simply failures of transitivity *for certain argument forms* and its like. The most interesting non-transitive logics retain transitivity for many argument forms.

Two applications of (T) would then yield that 'Graham Priest is wrong' does after all follow from 'The Strengthened-Liar sentence is true and the Strengthened-Liar sentence is not true'.

LOGICAL TRUTH Intuitively, '[Casey is a male and Casey is a sibling] iff it is not the case that [Casey is not a male or Casey is not a sibling]'[13] does not follow from 'Casey is a brother iff [Casey is a male and Casey is a sibling]'. No De Morgan Law should be entailed by an analysis of 'brother'. Yet:

(i′) 'Casey is a brother iff it is not the case that [Casey is not a male or Casey is not a sibling]' does seem to follow from 'Casey is a brother iff [Casey is a male and Casey is a sibling]' and '[Casey is a male and Casey is a sibling] iff it is not the case that [Casey is not a male or Casey is not a sibling]' (by transitivity of the biconditional), and so from the former only (by a version of suppression of logical truths);

(ii′) 'Casey is a brother iff [Casey is a male and Casey is a sibling]' does seem to follow from itself (by reflexivity);

(iii′) '[Casey is a male and Casey is a sibling] iff it is not the case that [Casey is not a male or Casey is not a sibling]' does seem to follow from 'Casey is a brother iff [Casey is a male and Casey is a sibling]' and 'Casey is a brother iff it is not the case that [Casey is not a male or Casey is not a sibling]' (by transitivity of the biconditional).

One application of (T) would then yield that '[Casey is a male and Casey is a sibling] iff it is not the case that [Casey is not a male or Casey is not a sibling]' does after all follow from 'Casey is a brother iff [Casey is a male and Casey is a sibling]'.

Some (Bolzano 1837—at least according to George 1983; 1986—and then Lewy 1958, 123–32; Geach 1958; Smiley 1959, 238–43; Walton 1979; Epstein 1979; Tennant 1987, 185–200, 253–65)[14] have taken these or similar intuitive judgements at face value and concluded that (T) does not unrestrictedly hold (see Lewy [1976, 126–31]; Routley et al. [1982, 74–8] for a critical discussion of this approach).[15] One possible way of elaborating the rationale for such judgements would be

[13] Throughout, I use square brackets to disambiguate English constituent structure.

[14] Sylvan [2000, 47–9, 98–9] intriguingly mentions some possible medieval and early-modern sources, but, to the best of my knowledge, a satisfactory investigation into the pre-Bolzanian history of this logical tradition has yet to be undertaken.

[15] While, as noted in n. 10, (T) in general implies versions of monotonicity, and while there are a variety of logics in which considerations of relevance lead to restrictions of monotonicity, it is worth noting that the particular applications of (T) involved in **CONTRADICTION** and **LOGICAL TRUTH** are unexceptionable in many such logics.

as follows (see von Wright 1957, 175, 177). Apart from no-premise and no-conclusion arguments, consequence should hold in virtue of some *genuine relation* between the *contents* of the premises and the *contents* of the conclusions—its holding should never be determined merely by the independent logical status of some premises (as logical falsities) or conclusions (as logical truths). Thus, apart from no-premise and no-conclusion arguments, consequence should never discriminate between logical truths and falsities on the one hand and logical contingencies on the other hand.

This intuitive constraint can then be made precise as the "filter condition" that an argument is valid iff it can be obtained by uniform substitution of sentences for atomic sentences from a classically valid argument none of whose premises or conclusions are logical truths or falsities (see Smiley 1959, 240).[16] The filter condition yields the desired results: it can easily be checked that each subargument of **CONTRADICTION** and **LOGICAL TRUTH** satisfies the filter condition, even though the overall arguments do not.

In particular, as for **CONTRADICTION**, notice that the two subarguments in (i) are valid, as they can be obtained from the valid argument 'Snow is white and grass is green. Therefore, snow is white (grass is green)' by substitution of 'The Strengthened-Liar sentence is true' for 'Snow is white' and of 'The Strengthened-Liar sentence is not true' for 'Grass is green'; the other subarguments satisfy the filter condition already in their present form. As for **LOGICAL TRUTH**, notice that the first subargument in (i′) is valid, as it can be obtained from the valid argument 'Snow is white iff grass is green; grass is green iff water is blue. Therefore, snow is white iff water is blue' by substitution of 'Casey is a brother' for 'Snow is white', of 'Casey is a male and Casey is a sibling' for 'Grass is green' and of 'It is not the case that [Casey is not a male or Casey is not a sibling]' for 'Water is blue'; the subargument in (iii′) is valid, as it can be obtained from the valid argument 'Snow is white iff grass is green; snow is white iff water is blue. Therefore, grass is green iff water is blue' by the same substitutions; the other subarguments satisfy the filter condition already in their present form. Given what counts as *meaning connection* in this framework, addition of (T) would lead to a gross overgeneration of meaning connections between sentences—the genuine intensional dependencies between the premises and conclusions of each subargument would overgenerate into the bogus intensional dependencies between the premises and conclusions of the overall arguments.

It is often claimed that the imposition of this and similar filter conditions on the consequence relation amounts to changing the subject matter of logic, and does

[16] The filter condition is so called as it acts as a filter on the classically valid arguments, selecting only the "good" ones (so that validity is in effect defined as classical validity plus something else).

not really engage with the traditional view according to which "*truth preservation*" is what consequence is all about. Notice that such a claim cannot be addressed in the particular case of **CONTRADICTION** by holding that some contradictions are true whilst not everything is true, so that the overall argument would fail to be truth preserving in the straightforward sense of having true premises and false conclusions. For the truth of only some contradictions would presumably invalidate subargument (iii) (*qua* not truth preserving in the straightforward sense) and therefore prevent a possible failure of transitivity (see Priest 2006, 110–22).

The claim is however highly dubious on other grounds. For it is plausible to assume that there is a notion of *implication*[17] (expressible in the language by $\rightarrow$) such that, if Δ, ψ follows from Γ, φ, then $\Delta, \varphi \rightarrow \psi$ follows from Γ. At least on some readings, 'If φ, then ψ' does presumably express such a notion in English. But, on such a reading, 'If the Strengthened-Liar sentence is true and the Strengthened-Liar sentence is not true, then Graham Priest is wrong' seems to be false (indeed false-only) if anything is. Under the present assumption, it would, however, be a logical truth if the overall argument of **CONTRADICTION** were valid. Even though, we may assume, truth preserving *in the straightforward sense* of not having true premises and false conclusions, such an argument would thus not be truth preserving *in the only slightly less straightforward sense* of being such that its validity would imply the validity of arguments which are not truth preserving in the straightforward sense. Under the present assumption about the behaviour of implication, there is thus a perfectly good sense in which, for someone attracted by the framework just sketched, licensing the validity of the overall argument of **CONTRADICTION** (T) would indeed lead to failures of truth preservation—what consequence is supposed to be all about (see section 8.3.2 for more on transitivity and truth preservation and Read [1981]; [2003] for a different reply on the issue of relevance and truth preservation).

8.2.2 *Tolerance*

Consider the three following premises:

(1) A man with 0 hairs is bald;
(2) A man with 1,000,000 hairs is not bald;
(3) If a man with i hairs is bald, so is a man with $i + 1$ hairs.

All these premises are intuitively true, and, presumably *a fortiori*, consistent. However, from (3) we have that, if a man with 0 hairs is bald, so is a man with 1 hair, which, together with (1), yields that a man with 1 hair is bald. Yet, from (3) we also

[17] Henceforth, by 'implication' and its like I will mean any operation expressed by some conditional or other.

have that, if a man with 1 hair is bald, so is a man with 2 hairs, which, together with the previous lemma that a man with 1 hair is bald, yields that a man with 2 hairs is bald. With another 999,997 structurally identical arguments, we reach the conclusion that a man with 999,999 hairs is bald. From (3) we also have that, if a man with 999,999 hairs is bald, so is a man with 1,000,000 hairs, which, together with the previous lemma that a man with 999,999 hairs is bald, yields that a man with 1,000,000 hairs is bald. 999,999 applications of (T) (and of monotonicity) would then yield that the contradictory of (2) does after all follow from (1) and (3).

In my works on vagueness referenced in section 8.1, I have taken these intuitive judgements at face value and concluded that (T) does not unrestrictedly hold (earlier, Weir [1998, 792–4]; Béziau [2006] had also briefly entertained this possibility).[18] One possible way of elaborating the rationale for these judgements would be

[18] If I believe that (T) does not unrestrictedly hold for vague discourses, am I not barred from accepting any of its applications involving vague expressions? That would be extremely problematic, since most of our reasoning involves vague expressions and such reasoning often seems to involve applications of (T) (in fact, the arguments in this chapter are no exception!). Before addressing this issue, it is important to realize that it does not constitute an *idiosyncratic* problem for my view on vagueness, but that it represents a *general* problem common to many other cases of logical deviance. The common problem can abstractly be described as the one of recovering what are, even from the point of view of a deviant logician, intuitively acceptable applications of a logical principle whose unrestricted validity she has rejected precisely for the discourses to which those applications belong. Thus, to take a different instance of the same problem, a deviant logician who, because of the semantic paradoxes, has rejected the unrestricted validity of the law of excluded middle for discourses about truth will still want to accept an instance of that law such as 'Either 'Socrates was in the agora on his 23rd birthday' is true or it is not'. (Indeed, there is arguably a *parallel* problem for transitivists who reject tolerance principles like (3), since they need to recover intuitively acceptable applications of (3) and its like, as when, for example, after noticing that one has lost 1 hair one infers, from the fact that one was not bald before, that one is still not bald.) The problem has at least two aspects: at the *philosophical* level, how to motivate the distinction between acceptable and unacceptable applications of a not unrestrictedly valid principle; at the *logical* level, how to recover the acceptable applications. In the case of tolerance-driven non-transitivism, both aspects would deserve a much more extended treatment than can be given in this chapter, but let me briefly sketch my own take on them (focussing on those applications of (T) that are indeed not formally valid—as I have already mentioned in n. 12, the most interesting non-transitive logics retain (T) for many argument forms). As for the first aspect, the only fault the tolerance-driven non-transitivist finds with (T) is that, by sorites reasoning, it allows one to use intuitively correct tolerance principles like (3) to go down the slippery slopes associated with vague predicates. She will thus regard an application of (T) as unacceptable only if it is in this sense "*soritical*" (while intuitive enough for the purposes of this footnote, in a fuller treatment the notion certainly could and should be made more precise). On this view, many applications of (T) remain acceptable even if they involve vague expressions, since such applications will not involve tolerance principles at all or, even if they do, will still be non-soritical. As for the second aspect, the tolerance-driven non-transitivist could simply take acceptable applications of (T) to be, although not *formally* valid, *materially* valid in virtue of the fact that the specific occurrences of non-logical expressions in them determine that they are non-soritical. Alternatively, and focussing on the tolerant logics advocated in my works on vagueness referenced in section 8.1, the tolerance-driven non-transitivist could add *further principles* to her theory which, together with the initial premises of a target acceptable application of (T), formally entail (in those tolerant logics) the final conclusions of that application (see Zardini [forthcoming a] for some details). Thanks to an anonymous referee for pressing me on these issues.

as follows. A major point of a vague predicate is to draw a difference in application between some cases which are far apart enough on a dimension of comparison relevant for the application of the predicate. The predicate should *discriminate* between some such cases. Hence, (1) and (2) must be enforced. Still, another major point of a vague predicate is not to draw any difference in application (from a true application to anything falling short of that) between any two cases which are close enough on a dimension of comparison relevant for the application of the predicate. The predicate should *not discriminate* between any two such cases. Hence, (3) must be enforced. Moreover, instances of (3) should allow *modus ponens*: what substance is there to the idea that there is no sharp boundary between $\mathtt{i}$ and $\mathtt{i}+1$ in matters of baldness if, given the premise that a man with $\mathtt{i}$ hairs is bald, it does not follow that a man with $\mathtt{i}+1$ hairs is bald? Given what counts as *indifference connection* in this framework, addition of (T) would lead to a gross overgeneration of indifference connections between sentences—the correct mandate of not drawing any distinction in matters of baldness between people who differ by 1 hair would overgenerate into the incorrect mandate of not drawing any distinction in matters of baldness between people who differ by 1,000,000 hairs.

8.2.3 *Evidence*

Finally, I would also like to put on the table a case of a non-deductive, *defeasible* consequence relation—that is, roughly, a relation which is supposed to hold between premises and conclusions iff the truth of some of the latter is *reasonable* by the lights of the state of information represented by the former, even if not *guaranteed* by their truth (as is supposed to be the case for a *deductive* consequence relation). For example, the inference of 'Al is a native speaker of Italian' from 'Al was born in Little Italy' is eminently reasonable (all else being equal), as well as the inference of 'Al was born in Italy' from 'Al is a native speaker of Italian'. However, the inference of 'Al was born in Italy' from 'Al was born in Little Italy' is eminently unreasonable. Under this intuitive understanding of defeasible consequence:

(I) 'Al is a native speaker of Italian' is a consequence of 'Al was born in Little Italy';
(II) 'Al was born in Italy' is a consequence of 'Al is a native speaker of Italian';
(III) 'Al was born in Italy' is not a consequence of 'Al was born in Little Italy',

whereas, given (I) and (II), (T) would rule out (III).

One possible way of elaborating the rationale for these judgements would be as follows. Whereas on all *probability functions* reasonable in the light of how things actually are the conditional probabilities of 'Al is a native speaker of Italian' and

'Al was born in Italy' on 'Al was born in Little Italy' and 'Al is a native speaker of Italian' respectively are both very high, the conditional probability of 'Al was born in Italy' on 'Al was born in Little Italy' is very low (if not $= 0$!).

The idea can be made more precise in different specific ways. Here is a fairly general recipe. Let a model $\mathfrak{M}$ of a language $\mathcal{L}$ be a probability function on the sentences of $\mathcal{L}$. Let the conditional probability functions be *totally* defined—assume to that effect a suitable probability calculus (for well-known examples, see Popper [1955]; Rényi [1955]). The probability function corresponding to $\mathfrak{M}$ will thus be in effect determined by the set of conditional probability functions—the unconditional probability in $\mathfrak{M}$ of φ being the conditional probability in $\mathfrak{M}$ of φ on ψ, for some logical truth ψ (assuming a suitable logic in the characterization of the probability calculus). Define then Δ to be a δ-*consequence* of Γ in $\mathfrak{M}$ iff the conditional probability in $\mathfrak{M}$ of the disjunction of all the coordinates of Δ on the conjunction of all the coordinates of Γ is $\geqslant \delta$. Finally, define Δ to be a δ-consequence of Γ iff, for every model $\mathfrak{M}$, Δ is a δ-consequence of Γ in $\mathfrak{M}$. Under natural assumptions, it's easy to check that, as defined, for every $\delta > 0$ δ-consequence is just classical consequence (for $\delta = 0$, δ-consequence is of course trivial). The desired supra-classical strength comes by restricting the range of admissible models to a set of (contextually determined) "reasonable" probability functions.[19]

No doubt this recipe still leaves a lot of leeway in the choice of the probability calculus and of the appropriate restrictions on models. However, it seems plausible that, whichever specific implementation is eventually chosen, the consequence relation so obtained will have a decent claim to codify at least in part the (contextually determined) canon of defeasible reasoning. If so, such a canon will not satisfy (T): given what counts as *probabilification connection* in this framework, addition of (T) would lead to a gross overgeneration of probabilification connections between sentences—the reasonable rules of thumb that people born in Little Italy (typically) speak Italian and that Italian speakers were (typically) born in Italy

[19] Literature on defeasible consequence usually discusses the *cumulative-transitivity* principle:

(CT) If, for every $\varphi \in \mathrm{ran}(\Theta)$, $\Gamma \vdash \Delta, \varphi$, and $\Gamma, \Theta \vdash \Delta, \Xi$, then $\Gamma \vdash \Delta, \Xi$

instead of (T) (for the reason I have mentioned in n. 10). Interestingly, although they constitute a counterexample to (T), (I)–(III) do not constitute a counterexample to (CT), for 'Al was born in Italy' is not a consequence of 'Al is a native speaker of Italian' *and* 'Al was born in Little Italy'. Indeed, while they fail to satisfy (T), many defeasible consequence relations do satisfy (CT). However, it is easy to see that the kind of defeasible probabilistic consequence relation defined in the text will fail to satisfy (CT) as well as (T) in many natural cases (see n. 47 for a concrete example). Thanks to Cian Dorr, Sven Rosenkranz, and Martin Smith for discussions of this issue.

would overgenerate into the unreasonable rule of thumb that people born in Little Italy were (typically) born in Italy.

It might legitimately be wondered what the point is of introducing a *defeasible* consequence relation in the course of an attempt at understanding the idea that *deductive* consequence is non-transitive. Yet, it will turn out that one of the gateways to this understanding is constituted by an appreciation of the *normative import of consequence on rational attitudes*. The non-transitivist can be made sense of as interpreting such import in a particular way. However, that import is not of course a privilege of deductive consequence, and so study of defeasible consequence relations with certain properties (such as the failure of transitivity) may well shed light on some features of deductive consequence relations with the same properties. Indeed, especially in section 8.4.3, I will argue for the claim that the normative force of non-transitive deductive consequence shares crucial features with the normative force of defeasible consequence (in particular, of defeasible probabilistic consequence).[20,21]

8.3 Two Objections to the Very Idea of Non-transitive Consequence

8.3.1 Consequence, inference, and derivation

I can see (and have encountered, in print and conversation) two main objections concerning the very idea of non-transitive consequence, objections which, if correct, would seem to doom from the start any interesting use of a non-transitive

[20] In fact, there seem to be some even deeper connections between probabilistic structure and at least some kinds of non-transitive deductive consequence relations. For example, the tolerant logics advocated in my works on vagueness referenced in section 8.1 fail to satisfy some principles (such as implication in the premises, reasoning by cases, conjunction in the conclusions, etc.) which are also invalid from a probabilistic point of view, even though they are valid on many non-probabilistic codifications of defeasible consequence, such as mainstream non-monotonic logics (see Makinson [2005] for a useful overview of these). I reserve to future work the exploration and discussion of these more specific connections.

[21] Defeasible consequence relations have many structural similarities with *conditionals*. Conditionals arguably fail to satisfy the relevant analogues of (T) and (what is seldom noticed) of (CT) (not to speak of monotonicity), for reasons very similar to those for which certain defeasible consequence relations fail to satisfy (T) and (CT). That is generally a pertinent observation given the tight connections between consequence and conditionals (indeed, it is attractive to see consequence as a special, limit case of implication, see section 8.3.2). The observation is even more pertinent given that the problems and views I will discuss in sections 8.3 and 8.4 find their analogues in the case of conditionals. For lack of space, in this chapter I will however have to focus on consequence relations and leave for another occasion the application of the framework developed in sections 8.3 and 8.4 to the case of non-transitivity of conditionals. Thanks to Sven Rosenkranz for impressing upon me the importance of this connection.

consequence relation. Their rebuttal will help to dispel some misunderstandings of what non-transitivity of consequence amounts to. A more positive characterization will be offered in section 8.4.

There should be an uncontroversial sense in which logic is oftentimes substantially *informative*. There should be an uncontroversial sense in which reasoning oftentimes leads to the *discovery* of new truths. For example, there should be an uncontroversial sense in which the derivation from a standard arithmetical axiom system of the conclusion that there are infinitely many prime numbers is rightly regarded as a substantial discovery about natural numbers, accomplished by purely logical means (under the assumption of the truth of the axioms), no matter what one's views are about the ultimate aptness of logical vocabulary to represent features of the world. However, as Quine among others has stressed (see e.g. Quine 1986, 80–94), the *elementary steps* of logic are, in a sense, *obvious*.[22] In what sense can then logic still be substantially informative?

Well, as the traditional thought about this puzzle has always been at pains to stress, a *series* of completely obvious elementary steps may well lead to a completely unobvious conclusion. In other words, the relation x-is-an-obvious-consequence-of-y is non-transitive. However, it is important to note that this observation by itself does not yet explain away the puzzle: a series of one-foot steps may well lead to cover a considerable distance, but no one has ever supposed this to show that one can take a much longer step than the length of one's legs would give reason to suppose. Why is then the fact that φ_i is reachable in i elementary steps from φ_0 supposed to show that there is some interesting connection between φ_i and φ_0? The crucial, implicit, auxiliary assumption must be that *consequence, as opposed to obvious consequence, is indeed transitive*, so that the path to an unobvious final conclusion can be obliterated, in the sense that the conclusion can be seen as already following from the initial premises. As Timothy Smiley so nicely put it, "the whole point of logic as an instrument, and the way in which it brings us new knowledge, lies in the contrast between the transitivity of 'entails' and the non-transitivity of 'obviously entails' " (Smiley 1959, 242). Thus, there seems to be little space for consequence not to be transitive, if logic is to preserve its function as a means of discovering new truths.

[22] Quine uses 'obvious' in such a way that an apparently deviant logician would be best translated non-homophonically (see also Quine 1960, 57–61). I find very dubious that even the elementary steps of logic are "obvious" in this sense and do not mean anything that strong—only that, once one has adopted a reasonable logic, its elementary steps look most straightforward and least informative, unlike the step from a standard arithmetical axiom system to Euclid's theorem. Thanks to Graham Priest and Stewart Shapiro for stimulating discussions of Quine's views on these matters.

I accept that logic is to preserve such a function, but, as it stands, I contest the very coherence of the previous objection. Suppose that ψ obviously follows from φ and that χ obviously follows from ψ but not so obviously from φ. If consequence is indeed transitive, then, *at least as far as consequence is concerned*, the step from φ to χ is no more mediated, and thus no less "elementary", than those from φ to ψ and from ψ to χ: χ just follows straight from φ as it follows straight from ψ (and as ψ follows straight from φ)—the idea that χ bears the more complex relation to φ of being reachable in two "elementary" steps from it but not in one is simply a non-transitivist illusion. If then the step from φ to χ is indeed unobvious, it is just not true, in a transitivist framework, that every "elementary" logical step is obvious. Thus, the solution offered to the puzzle (transitivity) simply denies one of the elements from which the puzzle arose. To repeat, in a transitive logic the step from φ to χ is "elementary" in the sense that, *at least as far as consequence is concerned*, it is no more mediated than those from φ to ψ and from ψ to χ. Indeed, in this sense, in a transitive logic *every* valid step—no matter how remotely connected the premises and the conclusions may in effect be—is an "elementary" step. The root of the trouble is then that the transitivity solution simply obliterates the *structure* presupposed by the puzzle: namely, the distinction between elementary logical steps and non-elementary ones, *as long as that structure is supposed to be generated by the relation of consequence* (a supposition without which, as we are about to see, the objection against the non-transitivist evaporates).

This strongly suggests that the elementary/non-elementary distinction has been mislocated by the objection. It is not a distinction to be drawn at the level of *consequence*—rather, it is a distinction to be drawn at the level of *inference*. For our purposes, the notion of an elementary inference can be left at an intuitive level: it is understood in such a way that the inference of φ from 'φ and ψ' is elementary, whereas that of '[φ or χ] and [ψ or υ]' from 'φ and ψ' is not. This is just as good, since the puzzle was an *epistemic* one and the drawing of an inference is one of the ways in which the validity of an argument can be *recognized* by a subject. The transitivist herself has thus to acknowledge that the puzzle should properly be stated as the puzzle of how any inference can be substantially informative given that every elementary inference is least informative. Her own solution to the puzzle may then be revised as follows. x-is-elementarily-inferrable-from-y is non-transitive, yet it entails x-is-inferrable-from-y. Furthermore, by the soundness of inferrability (with respect to consequence), x-is-inferrable-from-y entails x-is-a-consequence-of-y. The transitivity of the latter in turn ensures that an unobvious final conclusion already follows from the initial premises.

The revised argument is coherent and might seem to offer a satisfactory explanation of the revised puzzle from a transitivist perspective. However, if x-is-inferrable-from-y is itself transitive, the appeal to the transitivity of consequence is now exposed as a superfluous detour. For a simpler argument can run as follows. x-is-elementarily-inferrable-from-y is non-transitive, yet it entails x-is-inferrable-from-y. The transitivity of the latter in turn ensures that an unobvious final conclusion can already be inferred from the initial premises, and so, by the soundness of inferrability (with respect to consequence), that the unobvious final conclusion follows from the initial premises.

Now, it is true that, for the little that has been said so far, x-is-inferrable-from-y need not be transitive from a non-transitivist point of view. Yet, any theoretical employment of logic by a being whose cognitive architecture resembles that of humans is likely to require a *systematization* of pre-theoretical judgements of validity in terms of a small set of syntactic rules which are at least sound (and possibly complete) with respect to consequence. It is simply a fact that such a systematization provides one of the most effective methods a human (and anyone with a similar cognitive architecture) can employ in order to explore what follows from what (just as something along the lines of the standard rules for addition, subtraction, multiplication, and division provides one of the most effective methods a human (and anyone with a similar cognitive architecture) can employ in order to explore what gives what).

To stress, I don't think that there is an immediate connection between inference and syntactic rules: I think it's clear on reflection that one can draw an inference simply in virtue of one's appreciation of the logical concepts involved in it, even if one is not in possession of a set of syntactic rules which would allow the relevant derivation. Yet, as I have just noted, given the way human cognition works there is every theoretical reason for transitivists and non-transitivists alike to accept a systematization of our pre-theoretical judgements of validity in terms of syntactic rules. And, while there is no conceptual bar to the derivational system thus generated being itself non-transitive, it is of course a desirable epistemic feature also from the non-transitivist perspective that the derivational system used in the study of a non-transitive consequence relation be itself transitive, in the sense that *any new application of a syntactic rule preserves the property of being a correct derivation*. This feature is desirable exactly because non-transitive consequence can be just as unobvious as transitive consequence, whence the need arises for derivational techniques offering the epistemic gain flowing from the asymmetry between the non-transitivity of x-is-obviously-derivable-from-y (x-is-elementarily-derivable-from-y) and the transitivity of x-is-derivable-from-y (for an example of non-transitive logics with transitive sound and complete derivational systems, see Zardini [2013h]).[23] No further asymmetry between the non-transitivity of x-is-an-obvious-consequence-of-y and the alleged transitivity of x-is-a-consequence-of-y is required.

8.3.2 *Consequence and truth preservation*

We have already encountered in section 8.2.1 the (rather vague) claim that consequence is all about truth preservation, in the sense that whether the consequence relation holds between certain premises and conclusions is *wholly* determined by whether the conclusions preserve the truth of the premises. On the basis of an intuitive understanding of such notion of truth preservation, the following objection can be mounted.

The non-transitivist, we may assume, claims, for some φ, ψ, and χ, that ψ follows from φ and χ from ψ, but χ does not follow from φ (i.e. that (T) fails in some single-premise and single-conclusion case). So, since truth preservation is necessary for consequence, the non-transitivist should concede that ψ preserves the truth of φ and that χ preserves the truth of ψ. Yet, on the face of it, truth preservation is a transitive relation: if ψ preserves the truth of φ and χ preserves the truth of ψ, then it would seem that χ also preserves the truth of φ. For suppose that ψ preserves the truth of φ and that χ preserves the truth of ψ, and suppose that φ is true. Then, since ψ preserves the truth of φ, ψ should be true as well. But χ preserves the truth of ψ, and so, since ψ is true, χ should be true as well. Thus, under the supposition that φ is true (and that ψ preserves the truth of φ and that χ preserves the truth of ψ), we can infer that χ is true. Discharging that supposition, we can then infer (still under the suppositions that ψ preserves the truth of φ and that χ preserves the truth of ψ) that, if φ is true, so is χ, which might seem sufficient for χ's preserving φ's truth. Having thus argued that truth preservation is transitive, the objection is completed by noting that truth preservation suffices for consequence, so that, since χ preserves the truth of φ, χ follows from φ, contrary to what the non-transitivist claims.

[23] To get a concrete sense of how a transitive derivational system can be sound and complete with respect to a non-transitive consequence relation, just take your favourite transitive derivational system presented in sequent-calculus style which does not admit of cut elimination. Throw the cut rule out of the derivational system (if it was included in it) and take your consequence relation to be simply the set of sequents provable in the resulting derivational system. The resulting derivational system itself is obviously sound and complete with respect to that consequence relation, although the former will still be transitive (in the specific sense just explained in the text) while the latter will be non-transitive (in the sense of failing to satisfy (T)). Similar examples could be provided, for instance, for derivational systems presented in natural-deduction style by suitably complicating with provisos the derivational system's syntactic rules. Thanks to Marcus Rossberg and Moritz Schulz for discussions on this issue.

A proper assessment of the objection requires first an adequate explication of the underlying notion of truth preservation. I know of no better way of spelling out this notion than in terms of a *conditional* statement: certain conclusions preserve the truth of certain premises iff, *if* every premise is true, *then* some conclusion is true (where, for the purposes of the discussion to follow, we can afford to remain rather neutral as to the exact behaviour of the conditional). Henceforth, I will officially adopt this explication of the notion (save for briefly considering a possible alternative understanding at the end of this section). Of course, so stated in terms of an unadorned indicative conditional, truth preservation alone is not *sufficient* for consequence: if 'A match is struck' is true, so is 'A match will light', but the latter does not in any sense logically follow from the former. Many authors would be sceptical that any sufficient non-circular strengthening of the intensional force of the conditional is available, but I remain an optimist (see Zardini 2013f; 2014b). In the interest of generality, I will however assume that there is no interesting sense of 'truth preservation' in which truth preservation is sufficient for consequence, and show how the objection from truth preservation against the non-transitivist can be modified so as to need only the assumption that truth preservation is *necessary* for consequence (in any event, my reply to that modified objection will apply with equal force to the original one).

Clearly, if truth preservation is not sufficient for consequence, the original objection against the non-transitivist presented at the beginning of this section fails. In particular, the objection breaks down at the very last step 'since χ preserves the truth of φ, χ follows from φ', which assumes the sufficiency of truth preservation for consequence. Yet, a more sophisticated version of the objection could still be made to run against some applications that the non-transitivist envisions for her logics (for example, to the sorites paradox). To see this, let us classify three degrees of "*non-transitive involvement*":

- An application of non-transitivity which only requires a verdict of invalidity concerning a transitivistically valid target argument is *weak*;
- An application which in addition requires the rejection of a commitment to accepting the conclusions of a transitivistically valid target argument all of whose premises are accepted[24] is *intermediate*;
- An application which in addition requires the falsity of all the conclusions of a transitivistically valid target argument all of whose premises are true is *strong*.[25]

[24] For simplicity's sake, I will henceforth take sentences as the objects of acceptance and rejection. The whole discussion may be recast, more clumsily, in terms of propositions.

[25] Interestingly, our three examples of applications of non-transitivity in section 8.2 suffice to cover all the three degrees of non-transitive involvement: the application presented in section 8.2.1 is weak, the one presented in section 8.2.3 is intermediate and the one presented in section 8.2.2 is strong.

Focus then on strong applications of non-transitivity (an analogous discussion could be run for intermediate applications).[26] For some of these, the non-transitivist wishes to maintain that φ is true but χ false, even though ψ follows from φ and χ from ψ. However, without assuming the sufficiency of truth preservation for consequence, the objection from truth preservation presented at the beginning of this section can still be used to reach the conclusion that truth preservation is transitive, which, together with the necessity of truth preservation for consequence, yields that χ preserves the truth of φ—that is, if φ is true, so is χ.[27] And that conditional sits badly with the other commitments (truth of φ, falsity of χ) the non-transitivist would wish to undertake (exactly how badly it sits will depend of course on the details of the logic—for example, these claims are jointly inconsistent in the tolerant logics advocated in my works on vagueness referenced in section 8.1). Now, because of various considerations relating to *context dependence* and to *indeterminacy* which would lead us too far afield to rehearse here, I would be wary of principles stating that certain versions of truth preservation are necessary for validity (see Zardini 2012b; 2013b). Still, none of the considerations underlying that caution target the necessity of truth preservation for the kinds of arguments that are typically considered in strong applications of non-transitivity, so that I am willing to concede that, for those cases, truth preservation is indeed necessary for consequence.[28]

[26] Henceforth, I will mainly talk about non-transitive logics tailored to *intermediate* and *strong* applications of non-transitivity, as these are arguably the philosophically most problematic and interesting cases to be made sense of. This qualification must be understood as implicit in the following. However, although I cannot give a detailed treatment of the issue here, I hasten to add that this is very likely not a real limitation. For, in the case of *suppositional* acceptance if not in that of *flat-out* acceptance, weak applications will very likely have to replicate metaphysical and normative structures identical to those which I will argue are generated by intermediate and strong applications.

[27] In fact, at least one of the most prominent strong applications of non-transitivity (i.e. the one to the sorites paradox) presents some recalcitrance against being fitted into the mould used in the text. For that is an application with *multiple-premise* arguments, so that transitivity of truth preservation is not applicable to such chains of arguments in the direct way exploited by the modified objection in the text. One way to recover an application with multiple-premise arguments would be to modify the relevant arguments, so as to bring all the premises $\varphi_0, \varphi_1, \varphi_2 \ldots, \varphi_i$ used at some point or other in the chain up front, collect them together in a single, long "conjunction" (at least assuming that they are finite) and carry them over from conclusion to conclusion adding to the relevant intermediate conclusion the "conjuncts" $\varphi_0, \varphi_1, \varphi_2 \ldots, \varphi_i$ (scare quotes being used here since, for example in tolerant logics, standard conjunction has not the right properties to do the job and a new operation would have to be introduced instead). An alternative way to recover an application with multiple-premise arguments would be to use a generalized version of transitivity of truth preservation, which, given the relevant chain of arguments $a_0, a_1, a_2 \ldots, a_i$ and under the assumption that all the side premises in those arguments are true, says that, if each of the arguments is truth preserving, so is the argument whose premises are those of a_0 and whose conclusion is that of a_i.

[28] Some authors have recently started to talk as though the semantic paradoxes gave a reason to reject that truth preservation is necessary for consequence even in the cases that I myself would regard as unproblematic (see e.g. Field 2006). The situation is however more neutrally described in

Before giving my reply to the modified objection from truth preservation, I would like to undertake a brief digression on truth preservation which I hope will prove instructive for understanding some features of non-transitive logics. One might think that any worry about the connection between consequence and truth preservation is simply misplaced because of the fact that consequence is trivially guaranteed to be equivalent with the preservation of at least *some kind* of truth, even if not truth *simpliciter*. For consequence is often defined as preservation, for every model $\mathfrak{M}$, of truth in $\mathfrak{M}$. Consequence would then be trivially guaranteed to be equivalent with the preservation of truth in every model.

Setting aside the important question about what kind of truth "truth in a model" really is, what I want to remark on here is that already the identification of consequence with the *preservation* of something or other (in some class of structures or other) is unwarranted in our dialectic, as one can define well-behaved consequence relations without appealing to anything recognizable as preservation of something or other (in some class of structures or other).[29] Tolerant logics are an example of such a consequence relation developed in a non-transitive framework; an example driven by a completely different kind of consideration (not affecting transitivity) can be found in Martin and Meyer [1982]. The general structural point in these logics is that *the collection of designated values relevant for the premises is not identical with the collection of designated values relevant for the conclusions*: in tolerant logics, the former is in effect a proper *subcollection* of the latter, whereas the logic S of Martin and Meyer [1982] can be seen as a form of the dual position, in which the former is a proper *supercollection* of the latter (thereby leading to the failure not of transitivity, but of another property considered by Tarski [1930] essential for a consequence relation—namely, *reflexivity*).[30] Thus, in both tolerant logics and S, consequence is not equivalent with anything recognizable as preservation of something or other (in some class of structures or other); it is

terms of these authors' theories of truth being inconsistent with the principle that truth preservation is necessary for consequence, which would seem (at least to me) to be better taken as a reason for rejecting these authors' theories rather than as a reason for rejecting the principle (at least in the cases that I myself would regard as unproblematic), especially given that there are other theories of truth on the market which are perfectly consistent with the principle and even entail it (see, for example, the theory developed in my works on the semantic paradoxes referenced in section 8.1).

[29] Even granting an identification of consequence with truth preservation in some class of structures or other, it is well known that the standard set-theoretic notion of *model*, which replaces the generic notion of *structure*, has serious drawbacks in the analysis of the consequence relation of expressively rich languages. I won't go here into this further aspect of the complex relation between consequence and preservation of truth in a structure (see McGee [1992] for a good introduction to some of these issues).

[30] Thanks to the late Bob Meyer for a very helpful discussion of S and to Graham Priest for pointing out to me the duality connection.

rather equivalent with the *connection* of a certain collection of designated values with a certain *other* collection of designated values (in every model of the relevant class).

Let me be clear. It is not the case that a representation in terms of non-identity of the two collections of designated values *suffices* to ensure the non-transitivity of a logic (see Smith [2004] for an example of a transitive logic generated by such a representation). Nor is it the case that a representation in terms of non-identity of the two collections is *essential* for every non-transitive logic (see Smiley [1959] for an example of a non-transitive logic where no such natural representation seems to be forthcoming). Yet such a representation, where available, is a fruitful point of entry to at least one of the key thoughts behind the logic (and as such will be deployed in section 8.4.4). The representation also connects up neatly with usual representations of transitive logics (preservation of designated value in some class of structures or other), showing how non-transitivity arises from a very natural and straightforward generalization of the usual model-theoretic representation of a (transitive) consequence relation.

Moving on now to my reply to the modified objection from truth preservation, I reject that the non-transitivist is committed to χ's being true if φ is. For the most natural way in which this conclusion (and so the transitivity of truth preservation) might be thought to follow from the premises that ψ preserves the truth of φ and that χ preserves the truth of ψ requires the assumption that the consequence relation of the *metalanguage* (the language in which we talk about the truth of φ, ψ, and χ) is transitive (as against the non-transitivity of the consequence relation of the *object language* in which we talk about whatever φ, ψ, and χ talk about). More explicitly: the validity of the metalinguistic argument 'If 'φ' is true, then 'ψ' is true; if 'ψ' is true, then 'χ' is true. Therefore, if 'φ' is true, then 'χ' is true' (and so, under the current interpretation of truth preservation, the transitivity of truth preservation) boils down to the validity of the argument form 'If P_0, then P_1; if P_1, then P_2. Therefore, if P_0, then P_2', which is however invalid in many non-transitive logics (it is, for example, invalid in many tolerant logics).[31]

Moreover, virtually no deviance from classical logic worthy of this name should grant the assumption that the consequence relation of the metalanguage $\mathcal{M}$ talking about the truth and falsity of the sentences of the object language $\mathcal{O}$ over which

[31] I do not wish to suggest that the reply I am proposing is the only one available to the non-transitivist (although it is the kind of reply that, subject to the important refinement mentioned in the last point made in n. 36, I do endorse): I will present in section 8.4.1 a coherent kind of non-transitive position which would rather reject that truth preservation is necessary for consequence (in particular, while that position would accept that χ follows from ψ, it would reject that, if ψ is true, so is χ).

a deviation from classical logic is envisaged should itself be classical. For, given any decent theory of truth, this will be sufficient to reintroduce classical logic in 𝒪 itself. Consider, for example, a deviant intuitionist logician. Were she to accept 'Either 'φ' is true or 'φ' is not true' in ℳ (for φ belonging to 𝒪), the most natural theory of truth (namely one such that φ and ''φ' is true' are fully equivalent,[32] so that in particular ''φ' is not true' entails 'It is not the case that φ') would commit her to 'Either φ or it is not the case that φ'.

The point need not exploit the full equivalence between φ and ''φ' is true' which is induced by the *enquotation/disquotation* schema:

(ED) P iff 'P' is F.

We can produce similar results only, for example, with the right-to-left direction of (ED). If a metalinguistic *necessity* predicate were to behave classically, the intuitionist would have to accept in ℳ 'Either 'φ' is necessary or 'φ' is not necessary' (for φ belonging to 𝒪). Given that ''φ' is not necessary' is classically equivalent with ''It is not the case that φ' is possible', she would have to accept 'Either 'φ' is necessary or 'It is not the case that φ' is possible', whence, by substituting 'Either φ or it is not the case that φ' for 'φ', she would have to accept 'Either 'Either φ or it is not the case that φ' is necessary or 'It is not the case that either φ or it is not the case that φ' is possible', which entails in (any suitable modal extension of) intuitionist logic ''Either φ or it is not the case that φ' is necessary'. By right-to-left (ED), she would then have to accept 'Either φ or it is not the case that φ'.[33] Indeed, up to the very last step, the previous argument would go through also for necessity-like metalinguistic predicates[34] which fail to satisfy either direction of (ED), such as 'in

[32] Here and in what follows, someone (not me) might want to make a proviso for the semantic paradoxes. Even with that proviso, the argument in the text shows that all of 𝒪's *grounded* sentences behave classically—hardly a pleasing result for the intuitionist.

[33] The argument in the text should make it clear what exactly I mean when I say that a metalinguistic predicate Φ "behaves classically". What exactly I mean is that every instance of an argument form valid in classical first-order logic in which only Φ occurs as non-logical constant (treating quotation names as logical constants) is valid and, in addition, that, as is usual in the relevant extensions of classical first-order logic, Φ has a dual Ψ satisfying the principle ''φ' is not Φ iff 'It is not the case that φ' is Ψ' (in the argument in the text, 'possible' is naturally assumed to be such classical dual for 'necessary'). In particular, notice that, since 'a is F' is not an argument form valid in classical first-order logic, we cannot straightforwardly assume ''Either φ or it is not the case that φ' is necessary' (maybe on the grounds that 'Either φ or it is not the case that φ' is classically valid), whose most fine-grained form is precisely 'a is F'. Only the object-language logic—which in our example is intuitionist—is allowed, as it were, to "look inside" quotation names, whence it is crucial that, in the argument in the text, ''It is not the case that either φ or it is not the case that φ' is not possible' is valid in (any suitable modal extension of) intuitionist logic.

[34] A metalinguistic predicate Φ is *necessity-like* iff Φ is closed under *adjunction* (''φ' is Φ' and ''ψ' is Φ' entail ''φ and ψ' is Φ') and φ entails ''φ' is Φ' if φ is the *trivial truth* (i.e. the truth entailed by anything).

principle acceptable'. Thus, if a metalinguistic *in-principle-acceptability* predicate were to behave classically, the intuitionist would have to accept in $\mathcal{M}$ ‘‘Either φ or it is not the case that φ’ is in principle acceptable’ (for φ belonging to $\mathcal{O}$). (Notice that, although failing to satisfy either direction of (ED), ‘in principle acceptable’ and what is naturally assumed to be its classical dual ‘in principle non-rejectable’ still connect with the object language in a broadly “(ED)-style” fashion in the sense that, roughly, if one knows that φ is in principle acceptable, one should accept φ, and, if φ is inconsistent, φ is not in principle non-rejectable.)

The general observation implicit in the last paragraph is that a deviance from classical logic for a language $\mathcal{L}_0$ should equally apply to a language $\mathcal{L}_1$ as soon as some of the notions expressible in $\mathcal{L}_1$ are best thought of as exhibiting the same problematic properties which motivate a deviance from classical logic for $\mathcal{L}_0$, and that, in some cases, this may be so exactly because of some systematic *connections* that those notions bear to notions expressible in $\mathcal{L}_0$—for example, connections established by some broadly (ED)-style principle when $\mathcal{L}_1$ is the metalanguage of $\mathcal{L}_0$. Importantly, this general observation fatally affects also revisions of the objection from truth preservation which replace the notion of truth preservation with that (structurally identical) of *closure of knowledge or of other epistemic properties under logical consequence* (for example, ‘known’ still satisfies the right-to-left direction of (ED)).[35]

It is worth seeing in closing how the objection from truth preservation fares if the current (and standard) explication of the notion of truth preservation in terms of the notion of implication is rejected in favour of a *primitive* relation of truth preservation. While under the former explication there was a logical guarantee that the relation is transitive (assuming a transitive logic!), now that guarantee is lost and the alleged transitivity must be postulated as a specific law governing the relation. Pending further argument, there do not seem to be clear reasons for the non-transitivist to accept this postulation. But what is most important to note is that even the acceptance of a transitivity postulate for the relation would not by itself wreak havoc at least for a strong application of non-transitivity like the one to the sorites paradox, for what is really at issue there is not so much simple transitivity, but the stronger assumption that a finite chain of elements connected by a relation R is such that its first element bears R to its last element. This is in effect the property of relations that is sometimes called ‘*chain transitivity*’ (see Parikh 1983), which, in a non-transitive logic, is usually *stronger* than the

[35] Thanks to Stephen Schiffer for interesting discussions on this last kind of objection. Within a tolerant logic, the point made in the text can actually be developed into a surprising defence of certain closure principles for knowledge against some very influential objections (see Zardini 2013c).

property of transitivity (see Zardini [forthcoming a] for details; of course, under minimal assumptions, the two properties are equivalent in a transitive logic). Even though there may ultimately be no need for the non-transitivist to pursue this latter strategy in the case of a primitive truth-preservation relation, I think there is good reason for her to embrace it in other cases in which independent grounds support the transitivity of a particular relation, as in the case of the identity relation (see again Zardini [forthcoming a] for details).[36]

[36] Going back to the standard explication of the notion of truth preservation in terms of the notion of implication, I have focussed on the fact that the argument 'If 'φ' is true, then 'ψ' is true; if 'ψ' is true, then 'χ' is true. Therefore, if 'φ' is true, then 'χ' is true' is invalid in many non-transitive logics. That argument has in turn been supported in the objection from truth preservation by the argument ''φ' is true; if 'φ' is true, then 'ψ' is true; if 'ψ' is true, then 'χ' is true. Therefore, 'χ' is true', which is equally invalid in many non-transitive logics (while the step from the validity of the latter argument to the validity of the former argument—an application of the *deduction theorem*—is valid in many non-transitive logics and can be taken for granted for our purposes). One might object that, never mind what many non-transitive logics say about it, especially the latter argument (a "metalinguistic double-*modus-ponens* argument") seems *intuitively very compelling*, so that its rejection is a cost of the reply I have given on behalf of the non-transitivist. In response to this worry, I would like to make four points. First, as a preliminary clarification of the limited aims of the reply I have given on behalf of the non-transitivist, I should stress that the purpose of the reply is not to argue that there is absolutely nothing to the objection from truth preservation (or to any other objection against the non-transitivist), but only to show that the non-transitivist can take a *coherent* and (at least from her point of view) *very natural* position in reply to that objection *that is compatible with the principle that truth preservation is necessary for consequence*. To be sure, adoption of that position comes with its own costs—just as adoption of any other position in the puzzling areas for which non-transitive logics have usually been proposed does—and rejection of the metalinguistic double-*modus-ponens* argument may well be among those, but, contrary to the spirit with which the objection from truth preservation has often been raised to me, the non-transitivist is not forced to a rather improbable rejection of the principle that truth preservation is necessary for consequence. Second, it is actually unclear (at least to me) that the metalinguistic double-*modus-ponens* argument is intuitively so compelling that its rejection should be regarded as a cost of the reply I have given on behalf of the non-transitivist. What seems clear (at least to me) is that that argument is naturally analyzed as involving two applications of *modus ponens* and a transitivity step, and that both applications of *modus ponens* are indeed intuitively valid; but it is unclear (at least to me) whether chaining those applications together and thus *obliterating the role of the intermediate conclusion* (''ψ' is true') *as a premise* in yielding the final conclusion (''χ' is true') has much intuitive compellingness. Third, it would seem that, if, contrary to the doubt I have just raised, the *metalinguistic* double-*modus-ponens* argument is regarded as intuitively compelling, the simpler *object-language* double-*modus-ponens* argument 'φ; if φ, then ψ; if ψ, then χ. Therefore, χ' should be regarded as no less intuitively compelling. But that argument usually has to be rejected by strong applications of non-transitivity, so that the worry raised would ultimately have nothing to do *in particular* with metalinguistic double-*modus-ponens* arguments, and so nothing to do *in particular* with the metalinguistic principle that consequence requires truth preservation. That is of course not to say that the worry would not be real; only that, contrary to the objection I have discussed in this section, there would be no specific problem of compatibility between non-transitivism and the principle that truth preservation is necessary for consequence. Fourth, I should mention that, at least focussing on the strong application of non-transitivity made in my works on vagueness referenced in section 8.1, insofar as one finds double-*modus-ponens* arguments intuitively compelling there are on the one hand well-behaved tolerant logics (not those focussed on in said works) in which they are valid, and there are on the other hand prospects for upholding their validity also in other tolerant logics by switching to *non-classical* (in fact, *tolerant*) *metatheories in which it is vague what the values*

8.4 Making Sense of Non-transitivity

8.4.1 Non-logical/logical dualism and non-transitivism

We have only started to scratch the surface of the philosophical underpinnings of a non-transitive logic. The rebuttal of the two previous objections has helped to dispel some misunderstandings of what non-transitivity of consequence amounts to, but not much has yet been offered by way of a positive characterization of a conception of consequence as non-transitive.

I think more progress on this issue can be made by asking why consequence is usually assumed to be transitive. Consider the following natural picture (taking now sets to be the terms of the consequence relation and restricting our attention to single-conclusion arguments). The laws of logic can be seen as an *operation*[37]

relevant for consequence are. On either scheme, the result would be the analogue for implication of the move discussed in the text for relations: again, what is really at issue in the sorites paradox is not so much a simple-transitivity (i.e. double-*modus-ponens*) argument like 'A man with 0 hairs is bald; if a man with 0 hairs is bald, so is a man with 1 hair; if a man with 1 hair is bald, so is a man with 2 hairs. Therefore, a man with 2 hairs is bald', but a chain-transitivity (i.e. iple-*modus-ponens*) argument like 'A man with 0 hairs is bald; if a man with 0 hairs is bald, so is a man with 1 hair; if a man with 1 hair is bald, so is a man with 2 hairs; if a man with 2 hairs is bald, so is a man with 3 hairs . . . ; if a man with 999,999 hairs is bald, so is a man with 1,000,000 hairs. Therefore, a man with 1,000,000 hairs is bald'. Obviously, on either scheme, truth preservation, even if explicated in terms of implication, is now harmlessly transitive (but not chain transitive). Thus, interestingly, if a non-transitivist pursuing either scheme were to identify consequence with truth preservation, consequence would itself be governed by a non-classical (indeed, non-transitive) logic and would after all be transitive (but not chain transitive—which, again, is what is really at issue in the strong application of non-transitivity to the sorites paradox). Thanks to an anonymous referee for pushing the worry discussed in this footnote.

[37] The study of consequence as an *operation* rather than *relation* goes back at least as far as Tarski [1930] (see Wójcicki [1988] for a comprehensive study within this approach). Under the simplifying assumptions made at the beginning of this section, the properties of reflexivity, monotonicity, and transitivity of a consequence relation correspond to the following properties of a consequence operation cons:

(i) $X \subseteq \text{cons}(X)$ (*extensivity*);
(ii) If $X \subseteq Y$, $\text{cons}(X) \subseteq \text{cons}(Y)$ (*preservation of inclusion*);
(iii) $\text{cons}(\text{cons}(X) \cup Y) \subseteq \text{cons}(X \cup Y)$ (*union-adjoint subidempotency*),

where cons is an operation from sets of sentences to sets of sentences. An operation satisfying conditions (i)–(iii) is a *Tarski closure operation*. A generalization of closure operations for modelling multiple-conclusion consequence relations are *Scott closure operations* (see Scott 1974). Further generalizations dealing with collections more fine-grained than sets are possible (see Avron [1991] for a start). Closure operations have first been identified by Kuratowski [1922], who also included the conditions:

(iv) $\text{cons}(X \cup Y) \subseteq \text{cons}(X) \cup \text{cons}(Y)$ (*preservation of binary unions*);
(v) $\text{cons}(\varnothing) \subseteq \varnothing$ (*preservation of nullary unions*),

which, while making sense for *topological* closure operations, seem rather out of place for *consequence* closure operations, since they in effect amount, respectively, to obliterating the extra logical power given by the *combination of premises* and to forcing that no sentences are *logical truths*. It may be log which, given a set of facts S,[38] applies to it yielding with logical necessity another such set log(S) = T (possibly identical with S)—just like, at least in the deterministic case, the laws of nature can be seen as an operation which, given the state of a system at a certain time, applies to it yielding with natural necessity the states of the system at subsequent times. Importantly, as in the case of truth preservation, the sense in which, given a set of facts S, the laws of logic apply to it yielding with logical necessity T is again best explicated in terms of a *conditional* statement to the effect that, if all the members of S hold, so do with logical necessity all the members of T (where the necessity operator has wide scope over the conditional).

Given this natural picture, the thought in favour of transitivity goes as follows. Presumably, the laws of logic enjoy *universal applicability*: they apply to any set of facts V *whatsoever* yielding with logical necessity log(V). Given a set of facts S, the laws of logic apply to it yielding with logical necessity T, but, by their universal applicability, given the set of facts T = log(S) they should also apply to it yielding with logical necessity another set log(log(S)) = U (possibly identical with T).[39] Recalling the explication given in the last paragraph, we should then have that, if all the members of S hold, so do with logical necessity all the members of T and that, if all the members of T hold, so do with logical necessity all the members of U, from which it may seem to follow that, if all the members of S hold, so do with logical necessity all the members of U—that is, that the laws of logic, given the set of facts S, apply to it already yielding with logical necessity U. Supposing for the rest of this paragraph that S is indeed a set of facts, this would extremely plausibly imply that the result (U = log(log(S))) of the operation of the laws of logic on the result (T = log(S)) of the operation of the laws of logic on S exists and is included in the result (log(S)) of the operation of the laws of logic on S—log would be subidempotent (which would thus validate a form of transitivity for

worth mentioning that operations satisfying (i), (ii), (iv), and (v) but not necessarily (iii) (known as '*preclosure operations*' or '*Čech closure operations*') form a well-behaved, well-understood, and interesting topological kind (see Čech 1966).

[38] As usual, I assume that there are such abstract entities as *states-of-affairs* which are described by sentences and which may hold or fail to hold, and I identify *facts* with states-of-affairs that hold. Throughout, 'S', 'T', 'U', and 'V' (possibly with numerical subscripts) are used as variables ranging over the set of sets whose members are states-of-affairs (which may or may not be facts).

[39] Throughout, U is simply understood as {s : s is a state-of-affairs and, if all the members of T hold, so does with logical necessity s} (T is understood analogously). Thus, assuming that states-of-affairs always exist, U always exists. But, for every V_0 and V_1, $V_1 = \log(V_0)$ only if all the members of $V_0 \cup V_1$ hold (since log is an operation on sets of *facts*). Thus, in particular, U = log(log(S)) only if all the members of log(S) ∪ U hold (a condition which is extremely plausibly satisfied under the supposition that all the members of log(S) hold).

the correlative relation of consequence).[40] These implications would be multiply repugnant for many applications of non-transitivity. Let us assume again, as at the beginning of section 8.3.2, that the non-transitivist claims, for some φ, ψ, and χ, that ψ follows from φ and χ from ψ, but χ does not follow from φ, and let us also assume that φ describes a fact and S is the singleton of that fact (and that monotonicity holds). Then, one rather crude way to bring out the repugnancy is to notice that, since $\log(\log(S))$ would exist and χ belongs to it, the state-of-affairs described by χ would hold, contrary to what many applications of non-transitivity require (among which all intermediate and strong applications). A more subtle way to bring out the repugnancy is to notice that, since, by subidempotency, $U \subseteq T$, χ, which describes a state-of-affairs belonging to U, would also describe a state-of-affairs belonging to T, contrary to what many applications of non-transitivity require.

However, as in the case of the objection from truth preservation, the conclusion that the laws of logic, given the set of facts S, apply to it yielding with logical necessity U (that is, that, if all the members of S hold, so do with logical necessity all the members of U), in its characteristic obliteration of the logical role played by the implicit intermediate conclusion (that is, that all the members of T hold), implicitly relies on transitivity. Given that, in the sense explained in section 8.3.2, facts-talk is no less "logically penetrable" than truth-talk, the considerations developed in that section apply here: by the non-transitivist's lights, the repugnant conclusion that the laws of logic, given the set of facts S, apply to it already yielding with logical necessity U does not follow from the assumptions characterizing the natural picture, even when taken together with the further assumption that the laws of logic enjoy universal applicability. Thus, the non-transitivist should not be seen as committed to rejecting any of those assumptions.

The point about the compatibility of non-transitivism with the universal applicability of the laws of logic is crucial. Keeping fixed the natural picture of the laws of logic as an operation on facts, it is very tempting to try to make sense of the non-transitivist's position as precisely rejecting the universal applicability of the laws of logic. More specifically, it is very tempting to try to make sense of the non-transitivist's position as relying on a distinction between *two different kinds of facts*, the non-logical and the logical. *Non-logical facts* are those provided, as it were, by the world itself, such as the fact that snow is white; the fact that, if snow is white, it reflects light; the fact that every piece of snow is white (I cannot but emphasize, as

[40] An operation op is *subidempotent* on a set S and ordering $\leqslant$ iff, for every $x \in S$, $op(op(x)) \leqslant op(x)$. Under the simplifying assumptions made at the beginning of this section, in our case $\leqslant$ is simply subset inclusion. It's easy to check that, with conditions (i) and (ii) of n. 37 in place, subidempotency on subset inclusion implies union-adjoint subidempotency.

the last two examples make clear, that non-logical facts need not in any sense be "atomic"). *Logical facts* are those that hold in virtue of the application of the laws of logic to the non-logical facts, such as the fact that either snow is white or grass is blue (holding in virtue of the application of the laws of logic to the fact that snow is white); the fact that snow is white and grass is green (holding in virtue of the application of the laws of logic to the fact that snow is white and the fact that grass is green); the fact that something is white (holding in virtue of the application of the laws of logic to the fact that snow is white). Of course, much more would have to be said about how to draw exactly the non-logical/logical distinction, but I take it that we have an intuitive grasp of it (as witnessed by our intuitive judgements in the foregoing cases) which will be sufficient for our purposes.

The non-transitivist would then be seen as advocating what we may call '*non-logical/logical dualism*', consisting in *rejecting the universal applicability of the laws of logic to logical facts*. Thus, in our original case, since T contains the fact described by ψ and, we may assume, such fact is logical, it is open to the dualist non-transitivist to reject that the laws of logic apply to T, and thus to reject that they yield U (in particular, that they force the state-of-affairs described by χ to hold). As should be clear, I am understanding the dualist non-transitivist as someone who not only rejects the *conditional* that, if all the members of T hold, so do all the members of U, but also refuses to *infer* the conclusion that all the members of U hold from the premise that all the members of T hold (we will see in sections 8.4.2 and 8.4.3 that, as opposed to her rejection of the conditional, her rejection of the inference is what she shares with a non-dualist non-transitivist). Still, I want to understand the dualist non-transitivist's rejection of the inference as *grounded* in her rejection of the conditional, and more specifically as grounded in her idea that the antecedent of the conditional ('All the members of T hold') might [be the case while its consequent ('All the members of U hold') is not]. I also want to understand the dualist non-transitivist as taking an analogous position in the case of truth preservation.[41] Thus, in our original case, she would maintain that ψ might [be true while χ is not], and so would reject that, if ψ is true, so is χ, contrary to the position I have recommended to the non-transitivist in section 8.3.2 (see n. 31).

It is interesting to observe that, under a certain extremely plausible assumption, non-logical/logical dualism *by itself* requires restrictions on the transitivity

[41] As should be clear, I see a structural identity between what a non-transitivist can or should say about *truth* and what she can or should say about *facts*. All the points made about non-transitivism and truth are meant to apply just as well to non-transitivism and facts and *vice versa*, even though sometimes the point is more easily made about non-transitivism and truth and some other times about non-transitivism and facts.

of consequence. For, whenever a restriction of the application of the laws of logic to some non-empty set[42] of logically contingent facts described by the premises $\varphi_0, \varphi_1, \varphi_2 \ldots$ and logically necessary facts described by the premises $\psi_0, \psi_1, \psi_2 \ldots$ is envisaged (so that the inference to the conclusion χ, which follows from them, is rejected), for every k φ_k must itself follow from some non-empty set X_k whose members describe non-logical facts. This is so because, even if not every logically contingent fact is non-logical (consider for example the fact that something is white), it is extremely plausible to assume that every logically contingent fact is ultimately grounded in certain non-logical facts in such a way as to be yielded from them by an application of the laws of logic (here I won't try though to justify this assumption). If transitivity were then to hold unrestrictedly, χ would already follow from $X_0 \cup X_1 \cup X_2 \ldots$ Since, however, the members of this set all describe non-logical facts, there would be no dualist bar to the application of the laws of logic to it, and so the state-of-affairs described by χ would be forced to hold as well.

Interesting as non-logical/logical dualism may be, as I have already explained four paragraphs back a non-transitivist is not committed to it (and can indeed accept its negation). I should also remark, however, on an important point of *agreement* between the dualist and the non-dualist non-transitivist. For reflect that, in a non-transitive framework, the dualist non-transitivist can be seen as doing one thing by means of a quite different one. That is, to go back to our original case, she can be seen as rejecting a commitment to accepting a consequence (χ) of what she is committed to accepting (ψ) by rejecting that, if ψ is true, so is χ (I take it to be extremely plausible that the dualist non-transitivist is committed to accepting ψ, since it follows from φ which, we may assume, describes a non-logical fact).[43] Even if a non-transitivist can disagree with the latter, she cannot but agree with the former, since, as I will explain in sections 8.4.2 and 8.4.3, the rejection of being committed to accepting a consequence of what she is committed to accepting is arguably the *crux* of her disagreement with the transitivist, one of the places at which their different *metalinguistic judgements* about validity are finally reflected in a clash of *object-language attitudes* (in this case, the transitivist's acceptance of χ against the non-transitivist's non-acceptance of χ). A non-transitivist need not disagree with the transitivist as to whether, if ψ is

[42] If the set is empty, then it is vacuously the case that all its members are non-logical facts, and so there is no dualist bar to the application of the laws of logic to it.

[43] One might wonder how plausible this rejection on the part of the non-logical/logical dualist is. Given that χ follows from ψ, the conditional that, if ψ is true, so is χ follows by the *deduction theorem* (and what I take to be utterly uncontroversial instances of (ED) for truth). It would thus seem that the non-logical/logical dualist is committed to an implausible rejection of the deduction theorem.

true, so is χ, but she has to disagree with the transitivist's willingness to undertake a commitment to χ's being true on the sole basis of her logical commitment to ψ's being true.

It should by now be clear that, in opposition to the dualist non-transitivist, a non-dualist non-transitivist can legitimately insist on not explaining her position in terms of some *deviant conception of truth*—in particular, in terms of a denial of the principle that truth preservation is necessary for consequence (claiming that, although χ follows from ψ, it is not the case that, if ψ is true, so is χ). That is helpfully compared with the fact that someone who rejects the *law of excluded middle* need not explain her position in terms of a deviant (e.g. *gappist*) conception of the relation between the truth of a sentence and the truth of its negation (e.g. such that it might be the case that neither a sentence nor its negation are true; see Field [2008] for a non-gappist rejection of the law of excluded middle), or with the fact that someone who rejects the rule of *disjunctive syllogism* need not explain her position in terms of a deviant (e.g. *dialetheist*) conception of the relation between the truth of a sentence and the truth of its negation (e.g. such that it might be the case that both a sentence and its negation are true; see Read [1988] for a non-dialetheist rejection of the rule of disjunctive syllogism). As in all those other cases, the heart of the logical deviance being proposed by the non-dualist non-transitivist is a certain conception of what counts as a *correct pattern of reasoning* rather than a certain deviant conception of what truth is.

8.4.2 *Logical nihilism and non-transitivism*

Still, regarded now as a proposal as to what counts as a correct pattern of reasoning, non-transitivism may look perilously close to a *logical nihilism* which rejects the universal validity of all the argument forms that trigger failures of (T) when satisfying the second conjunct of its antecedent.[44] For take without loss of generality single-conclusion arguments $a_0, a_1, a_2 \ldots, a_{i-1}$ of forms $\mathbb{F}_0$, $\mathbb{F}_1$, $\mathbb{F}_2 \ldots$, $\mathbb{F}_{i-1}$ and a single-conclusion argument a_i of form $\mathbb{F}_i$, and suppose that $a_0, a_1, a_2 \ldots, a_{i-1}$ and a_i satisfy the first conjunct and the second conjunct of (T)'s antecedent respectively. Consider a non-transitivist according to whom the resulting instance of (T) fails, and who accepts on non-logical grounds all the premises Γ of $a_0, a_1, a_2 \ldots, a_{i-1}$ and accepts (on non-logical or logical grounds)

[44] Of course, even in a non-transitive logic not *every* argument form is usually such, and so the label 'nihilist' might in the usual cases be a bit of an exaggeration; still, I will use it indiscriminately as it is even literally correct in some other cases, and I think conveys the right tones even in those cases in which it is literally speaking an exaggeration.

the premises Λ of a_i which are not conclusions of any of $a_0, a_1, a_2 \ldots, a_{i-1}$. Such non-transitivist would then have to accept the premises Λ, Θ of a_i (this seems extremely plausible; I will justify it more fully in section 8.4.3). Yet, given the foregoing explanation of what the non-transitivist regards as a correct pattern of reasoning, she will not regard herself as committed to accepting the conclusion Ξ of a_i. How could she then still maintain that a_i is *valid*, given that *she accepts its premises but refuses to infer its conclusion*? Is her refusal to infer the conclusion not an implicit admission that she does not regard a_i as valid (and thus that she does not regard $\mathbb{F}_i$, which is instantiated by a_i, as a universally valid argument form)? Notice how, in a certain respect, these questions are particularly pressing for a *non-dualist* non-transitivist, since she cannot help herself to the dualist rejection that, if every premise of a_i is true, so is the conclusion of a_i, rejection which would certainly go some way towards explaining the refusal to infer the conclusion of a_i.

Before addressing these urgent questions, notice that logical nihilism, as against non-logical/logical dualism, is actually not a possible option for a non-transitivist, at least in the following sense. It might well be that all of the logical nihilist, the non-logical/logical dualist, and the non-dualist non-transitivist accept the premises Γ of a classically valid argument a of form $\mathbb{F}$ while refusing to infer its conclusions Δ. The non-logical/logical dualist might do this because, even though she recognizes a as valid, she regards some coordinate of Γ as describing a logical fact, and so she rejects that, if every coordinate of Γ is true, so is some coordinate of Δ (and on these grounds refuses to infer Δ from Γ). The non-dualist non-transitivist might do this because, even though she recognizes a as valid and accepts that, if every coordinate of Γ is true, so is some coordinate of Δ, she still refuses to infer Δ from Γ (on grounds which we will explore shortly). The logical nihilist, however, refuses to infer Δ on the very simple grounds that she does not regard $\mathbb{F}$ as a universally valid argument form and that, in particular, she regards its instance a as invalid. She thus cannot regard a as involved in a possible failure of the transitivity of consequence, as the dualist and the non-dualist non-transitivists do. Logical nihilism is an *alternative* to non-transitivism, not one of its species.

To come back to the question as to how a (non-dualist) non-transitivist can recognize as valid an argument whose premises she accepts and whose conclusions she does not accept, we must observe that the connection, presupposed by this question, between recognizing an argument as valid and inferring the conclusions if one also accepts its premises is much less straightforward than it might seem at first glance. One can recognize an argument as valid and accept its premises while still not inferring its conclusions because one is somehow prevented by external circumstances from doing so (by a threat, a psychological breakdown, a sudden death, etc.). Or because one fails to recognize, maybe on account of their syntactic complexity, that the premises (conclusions) are indeed premises (conclusions) of an argument one recognizes as valid. Or because one has a general policy of not inferring conclusions, maybe for the reason that one has been told by one's guru that every inference is sacrilegious. Or because one simply cannot be required to infer all the conclusions of all the arguments one recognizes as valid and whose premises one accepts—this is arguably not a requirement on resource-bounded rationality, and, unless each and every single truth is an aim of belief, it is not clear why it should even be a requirement on resource-unbounded rationality.

The previous counterexamples may seem to trade on "*deviant*" cases. Still, unless a plausible independent characterization of "deviancy" is provided (a highly non-trivial task), the objection against the non-transitivist would seem to lose much of its force—why should non-transitivism be itself classified as one of the "deviant" cases (which we know from the foregoing counterexamples generally to exist)? Be that as it may, stronger, if more controversial counterexamples can be given where there is actually *epistemic force against the inference's being drawn*:

> **MO** Mo may be told by a source she is justified to trust that, if Mo's initial is 'M', then Mo is a horribly bad *modus-ponens* inferrer (which does not imply that Mo is horribly bad at recognizing the validity of *modus-ponens* arguments). Mo may also know that her initial is 'M'. Mo recognizes the validity of the relevant instance of the rule of *modus ponens*, yet, given that it would be self-defeating to believe that one is a horribly bad *modus-ponens* inferrer exactly *via* a *modus-ponens* inference, there is epistemic force against Mo's inferring the conclusion that she is a horribly bad *modus-ponens* inferrer.[45]
>
> **DAVE** Sincere Dave might believe of each of the 1,000,000 substantial and independent statements of his new history book that that statement is true—if Dave didn't really believe a statement to be true, why would he have put it in the book in the first place? Together with the certainly true assumption that those are all the statements in his book, it follows that all the statements in Dave's book are true. Dave recognizes the argument as valid, yet, given Dave's fallibility as a

[45] Thanks to Daniele Sgaravatti, Martin Smith, and Crispin Wright for discussion of this and similar examples.

historian, there is epistemic force against modest Dave's inferring the conclusion that all the statements in his book are true.[46,47]

As the reader will have spotted, these two candidate counterexamples can also be turned into candidate counterexamples (a new one and an old one respectively) to many *closure* principles for knowledge and justification. And, in fact, unsurprisingly all the main candidate counterexamples to closure principles I know of can conversely be turned into candidate cases in which one recognizes an argument as valid and accepts its premises while still not inferring its conclusions and in which there is actually epistemic force against the inference's being drawn.[48]

[46] This counterexample is of course a version of the preface paradox (see Makinson [1965]; see Christensen [2004] for a recent congenial discussion). The counterexample is even more telling given the relationships between probabilistic structure and some kinds of non-transitive deductive consequence relations remarked upon in n. 20. As Crispin Wright has emphasized to me in conversation, there is an important asymmetry between the counterexample, which crucially relies on the fact that there is epistemic force against accepting the conjunction of all the premises, and some application of non-transitivity, where one would wish not just to accept all the premises, but also their conjunction (more or less equivalently, one would wish not only to accept of every premise that it is true, but also to accept that every premise is true). The point of the counterexample is, however, only to show that there can be epistemic force against an inference's being drawn even if the argument is recognized as valid and all its premises are accepted. The *source* of this epistemic force in **DAVE** is such that it only applies to multiple-premise arguments, whereas the source of the force in some application of non-transitivity will evidently not carry this restriction, as it also happens in **MO** and in the other cases mentioned in n. 48 (Harman [1986, 11–24] is the *locus classicus* for the problematization of the connection between validity and inference).

[47] To connect with the issue mentioned in n. 19, the argument from β (a sentence expressing Dave's overall body of evidence relevant to the topics treated in his book) to γ_1 (where γ_i expresses the proposition that the ith statement in Dave's book is true) is defeasibly valid, as is the argument from β, γ_1 to 'γ_1 and γ_2'. Thus, by (CT), the argument from β to 'γ_1 and γ_2' would also be defeasibly valid. Yet, the argument from β, 'γ_1 and γ_2' to 'γ_1, γ_2 and γ_3' is also defeasibly valid. Thus, by (CT), the argument from β to 'γ_1, γ_2 and γ_3' would also be defeasibly valid. With another 999,997 structurally identical reasonings, we reach the conclusion that the argument from β to '$\gamma_1, \gamma_2, \gamma_3$...and $\gamma_{1,000,000}$' is also defeasibly valid. Since the last argument is not defeasibly valid, we can assume that, in at least one of those 999,999 reasonings, (CT) (and not just (T)) fails for the defeasible consequence relation in question. Notice that an interesting feature of cases like this in which it is also (CT) (and not just (T)) that fails is that, contrary to cases like that presented in section 8.2.3 in which it is only (T) that fails, there is no interesting sense in which, in any of the 999,999 reasonings in question, the state of information represented by the initial premise β *defeats* (or even makes less reasonable) the inference from the initial conclusion '$\gamma_1, \gamma_2, \gamma_3$...and γ_i' to the final conclusion '$\gamma_1, \gamma_2, \gamma_3..., \gamma_i$ and γ_{i+1}' (and so, in this respect, it is a bit misleading to call this kind of non-deductive consequence relations 'defeasible'). The problem seems rather that, in at least one of the 999,999 reasonings in question, *the way in which* Dave has arrived at the initial conclusion '$\gamma_1, \gamma_2, \gamma_3$...and γ_i' makes it unreasonable for him to draw the further inference to the final conclusion '$\gamma_1, \gamma_2, \gamma_3..., \gamma_i$ and γ_{i+1}' (contrast with the case in which Dave, who still has β, has arrived at '$\gamma_1, \gamma_2, \gamma_3$...and γ_i' not *via* successive defeasible inferences, but simply because a trustworthy source has revealed to him that that sentence is true: then, the further inference from β,'$\gamma_1, \gamma_2, \gamma_3$...and γ_i' to '$\gamma_1, \gamma_2, \gamma_3..., \gamma_i$ and γ_{i+1}' would seem eminently reasonable).

[48] Thus, at some point in the first performance of a 1,000,000-step proof of Fermat's Last Theorem (from φ_0 to $\varphi_{999,999}$), for a few i René accepts φ_i and recognizes as valid the

These counterexamples are certainly sufficient to open up conceptual space for a genuinely non-transitivist (rather than logical nihilist) position. Still, more needs to be said by way of a positive explanation of the refusal to draw an inference when an argument is recognized as valid and all its premises are accepted (at least, more needs to be said if one wants to avoid giving the explanation that the non-logical/logical dualist gives). Moreover, a doubt now arises as to the very point of non-transitivism: if a story needs to be told anyway in order to vindicate the rationality of accepting the premises of an argument recognized as valid while refusing to infer its conclusions, could such a story not be applied in a transitive framework, recognizing, say, that, since, [for every $\varphi \in \mathrm{ran}(\Theta)$, Γ entails Δ, φ, and Λ, Θ entails Ξ], Λ, Γ does entail Δ, Ξ, and accepting Λ, Γ, while refusing to infer Δ, Ξ? Such a doubt would certainly be pressing at least for the distinctive punch that intermediate and strong applications of non-transitivity are supposed to have.

The situation which seems to be emerging is this. The non-transitivist needs to differentiate between the *acceptance-related normative property* of being N_0 (which triggers the normative force of consequence once all the premises of an argument recognized as valid are N_0) and the acceptance-related normative property of being N_1 (which, while not sufficient for triggering such force, is nevertheless sufficient for generating a commitment to accepting a sentence that is N_1). Such a distinction between acceptance-related normative properties would allow the non-transitivist to insist that, in the cases in which she accepts the premises of an argument recognized as valid while refusing to infer its conclusions, this is so because the premises are only N_1 and not N_0. And, together with the additional claim that, if the premises of an argument recognized as valid are N_0, its conclusions are N_1, such a distinction between acceptance-related normative properties would also allow the non-transitivist to reply to the awkward question of the last paragraph by saying that, since Λ, Γ is N_0, if consequence were transitive,

argument from φ_i to φ_{i+1}, yet a case can be made that, given René's fallibility as an inferrer, for some such i there is epistemic force against René's drawing the relevant inference (see Hume 1739, Book I, Part IV, Section I); judging whether, in a series of 1,000,000 pairwise indiscriminable trees going from 100 ft tall to 1 ft tall she's looking at from the distance, the ith tree is at least 50 ft tall, for a few i and for some j Sherry accepts 'I know that I know that I know (j times) that the ith tree is at least 50 ft tall' and, given her limited powers of discrimination, recognizes as valid the argument from 'I know that I know that I know (j times) that the ith tree is at least 50 ft tall' to 'I don't know that I don't know that I know that I know that I know ($j-2$ times) that the $i+1$th tree is at least 50 ft tall', yet a case can be made that, given what Sherry knows about the last tree in the series, for some such i there is epistemic force against Sherry's drawing the relevant inference (see Zardini 2013g); looking at the zebras in the zoo, Fred accepts 'Those animals are zebras' and recognizes as valid the argument from 'Those animals are zebras' to 'Those animals are not cleverly disguised mules', yet a case can be made that, given Fred's ignorance of zoos' policies, there is epistemic force against Fred's drawing the relevant inference (see Dretske 1970).

Δ, Ξ would have to be N_1, and so she would after all be committed to accepting it. To the identification of the acceptance-related normative properties of being N_0 and of being N_1 we must now turn.

8.4.3 *The normativity of consequence*

Once again, let us go back to the case in which the non-transitivist accepts φ, we may assume, for non-inferential reasons,[49] recognizes both that ψ follows from φ and that χ follows from ψ but does not accept that χ follows from φ. I say that, in such a situation, her non-transitivism is sufficient to save her from a commitment to accepting χ. To see this, consider the highly plausible general normative principle to the effect that one has inferential reasons[50] to accept all *but no more* than the consequences of what one has *non-inferential* reasons to accept (insofar as one does have such non-inferential reasons).[51] Now, we may assume, the non-transitivist does not have non-inferential reasons to accept ψ. Since ψ is indeed a consequence of φ, and since she has non-inferential reasons to accept φ, by the general normative principle just introduced she has indeed inferential reasons to accept ψ, and so she is indeed committed to accepting ψ.[52] However, on her view, χ is not a consequence of φ, and so she is not committed to accepting χ simply because she has non-inferential reasons to accept φ—indeed, since in a non-transitive framework χ need not follow from anything she has non-inferential

[49] Henceforth, by 'having *non-inferential reasons* to accept φ' and its like I will really just mean 'having a subjective basis for accepting or in fact accepting φ not because it is the conclusion of a valid argument all of whose premises one has reason to accept or accepts'—acceptance for non-inferential "reasons" need not be objectively well grounded at all. My use of 'reason' and its like can thus be understood as roughly synonymous with '*subjective* reason' and its like.

[50] Henceforth, by 'having *inferential reasons* to accept φ' and its like I will really just mean 'having a subjective basis for accepting φ because it is the conclusion of a valid argument all of whose premises one has reasons to accept'. Notice that, contrary to my own focus up to section 8.4.2, this notion of an inferential reason is in some important respect *external*, as the possession of an inferential reason in this sense does not depend on one's recognition of the validity of the argument. Having said that, I stress that I am switching to the more external notion only to argue that the non-dualist non-transitivist can escape unwanted commitments even in such external sense: points analogous to the ones to be made will apply for a more internal notion that requires one's recognition of the validity of the argument.

[51] The 'insofar'-qualification will henceforth be implicitly understood; it is supposed to take care of the point forcefully made by Harman [1986, 11–12] to the effect that the fact that the conclusion is unacceptable may *defeat* the reasons one had for accepting the premises.

[52] This last step should make clear that, to simplify the discussion, I am assuming that the relevant (non-inferential or inferential) reasons to accept a certain sentence are always so strong as to imply a commitment to accepting that sentence. Conversely, given my understanding of 'reason' (see nn. 49 and 50), commitment to accepting a certain sentence is a reason (non-inferential or inferential, as the case may be) to accept that sentence. In view of this equivalence, I will often put points concerning one's having (non-inferential or inferential) reasons to accept a sentence in terms of one's being committed to accepting that sentence.

reasons to accept, by the general normative principle just introduced she has no reasons at all to accept χ (even though she is committed to accepting ψ and recognizes that χ follows from ψ!).

Drawing on the foregoing distinction between *having non-inferential reasons to accept* and *having inferential reasons to accept*, I propose that, roughly, we identify the acceptance-related normative properties of x-is-N_0 and x-is-N_1 with one-has-non-inferential-reasons-to-accept-x and one-has-inferential-reasons-to-accept-x respectively.[53] I take it that, as it has been introduced at the end of section 8.4.2, the distinction between x-is-N_0 and x-is-N_1 has some very intuitive appeal and import in our *ordinary evaluation of reasons*: we would ordinarily distinguish between one's reasons to accept a certain sentence being so strong as to permit (or even mandate) acceptance of whatever follows from that sentence and one's reasons to accept a certain sentence being simply strong enough as to permit (or even mandate) acceptance of that sentence. In the former case, one's reasons allow (or even mandate) one to take the sentence as an *initial point for further reasoning*, whereas in the latter case they only allow (or even mandate) one to take the sentence as a *terminal point of acceptance* (see Smith [2004, 196–9] for a defence of this distinction within a transitive framework). In the lights of the remarks already made in n. 20 and in connection with **DAVE**, it should go without saying that the distinction also makes perfectly good *probabilistic* and, more generally, *defeasible* sense. Moreover, I also take it that the proposed identification of that distinction with the distinction between one-has-non-inferential-reasons-to-accept-x and one-has-inferential-reasons-to-accept-x has some very intuitive appeal and import in our ordinary evaluation of reasons as well: we would ordinarily think that *the strength of our reasons decreases with further inferences* (and, contrary to standard assumptions in formal models of uncertainty, that this is so even if the arguments relied on in such inferences are deductively valid, *single-premise* arguments, as evidenced most clearly by the first two cases mentioned in n. 48).[54]

In such a distinction between acceptance-related normative properties, we see how a trace of non-logical/logical dualism does necessarily remain in the

[53] 'Roughly' because **DAVE** (and possibly other counterexamples too) shows that, in some cases, even if one has non-inferential reasons to accept the premises these are not N_0 (i.e. they do not trigger the normative force of consequence, at least with respect to certain conclusions). In what follows, I will ignore this further complexity.

[54] It is perhaps worth noting that neither acceptance-related normative property implies actual acceptance, whilst being actually accepted implies both being N_0 (see n. 49) and being N_1 (from n. 49, reflexivity of consequence and (NI) (see below)—indeed, by reflexivity of consequence and (NI), being N_0 itself implies being N_1). This is so because both acceptance-related normative properties pertain to what (subjectively) *ought to be* the case rather than to what *is* the case.

non-dualist non-transitivist's position: only, the distinction is not between two different kinds of *facts*, but between two different kinds of *reasons for accepting a sentence*—either non-inferential or inferential. In view of this distinction, the non-dualist non-transitivist can be seen not as endorsing the rather exotic restriction to non-logical facts of the application of the laws of logic, but as adhering unswervingly both to the verdicts of validity and invalidity issued by her non-transitive logic and to the general normative principle introduced two paragraphs back that the commitments generated by the laws of logic on a certain *position* (a sequence of sentences accepted for non-inferential reasons) coincide with the logical consequences of that position. That is, if one has non-inferential reasons to accept a certain sequence of sentences, one is indeed committed by logic and those very same reasons to accepting each and every consequence of them, but also committed only to that (at least by logic and those very same reasons), so that, if logic is non-transitive, one is not committed by logic and those very same reasons to accepting a consequence of a consequence of one's position which is not already a consequence of one's position, even if it is a consequence of something one is committed to accepting and even accepts (such "consequences at one remove," as it were, do not exist, of course, if logic is transitive, but they do if it isn't).

It is crucial to see that this general principle governing consequence can be accepted *by non-transitivists and transitivists alike as exhausting its normative force* (at least as far as the aspects we are concerned with here go). For our purposes, we can focus on its positive component, which can be fleshed out a bit more generally and precisely as the following principle of *connection between having non-inferential reasons to accept and having inferential reasons to accept*:

(NI) If one has non-inferential reasons to accept Γ, and Γ entails Δ, one has inferential reasons to accept Δ.

Notice that, *pending any further specification of the properties of the consequence relation*, (NI) does *not* imply that consequence in turn maps inferential reasons to accept a sequence of sentences onto inferential reasons to accept a sequence of sentences. In other words, pending any further specification of the properties of the consequence relation, while (NI) does imply that the normative force of consequence applies to premises one has non-inferential reasons to accept, producing inferential reasons to accept (and so commitments to accepting) the conclusions, it does not imply that such force applies to premises one has simply inferential reasons to accept. Indeed, the non-transitivist can be seen as exploiting exactly the fact that, by itself, (NI) does not imply the stronger principle of *preservation of inferential reasons to accept*:

(II) If one has inferential reasons to accept Γ, and Γ entails Δ, one has inferential reasons to accept Δ.

In particular, the non-transitivist can be seen as accepting (NI) while rejecting (II): on her view, one need not be committed to accepting consequences of commitments generated by logic.

It seems to me that, in her joint acceptance of (NI) and rejection of (II), the non-transitivist is occupying a reasonable position, given that the following theorem holds (under certain very plausible additional assumptions which I'll make explicit at the relevant stages in the proof):

Theorem 1. (NI) *implies* (II) *iff the consequence relation is transitive.*

Proof.

- *Left-to-right.* We prove the contrapositive. Take a non-transitive consequence relation **L** such that, for every $\varphi \in \text{ran}(\Theta)$, $\Gamma \vdash_L \Delta, \varphi$, and $\Lambda, \Theta \vdash_L \Xi$, but $\Lambda, \Gamma \nvdash_L \Delta, \Xi$, and consider an intermediate application of **L** by a subject s having non-inferential reasons only to accept Γ and Λ.[55] s's intermediate application of **L** is such that s does not accept Δ, Ξ, even though she does accept, for every $\varphi \in \text{ran}(\Theta), \Delta, \varphi$. In such a situation, (NI) only requires from s that she accept, for every $\varphi \in \text{ran}(\Theta)$, Δ, φ (since, for every $\varphi \in \text{ran}(\Theta)$, $\Gamma \vdash_L \Delta, \varphi$)—it does not require from s that she accept Δ, Ξ (since $\Lambda, \Gamma \nvdash_L \Delta, \Xi$). Therefore, s satisfies (NI). However, in such a situation, given plausible additional principles (II) does require from s that she accept Δ, Ξ. For, having non-inferential reasons to accept Γ, by (NI) s has inferential reasons to accept (and so is committed to accepting), for every $\varphi \in \text{ran}(\Theta)$, Δ, φ (since, for every $\varphi \in \text{ran}(\Theta)$, $\Gamma \vdash_L \Delta, \varphi$). By the additional principle of *semicolon-agglomeration of commitments to accepting "disjunctively"*:

  (SACAD) If, for some function seq from ordinals to sequences, for every $\alpha \in \text{dom}(\Pi)$ (with $\Pi = \psi_0, \psi_1, \psi_2 \ldots$),[56] one is committed to accepting "disjunctively" $\text{seq}(\alpha), \psi_\alpha$, then one is committed to accepting "disjunctively" $\text{seq}(0), \text{seq}(1), \text{seq}(2) \ldots, \Pi^{;}$

  (where ';', unlike ',', denotes a right-conjunctive structural punctuation mark[57] and $\Pi^{\dagger}$ the result of substituting $\dagger$ throughout as the structural punctuation

[55] Throughout this proof, in order to avoid excessive verbal clutter, I will sometimes let context disambiguate whether a sequence is accepted "conjunctively" (in the fashion of premises) or "disjunctively" (in the fashion of conclusions).

[56] Throughout, $\text{dom}(\Gamma)$ is the domain of Γ.

[57] For our purposes, a structural punctuation mark $\dagger$ is *right-conjunctive* iff $[\Gamma \vdash \Delta \dagger \varphi$ iff $[\Gamma \vdash \Delta$ and $\Gamma \vdash \varphi]]$. Needless to say, the usefulness of a right-conjunctive structural punctuation mark

mark of Π), s is committed to accepting "disjunctively" $\Delta, \Delta, \Delta \ldots, \Theta^{;}$. Hence, by contraction, s is committed to accepting "disjunctively" $\Delta, \Theta^{;}$.[58] Since s is also committed to accepting "conjunctively" Λ (having non-inferential reasons to accept "conjunctively" Λ), by the additional principle of *quasi-monotonicity of commitment to accepting "disjunctively" over implication of commitment to accepting*:

(QCADICA) If commitment to accepting "conjunctively" Π_0, Π_1 implies commitment to accepting "disjunctively" Σ_0, then commitment to accepting "conjunctively" Π_0 implies that commitment to accepting "disjunctively" $\Sigma_1, \Pi_1^{;}$ implies commitment to accepting "disjunctively" Σ_1, Σ_0,

we have that s is committed to accepting "disjunctively" Δ, Ξ (since the fact that $\Lambda, \Theta \vdash_L \Xi$ together with (NI) and (II) allows us to detach the main consequent of the relevant instance of (QCADICA)). Therefore, (II) has a consequence (i.e. that s is committed to accepting "disjunctively" Δ, Ξ) that (NI) does not have (and, if s were not committed to accepting "disjunctively" Δ, Ξ, (II) would not hold while (NI) might still hold).

- *Right-to-left*. By cases. Take a transitive consequence relation L and suppose that, for every $\varphi \in \text{ran}(\Theta_0)$, a subject s has inferential reasons to accept φ. This can be so:

  (i) Either because $\Gamma \vdash_L \varphi$, where $\Gamma \neq \langle . \rangle$ and s has non-inferential reasons to accept Γ (so that, by (NI), s has inferential reasons to accept φ);
  (ii) Or because $\langle . \rangle \vdash_L \varphi$ (so that, by (NI), s has inferential reasons to accept φ).

  Notice that, by the additional principle of *well-foundedness of inferential reasons over logically contingent sentences*:

  (WIRLCS) The relation x-is-a-coordinate-of-a-sequence-which-gives-s-inferential-reasons-for-y-and-x-is-not-y is well founded on the field of logically contingent sentences,

  if s has inferential reasons to accept φ because $\Gamma \vdash_L \varphi$, where $\Gamma \neq \langle . \rangle$, it is not possible that [Γ is such that, for some of its logically contingent coordinates φ_0, s has only inferential reasons to accept φ_0, namely that φ_0

derives from its ability to allow us to mimic conjunctive operations over sentences of a language which may well lack a conjunctive operator.

[58] The use of contraction is only needed in the proof because, as observed in n. 10, (T) is not a pure transitivity principle and does in fact imply contraction. Contraction would not be needed in the proof if we worked with one of the more convoluted transitivity principles mentioned in n. 10.

follows from premises Γ_0, and Γ_0 is in turn such that, for some of its logically contingent coordinates φ_1, s has only inferential reasons to accept φ_1, namely that φ_1 follows from premises Γ_1, and Γ_1 is in turn such that, for some of its logically contingent coordinates φ_2, s has only inferential reasons to accept φ_2, namely that φ_2 follows from premises Γ_2...]. This yields that, if $\langle . \rangle \nvdash_L \varphi$, each inferential reason s has to accept φ must ultimately be traceable back to a combination Γ_* of (possibly infinitely many, possibly infinitely long) sequences, each of which can be reached in a finite number of steps and each of which s has non-inferential reasons to accept. Hence, by (monotonicity and) (T), we have that $\Gamma_* \vdash_L \varphi$.[59] Letting $\Gamma = \Gamma_*$, this finite-chain case is reduced to case (i). Now, suppose also that $\Theta_0 \vdash_L \Xi$. Let Θ_1 be the sequence obtained from Θ_0 by replacing each coordinate falling under case (i) with its associated Γ and by deleting each coordinate falling under case (ii). Then, s has non-inferential reasons to accept Θ_1 and, by (monotonicity and) (T), $\Theta_1 \vdash_L \Xi$ just as well. Hence, by (NI), s has inferential reasons to accept Ξ. □

Indeed, given that (II) straightforwardly implies (NI) no matter whether the consequence relation is transitive or not as long as it is reflexive (since, as we have seen in n. 54, non-inferential reasons to accept will then imply inferential reasons to accept),[60] theorem 1 can be reformulated to the effect that [(NI) is *equivalent* with (II) iff the consequence relation is transitive]. In view of this, it seems to me that the non-transitivist can reasonably insist that the *pure principle*, free of any relevant assumption concerning formal properties of consequence, which encodes its normativity (principle which is endorsed by non-transitivists and transitivists alike) is (NI) rather than (II), (II) being associated with such normativity only because equivalent with (NI) under the (rejected) assumption of transitivity. Hence, even though a non-transitive logic is naturally hospitable to a certain "softening" of the normative force of consequence, this does not mean that the core of that force is not preserved (let alone that no important requirement is placed by non-transitive consequence on rational beings).

Before proceeding further, an absolutely essential feature of this dialectic must be made clear. (II) is in effect a principle of *closure of having inferential reasons to accept* under logical consequence. Why then cannot one apply in this case a strategy analogous to the one we used in order to uphold in a non-transitive

[59] As noted in n. 10, (T) in fact implies monotonicity. Monotonicity would not even *prima facie* be needed in the proof if we worked with one of the more convoluted transitivity principles mentioned in n. 10.

[60] The assumption of reflexivity will henceforth be implicitly understood.

framework the necessity of truth preservation for consequence, which is in effect a principle of *closure of truth* under logical consequence? In the case of truth and of the other properties we have considered in section 8.3.2, it has been observed that there seem to be bridge principles linking the languages talking about these properties and the original language $\mathcal{L}_0$ claimed to be non-classical, principles which force the logic of the former languages to be itself non-classical. No such principle seems to govern properties like one-has-inferential-reasons-to-accept-x and, more generally, there does not seem to be any reason why the language $\mathcal{L}_1$ talking about this property should exhibit the same problematic features which motivated a deviance from classical logic for $\mathcal{L}_0$. To give one example, for an intuitionist the language talking about which sentences of a standard quantified arithmetical language a mathematician has inferential reasons to accept may well lack the *unsurveyability* characteristic of the standard quantified arithmetical language, and so may well be classical (it may well be a surveyable matter which sentences the mathematician has non-inferential reasons to accept, and even an intuitionist would typically take it to be a classical matter which sentences follow from these in intuitionist logic). Or, to give another example more germane to our concerns, for a tolerant logician, the language talking about which sentences of an ordinary language a competent speaker has inferential reasons to accept may well lack the *vagueness* characteristic of the ordinary language, and so may well be classical (it may well be a precise matter which sentences the speaker has non-inferential reasons to accept, and even a tolerant logician would typically take it to be a classical matter which sentences follow from these in a tolerant logic). In all such cases, it is the respects in which the application of 'Subject s has inferential reasons to accept 'φ'' (as against the application of φ, ''φ' is knowable', ''φ' is justifiedly believable', etc.) depends on *brutely factual subjective features* (see nn. 49 and 50) that bring in a crucial *mismatch* between the truth conditions of the former sentence and those of the latter sentences, so that the problematic features which motivated a deviance from classical logic for the latter sentences need not be present in the former sentence.

Moreover, even if, contrary to what I have just argued, the logic of $\mathcal{L}_1$ were non-transitive, (II) would still be unacceptable for most non-transitivists. Once more, suppose that a subject s has non-inferential reasons to accept φ, and that ψ follows from φ and χ from ψ. Suppose also that the argument from φ to χ is non-transitivistically invalid, so that, *qua* non-transitivists, we would wish to avoid imputing to s any commitment to accepting χ, despite her having non-inferential reasons to accept φ. Now, by (NI) and ψ's following from φ, we can infer that s has inferential reasons to accept (and so is committed to accepting) ψ. But reflect that the satisfaction of s's commitment to accepting ψ requires s to *treat* ψ *in a certain way* (at least, very roughly, to assent to ψ on the basis of an inference if queried under normal circumstances). Crucially, such a way is also sufficient to establish s's having inferential reasons to accept ψ independently of s's having non-inferential reasons to accept φ together with ψ's following from φ and (NI). This means that there will typically be not only inferential reasons to accept that s has inferential reasons to accept ψ (namely, those provided by s's having non-inferential reasons to accept φ together with ψ's following from φ and (NI)), but also non-inferential reasons to accept that (namely, those provided by the observation of s's behaviour). More precisely, there will be such reasons whenever s in fact satisfies her commitment to accepting ψ. Having non-inferential reasons both to accept that s has inferential reasons to accept ψ and to accept (II), we would thus be committed to accepting that s has inferential reasons to accept (and so is committed to accepting) χ! (Of course, the argument propagates forward to any conclusion connected with χ through a chain of valid arguments.)

A very general lesson can be extracted from the main turn of the previous argument. Say that a sentence is *non-inferentially inaccessible* for a subject s at a time t iff its supposed truth does not imply that s at t has any non-inferential reason to accept it. What the previous argument shows is that an at least intermediate or strong application of non-transitivity with respect to the single-conclusion argument a_0 and the argument a_1 made by a subject s at a time t requires the conclusion of a_0 to be non-inferentially inaccessible for s at t. This squares nicely with the joint acceptance of (NI) and rejection of (II) typical of at least intermediate and strong applications of non-transitivity: for what these imply is that whether or not the normative force of consequence applies to a subject's acceptance of a sentence should depend on the sentence's *pedigree* (non-inferential or inferential) in the subject's cognitive history, whereas lack of non-inferential inaccessibility precisely obliterates any distinction which might be drawn at that level.

Thus, sentences accepted *qua* conclusions of valid arguments all of whose premises are accepted may just not have the right pedigree to enter in turn as premises into further valid arguments possessing normative force. It is in this *relevance of the premises' pedigree to the normative force of a valid argument* that deductive non-transitive reasoning comes close to defeasible reasoning. For example, consider the defeasible validity of the argument from 'Al is a native speaker of Italian' to 'Al was born in Italy' (see section 8.2.3) and suppose that one accepts its premise. Is one committed to accepting its conclusion? *That will depend on the reason why one accepts its premise*: if one accepts the premise because one has heard Al's fluent Italian speech (and one has no evidence against the conclusion), one is committed to accepting the conclusion, but, if one accepts the premise because one has in turn inferred it from 'Al was born in Little Italy', one will not be so committed. Contrast this relevance of the premises' pedigree to the normative force of a valid argument in the case of deductive non-transitive reasoning and defeasible reasoning with the irrelevance of the premises' pedigree to the normative force of a valid argument in the case of deductive transitive reasoning. For example, consider the transitive validity of the argument from 'Al is a native speaker of Italian' to 'Al is a native speaker of Italian or Al was born in Italy' and suppose that one accepts its premise. Then, one is committed to accepting its conclusion, *no matter for what reason one accepts its premise*. Non-dualist non-transitivism can thus be seen as an interesting hybrid joining important aspects of the *metaphysics* characteristic of deductive consequence relations as opposed to defeasible ones (i.e. the guarantee of truth preservation) with important aspects of the *normativity* characteristic of defeasible consequence relations as opposed to deductive ones (i.e. the relevance of the premises' pedigree to the normative force of a valid argument).

We can go a bit deeper. What underlies the relevant aspects of the normativity characteristic of defeasible consequence relations is the fact that the conclusion of a defeasibly valid argument can (in constrained ways) be *stronger* than its premise. Here, 'x is stronger than y' is of course not meant in the rather definitional sense of y's following from x but not *vice versa* (according to the relevant consequence relation); rather, it is meant in the intuitive sense of x's "*being weightier than*" y, with the more precise and specific consequences that x *is (epistemically) less likely than* y and that x *has consequences that* y *does not have* (according to the relevant consequence relation). Now, *exactly the same holds for deductive non-transitive consequence relations*: the only relevant difference with defeasible consequence relations is that, while in the latter consequence relations the conclusion being stronger than the premise goes together with (and, one may wish to add, is grounded in) the possibility of the premise's being true while the conclusion is not, this is not so in the former consequence relations. Hence, even for deductive non-transitive consequence relations, it is the case that, if χ is stronger than ψ, it might be that one has reasons to accept ψ but no reasons to accept χ even if χ follows from ψ—whether one does have the latter reasons will depend on the specifics of one's reasons to accept ψ (and on the specifics of the constrained ways in which χ is stronger than ψ).[61] In terms of the insightful taxonomy of

[61] Of course, it will be particularly problematic to suppose that one has reasons to accept χ if one's reasons for accepting ψ consist in reasons for accepting φ and in ψ's following from φ (for ψ might in turn be stronger than φ). Notice that, in addition to explaining the relevant aspects of the *normativity* characteristic of deductive non-transitive consequence relations, this conception also

Salmon [1967, 5–11], a non-transitivistically deductively valid inference can thus be "*ampliative*" while remaining "*demonstrative*", and be so in a way that goes beyond what Salmon and many others seem to have been able to conceive, since such demonstrative inference can be ampliative not only in the broadly *semantic* sense of the conclusion's not being implicit in the premise, but also in the broadly *epistemic* sense of the conclusion's being less likely than the premise. Coming back to a theme emerged already in section 8.3.1, we can then see how non-transitive logics vindicate a radical sense in which *logic* can be *informative*.

8.4.4 Non-transitivity and asymmetries between premises and conclusions

Let us draw together some of the contrasts explored so far. Exploiting the distinction just unearthed, the left-hand and right-hand sides of a sequent can be interpreted in a non-transitive framework as expressing a *connection* between what we have non-inferential reasons to accept and what we have inferential reasons to accept (which does not of course rule out that, independently, we also have non-inferential reasons to accept it: reasons-based acceptance can be overdetermined). This distinction has substance *for transitivists and non-transitivists alike*: there is a perfectly good sense in which I have non-inferential reasons to accept that snow is white but have simply inferential reasons to accept that either snow is white or grass is blue. It should thus actually be *common ground* that consequence can indeed fail to preserve non-inferential reasons to accept, and that it only guarantees that, if one has non-inferential reasons to accept the premises, one has inferential reasons to accept the conclusions.

Within this common ground, the debate on transitivity can then be understood as follows. The non-transitivist can be seen as thinking that consequence can also fail to preserve inferential reasons to accept. She can thus be seen as focussing on the *common-ground* property of consequence of *guaranteeing* inferential reasons to accept *given* non-inferential reasons to accept, whilst the transitivist can be seen as focussing on the *further* (alleged) property of consequence of *preserving* inferential reasons to accept (which, as shown by theorem 1, follows from the common ground-property iff consequence is transitive). The non-transitivist thinks that consequence *only guarantees a connection* between the *two different* acceptance-related normative properties of being N_0 and of being N_1, whilst the

accounts for the very *non-transitivity* of such consequence relations. Suppose that ψ follows from φ: if ψ is less likely than φ and χ is less likely than ψ, the difference in likelihood between χ and ψ might be small enough as to be compatible with χ's following from ψ, while the difference in likelihood between χ and φ might be large enough as to be incompatible with χ's following from φ; even more straightforwardly, let χ be one of the consequences that ψ has but φ does not have.

transitivist thinks that it *also preserves* the *single* acceptance-related normative property of being N_1.

It is worth noting that, parallel to the *normative* distinction between having non-inferential reasons to accept and having inferential reasons to accept—which can be used to mark a major point of disagreement between the *non-transitivist* and the *transitivist*—a similar distinction can be drawn at the *metaphysical* level between truth simply in virtue of how the world is and truth in virtue of how the world is and the laws of logic—which can be used to mark a major point of disagreement between the *dualist* and the *non-dualist*. The left-hand and right-hand sides of a sequent can equally legitimately be interpreted in a non-transitive framework as expressing a *connection* between what is true simply in virtue of how the world is and what is true in virtue of how the world is and the laws of logic (which does not of course rule out that, independently, it is also true simply in virtue of how the world is: truth can be overdetermined). This distinction too has substance *for dualists and non-dualists alike*: there is a perfectly good sense in which it is true simply in virtue of how the world is that snow is white, but it is true in virtue of how the world is and the laws of logic that either snow is white or grass is blue. It should thus actually be *common ground* that consequence can indeed fail to preserve truth simply in virtue of how the world is, and that it only guarantees that, if all the premises are true simply in virtue of how the world is, some conclusion is true in virtue of how the world is and the laws of logic.

The dualist can then be seen as thinking that consequence can also fail to preserve truth in virtue of how the world is and the laws of logic. She can thus be seen as focussing on the *common-ground* property of consequence of *guaranteeing* truth in virtue of how the world is and the laws of logic *given* truth simply in virtue of how the world is, whilst the non-dualist (either transitivist or non-transitivist) can be seen as focussing on the *further* (alleged) property of consequence of *preserving* truth in virtue of how the world is and the laws of logic (which, as indicated in n. 43, does follow if the deduction theorem holds). The dualist thinks that consequence *only guarantees a connection* between *two different* metaphysical properties, whilst the non-dualist thinks that it *also preserves* a *single* metaphysical property.

8.4.5 *The locality of non-transitive consequence*

In the peculiar joint acceptance of (NI) and rejection of (II) we see one sense in which the requirements placed by a non-transitive logic might typically be *local*, in this case extending only so far as the consequences of sentences accepted for non-inferential reasons go rather than stretching to cover every consequence of any sentences one is committed to accepting for whichever reasons. Indeed, the idea that non-transitive consequence has distinctively "*localist*" features has been lurking behind much of our discussion in sections 8.4.1–8.4.4, and it is now time to make it emerge more clearly.

To start with a suggestive spatial metaphor,[62] the non-transitivist's picture of logical space is the rather unusual one which sees the space of consequences of a given point (that is, of a given sequence of sentences) as being bounded by a *horizon*: there are bounds to what is logically necessary in relation to a given point which are not generally such in relation to themselves and to some other points included in the same original horizon, so that *a movement inside the horizon can result in a movement of the horizon itself.* Needless to say, this picture goes precisely against the more traditional picture of logical space (inspiring the view of consequence as a *closure* operation introduced in n. 37) which sees the space of consequences of a given point as being bounded by a *frame*: there are bounds to what is logically necessary in relation to a given point which stretch so far as to be such also in relation to themselves and to all other points included in the same original frame, so that *no movement inside the frame can result in a movement of the frame itself.*

Throughout sections 8.4.1–8.4.4, we have explored at some length two alternative ways (dualist and non-dualist) in which this horizon-like picture of logical space can be made better sense of with what has in fact been the general idea that the effects of consequence are peculiarly local—that *consequence manages to constrain the truth value of sentences only at one remove from where one starts.* As we have seen, this may be because consequence in turn fails to impose a truth-preservation constraint on the sentences whose truth values have been so constrained (if one is a dualist), or it may be instead because such a constraint can still be rationally not accepted to have certain effects (if one is a non-dualist). Either way, the upshot is that, although consequence does constrain the truth value of sentences at one remove from where one starts, it does not constrain the truth value of sentences at further removes—on longer distances, its constraining force gives out.

In conformity with the focus of our discussion in sections 8.4.1–8.4.4 (see n. 26), this point about the sensitivity of the constraining force of consequence to the distance from where one starts applies straightforwardly only to intermediate and strong applications of non-transitivity. But such sensitivity can in turn be seen to arise as a special case from the more general non-transitivist thought that a set of sentences may exhibit a *structure of merely local connections*, with consequence

[62] Suggested to me in another context by Josh Parsons.

playing the natural role of a "one-stop inference ticket" that reflects such structure. Thus, to go back to the examples of applications of non-transitivity of section 8.2, in the case of relevance consequence is supposed to reflect merely locally holding *meaning* connections; in the case of tolerance, consequence is supposed to reflect merely locally holding *indifference* connections; in the case of evidence, consequence is supposed to reflect merely locally holding *probabilification* connections.

Accordingly, on this localist picture, transitivity can be seen to fail because, generally speaking, it would create *spurious connections*. To go back again to the specific examples of applications of non-transitivity of section 8.2, in the case of relevance, transitivity would create a spurious meaning connection between sentences which have none (even though they are the opposite extremes of a chain of genuinely connected sentences); in the case of tolerance, transitivity would create a spurious indifference connection between sentences which have none (even though they are the opposite extremes of a chain of genuinely connected sentences); in the case of evidence, transitivity would create a spurious probabilification connection between sentences which have none (even though they are the opposite extremes of a chain of genuinely connected sentences). In all these cases, transitivity would obliterate a *non-trivial distance structure* which non-transitivists think is exhibited by a set of sentences, inflating local connections between these into global ones.

8.4.6 *Non-transitivist theories, situations, and worlds*

It is in view of what has been said that, in a non-transitive framework and under a certain assumption to be introduced shortly, two different readings of theoretical notions defined using the notion of *closure under logical consequence* must be sharply distinguished as being non-equivalent (in this section, for simplicity's sake, we assume monotonicity and return to taking sets to be the terms of the consequence relation and to restricting our attention to single-conclusion arguments). Notions so defined can be seen as having at their core the notion of a *theory*, straightforwardly defined as being any set of sentences closed under logical consequence. The theoretical interest of a notion which, in virtue of the closure clause, outruns that of a mere set of sentences should be evident in view of the normativity of consequence. One has non-inferential reasons to accept a set of sentences X. By (NI), one has thereby inferential reasons to accept (and so is committed to accepting) not only X, but also the set of all the consequences of X. The *objects of commitment* are thus always closed under logical consequence in the specified sense. And such objects well deserve to be called 'theories', for theories are traditionally thought precisely to be the objects of commitment: theories are what people are traditionally thought to hold, defend, attack, revise, try to confirm, etc.

Under the simplifying assumptions made at the beginning of this section, we can identify positions (introduced in section 8.4.3) with arbitrary sets of sentences; the theory of a position is then the closure under logical consequence of that position (let us denote by 'thr_{L}' the function from positions to theories under the consequence relation L). A theory $\mathcal{T}$ is *prime* iff, whenever 'φ or ψ' belongs to $\mathcal{T}$, either φ or ψ belongs to $\mathcal{T}$: primeness is the property of a theory to provide a witness for each of its assertions. A theory $\mathcal{T}$ is *maximal* iff, for every φ, either φ or 'It is not the case that φ' belongs to $\mathcal{T}$: maximality is the property of a theory to settle every question. A prime theory represents a *situation*, a maximal theory a *world*; φ is *true* in a situation* (*true* in a world*) iff φ belongs to the theory representing that situation (world). Call the logic under which a theory is closed '*target logic*' (given our purposes, this will almost always be a non-transitive logic), the logic of the language talking about theories '*background logic*'.

What do non-transitivist theories, situations, and worlds look like? There is no reason to think that the language talking about the logical consequences of sets of sentences of a language $\mathcal{L}$ should exhibit the same problematic features which motivate a deviance from classical logic for $\mathcal{L}$. It behoves us then to consider the case where the background logic is classical. Under this assumption, two readings of the definition 'The theory of a position is the closure under logical consequence of that position' must be sharply distinguished as being non-equivalent:

(i) The theory of a position P is the set of the logical consequences of P;
(ii) The theory of a position P is the smallest set $\mathcal{T}$ such that:
  (a) $\mathcal{T}$ contains all the logical consequences of P;
  (b) If φ is a consequence of $\mathcal{T}$, $\varphi \in \mathcal{T}$.[63]

Under the assumption of transitivity of the background logic, the notion delivered by reading (ii) is clearly too strong at least for intermediate and strong applications of non-transitivity, since it will force the theory of a position—that to which one is committed—to contain sentences which, according to the non-transitive target logic, are not logical consequences of the position.

Again, as in the case of (NI) and (II), it seems to me that, in her use of reading (i) rather than reading (ii), the non-transitivist is occupying a reasonable position, given that the following theorem holds:

[63] That is, the greatest lower bound under $\subseteq$ of the class of sets satisfying (a) and (b): that is, the set $\mathcal{T}_0$ such that $\varphi \in \mathcal{T}_0$ iff, for every $\mathcal{T}_1$ satisfying (a) and (b), $\varphi \in \mathcal{T}_1$.

Theorem 2. *Let $thr(i)_L$ and $thr(ii)_L$ be the functions corresponding respectively to readings (i) and (ii). Then, [for every* P, $\varphi \in thr(ii)_L(P)$ *only if* $\varphi \in thr(i)_L(P)$*] iff* L *is transitive.*

Proof.

- *Left-to-right.* We prove the contrapositive. Take a non-transitive consequence relation L such that, for every $\varphi \in Y, X \vdash_L \varphi$ and $Z \cup Y \vdash_L \psi$, but $Z \cup X \nvdash_L \psi$, and consider a position $P = Z \cup X$. Then, $\psi \notin thr(i)_L(P)$ (since $Z \cup X \nvdash_L \psi$), whereas $\psi \in thr(ii)_L(P)$ (since $Z \subseteq thr(ii)_L(P)$ by reflexivity of L and $Y \subseteq thr(ii)_L(P)$, while $Z \cup Y \vdash_L \psi$).
- *Right-to-left.* Take a transitive consequence relation L and consider an arbitrary position P. Suppose that $\varphi \in thr(ii)_L(P)$. We aim to prove that $P \vdash_L \varphi$. We do so by first defining by transfinite recursion the following hierarchy of positions:

  (i′) $P_0 = \{\varphi : P \vdash_L \varphi\}$;
  (ii′) $P_{\alpha+1} = P_\alpha \cup \{\varphi : P_\alpha \vdash_L \varphi\}$;
  (iii′) $P_\lambda = \bigcup(\{P_\alpha : \alpha < \lambda\})$.

  Notice that, since the language is finitary, by the well-ordering of the ordinals there will be a (not very big) first ordinal κ at which the process stabilizes and no new sentences are admitted as consequences—that is, for every $\alpha \geqslant \kappa$, $P_\alpha = P_\kappa$. Consider then the set P_κ. P_κ satisfies (a), since $P_0 \subseteq P_\kappa$. P_κ also satisfies (b), since $P_{\kappa+1} = P_\kappa$. Moreover, P_κ is the smallest set to do so. For suppose that $\mathcal{T}$ satisfies (a) and (b). We prove by transfinite induction that, for every $\alpha \leqslant \kappa$, $P_\alpha \subseteq \mathcal{T}$:

  (i″) Since, by (a), $\{\varphi : P \vdash_L \varphi\} \subseteq \mathcal{T}$, $P_0 \subseteq \mathcal{T}$;
  (ii″) If $P_\alpha \subseteq \mathcal{T}$, since, by (b), $\{\varphi : P_\alpha \vdash_L \varphi\} \subseteq \mathcal{T}$ as well, $P_{\alpha+1} \subseteq \mathcal{T}$;
  (iii″) If, for every $\alpha < \lambda$, $P_\alpha \subseteq \mathcal{T}$, then $\bigcup(\{P_\alpha : \alpha < \lambda\}) \subseteq \mathcal{T}$ as well.

  Thus, if $\varphi \in P_\kappa$, $\varphi \in \mathcal{T}$, and so P_κ is the smallest set to satisfy (a) and (b). It then follows that $P_\kappa = thr(ii)_L(P)$. We can now prove by transfinite induction that, given the transitivity of L, if $\varphi \in thr(ii)_L(P)$ ($\varphi \in P_\kappa$), then $P \vdash_L \varphi$. We do so by proving by transfinite induction that, for every $\alpha \leqslant \kappa$, if $\varphi \in P_\alpha$, then $P \vdash_L \varphi$:

  (i‴) If $\varphi \in P_0$, then $P \vdash_L \varphi$;
  (ii‴) If $\varphi \in P_{\alpha+1}$, then $\varphi \in P_\alpha \cup \{\varphi : P_\alpha \vdash_L \varphi\}$, and so:
    (a′) Either $\varphi \in P_\alpha$, in which case, by the induction hypothesis, $P \vdash_L \varphi$;

(b′) Or $\varphi \in \{\varphi : P_\alpha \vdash_L \varphi\}$, in which case, since, by the induction hypothesis, for every $\psi \in P_\alpha$, $P \vdash_L \psi$, by (T) $P \vdash_L \varphi$ as well;

(iii‴) If $\varphi \in P_\lambda$, then, for some $\alpha < \lambda$, $\varphi \in P_\alpha$, and so, by the induction hypothesis, $P \vdash_L \varphi$.

The proof is completed by observing that, since $P \vdash_L \varphi$, $\varphi \in \text{thr(i)}_L(P)$ as well.

□

Indeed, given that it is straightforward that, [for every P, $\varphi \in \text{thr(i)}_L(P)$ only if $\varphi \in \text{thr(ii)}_L(P)$] no matter whether L is transitive or not (because of (a)), theorem 2 can be reformulated to the effect that, [[for every P, $\varphi \in \text{thr(i)}_L(P)$ *iff* $\varphi \in \text{thr(ii)}_L(P)$] iff L is transitive]. In view of this, it seems to me that the non-transitivist can reasonably insist that the *pure notion*, free of any relevant assumption concerning formal properties of consequence, which plays the complex role usually assigned to the notion of a theory (role which is the same for non-transitivists and transitivists alike) is the one expressed by reading (i) rather than the one expressed by reading (ii), the latter being associated with such role only because reading (ii) is equivalent with (i) under the (rejected) assumption of transitivity.

Once reading (i) is distinguished as the appropriate notion to use when the target logic is non-transitive, the theory of theories can proceed very much as before (see e.g. Barwise and Perry 1983, 49–116). For our purposes, we only need to note the following concerning situations. A prime theory $\mathcal{N}$ with a non-transitive target logic L representing a certain situation may be such that $\psi \in \mathcal{N}$, $\psi \to \chi \in \mathcal{N}$ but $\chi \notin \mathcal{N}$ (such will be the case for example in tolerant logics if only φ, $\varphi \to \psi$, and $\psi \to \chi$ belong to the position $\mathcal{N}$ is a theory of). The situation thus represented would be one where a conditional ($\psi \to \chi$) and its antecedent (ψ) are true*, but the consequent (χ) is not. At first glance, this may of course look like an unwelcome consequence, since L may actually be such that the rule of *modus ponens* is unrestrictedly valid in it (as it is in tolerant logics). Even worse, if the law of excluded middle also holds unrestrictedly in L (as it does in the tolerant logics advocated in my works on vagueness referenced in section 8.1), given $\mathcal{N}$'s primeness we would have that $\neg\chi \in \mathcal{N}$, and so that $\neg\chi$ is true* in the situation represented by $\mathcal{N}$! However, it should by now be clear that, analogously to the cases discussed in section 8.3.2, these consequences are due to the choice of adopting a *transitive background logic* in the theory of theories. This choice imposes that the link between the technical notion of truth* in a situation (and of truth* in a world)—which is then governed by a transitive logic—and the informal and philosophical notion of truth in a situation (and of truth in a world)—which, just as the notion of truth, is sensitive to the failures of transitivity in the logic of the object language—be at best very complex and mediated.

Again, the generality of the point can be illustrated with reference to more well-known deviations from classical logic. Consider a classical theory $\mathcal{C}$ of [a prime intuitionist theory $\mathcal{I}$ formulated in language $\mathcal{L}$]. $\mathcal{C}$ entails that the situation $\mathfrak{i}$ represented by $\mathcal{I}$ is actually such that, for every $\varphi \in \mathcal{L}$, either φ is true* in $\mathfrak{i}$ or φ is not true* in $\mathfrak{i}$ (even though, of course, being the target logic intuitionist it need not be the case that, for every $\varphi \in \mathcal{L}$, either $\varphi \in \mathcal{I}$ or $\neg\varphi \in \mathcal{I}$). Such a conclusion would be repugnant given what would seem to be the most natural theory of truth in a situation (in the informal and philosophical sense) available to an intuitionist (namely one such that φ's being not true in a situation implies that φ is false in that situation and so that $\neg\varphi$ is true in that situation).

If a non-transitive logic is adopted as background logic, however, things change drastically and reading (ii) becomes again a viable option for intermediate and strong applications of non-transitivity. For example, going back to the theory $\mathcal{N}$ discussed two paragraphs back, reading (ii) no longer has the unwelcome consequence of declaring that $\chi \in \mathcal{N}$. For, in a non-transitive background logic, that $\chi \in \mathcal{N}$ does not follow from (set theory and) reading (ii), φ's belonging to $\mathcal{N}$, $\varphi \rightarrow \psi$'s belonging to $\mathcal{N}$, and $\psi \rightarrow \chi$'s belonging to $\mathcal{N}$. Indeed, there is now at least one reason to prefer reading (ii) over reading (i), since, if the background logic $\mathbf{L_0}$ is non-transitive (but not so as to affect definitional reasoning), reading (ii) allows for the construction of a theory $\mathcal{T}$ with a non-transitive target logic $\mathbf{L_1}$ representing a situation s such that, if $X \vdash_{\mathbf{L_1}} \varphi$, then *truth* in* s *is preserved* from X to φ (as befits the informal and philosophical notion of truth in a situation). For suppose that $X \vdash_{\mathbf{L_1}} \varphi$ and that every member of X is true* in s. Then, by definition of truth* in s, it follows that $X \subseteq \mathcal{T}$. From this, $X \vdash_{\mathbf{L_1}} \varphi$, and reading (ii), it follows that $\varphi \in \mathcal{T}$, which in turn entails, by definition of truth* in s, that φ is true* in s as well (notice how this argument only requires transitivity for definitional reasoning in $\mathbf{L_0}$).

8.5 Conclusion

Appealing as they may be, applications of non-transitivity require an extended philosophical discussion of what sense there is to be made of a rational being who reasons using a non-transitive logic. This chapter has no doubt been only a first stab at meeting that pressing request in its generality. We have seen how this task involves dealing with some of the hardest problems at the interface between the philosophy of logic and the theory of rationality: the connection between structural rules and the very nature of the premises and conclusions of an argument (and of their acceptance or rejection); the relation between deductive and defeasible consequence relations; the relation between consequence and inference; the relation between consequence, preservation of truth, and preservation of other epistemic properties; the normativity of consequence; the relation between premises and conclusion of a valid argument; the nature of the objects of logical commitment; the problem of theorizing about how all this behaves in a certain logic by using a different one, etc. From a wider perspective, I hope that the foregoing investigations about how non-transitivity impinges on these and other issues have helped to show them in a new, more general light, and that the conceptualizations made and the distinctions drawn will prove fruitful also in the examination of the philosophical foundations of other logics.[64]

References

Avron, Arnon. 1991. "Simple consequence relations." *Information and Computation* 92:105–40.

Barwise, Jon and Perry, John. 1983. *Situations and Attitudes*. Cambridge, MA: MIT Press.

Béziau, Jean-Yves. 2006. "Transitivity and paradoxes." In Jurğis Šķilters, Matti Eklund, Ólafur Jónsson, and Olav Wiegand (eds.), *Paradox: Logical, Cognitive and Communicative Aspects*, volume 1 of *The Baltic International Yearbook of Cognition, Logic and Communication*, 87–92. Riga: University of Latvia Press.

Bolzano, Bernhard. 1837. *Wissenschaftslehre*. Sulzbach: Seidel.

[64] Earlier versions of the material in this chapter were presented in 2008 at the 3rd International Conference on Philosophy (ATINER, in Athens) and at the Arché Philosophy of Logic Seminar (University of St Andrews); in 2009, at the Formal Philosophy Research Seminar (Catholic University of Leuven), at a research seminar at the University of Aarhus, and at the 7th GAP Conference on *Reflections and Projections—Challenges to Philosophy* (University of Bremen); in 2010, at the Workshop on *Vagueness and Self-Reference* (New University of Lisbon/University of Évora, in Lisbon); in 2011, at the 4th Workshop on *Vagueness and Physics, Metaphysics, and Metametaphysics* (University of Barcelona). I would like to thank all these audiences for very stimulating comments and discussions. Special thanks go to Cian Dorr, Hartry Field, Patrick Greenough, Richard Heck, Dan López de Sa, the late Bob Meyer, Josh Parsons, Graham Priest, Steve Read, Sven Rosenkranz, Marcus Rossberg, Ricardo Santos, Kevin Scharp, Stephen Schiffer, Moritz Schulz, Johanna Seibt, Daniele Sgaravatti, Stewart Shapiro, Martin Smith, Robbie Williams, Crispin Wright, and an anonymous referee. I am also grateful to the editors Colin Caret and Ole Hjortland for inviting me to contribute to this volume and for their support and patience throughout the process. At different stages during the writing of this chapter, I have benefitted from a RIP Jacobsen Fellowship, an AHRC Postdoctoral Research Fellowship, a UNAM Postdoctoral Research Fellowship, and the FP7 Marie Curie Intra-European Research Fellowship 301493, as well as from partial funds from the project FFI2008-06153 of the Spanish Ministry of Science and Innovation on *Vagueness and Physics, Metaphysics, and Metametaphysics*, from the project FFI2011-25626 of the Spanish Ministry of Science and Innovation on *Reference, Self-Reference and Empirical Data*, from the project CONSOLIDER-INGENIO 2010 CSD2009-00056 of the Spanish Ministry of Science and Innovation on *Philosophy of Perspectival Thoughts and Facts* (PERSP), and from the FP7 Marie Curie Initial Training Network 238128 on *Perspectival Thoughts and Facts* (PETAF).

Čech, Eduard. 1966. *Topological Spaces*. Prague: Academia, revised edition.

Christensen, David. 2004. *Putting Logic in Its Place*. Oxford: Oxford University Press.

Cobreros, Pablo, Égré, Paul, Ripley, David, and van Rooij, Robert. 2012. "Tolerant, classical, strict." *Journal of Philosophical Logic* 41:347–85.

Dretske, Fred. 1970. "Epistemic operators." *Journal of Philosophy* 67:1007–23.

Epstein, Richard. 1979. "Relatedness and implication." *Philosophical Studies* 36:137–73.

Evans, Gareth. 1976. "Semantic structure and logical form." In Gareth Evans and John McDowell (eds.), *Truth and Meaning*, 199–222. Oxford: Oxford University Press.

Field, Hartry. 2006. "Truth and the unprovability of consistency." *Mind* 115:567–605.

Field, Hartry. 2008. *Saving Truth from Paradox*. Oxford: Oxford University Press.

García-Carpintero, Manuel. 1993. "The grounds for the model-theoretic account of the logical properties." *Notre Dame Journal of Formal Logic* 34:107–31.

Geach, Peter. 1958. "Entailment." *Proceedings of the Aristotelian Society Supplementary Volume* 32:157–72.

George, Rolf. 1983. "Bolzano's consequence, relevance, and enthymemes." *Journal of Philosophical Logic* 12:299–318.

George, Rolf. 1986. "Bolzano's concept of consequence." *Journal of Philosophy* 83:558–64.

Harman, Gilbert. 1986. *Change in View*. Cambridge, MA: MIT Press.

Hume, David. 1739. *A Treatise of Human Nature*. London: Noon and Longman.

Kuratowski, Kazimierz. 1922. "Sur l'operation $\overline{A}$ de l'analysis situs." *Fundamenta Mathematicae* 3:182–99.

Lewy, Casimir. 1958. "Entailment." *Proceedings of the Aristotelian Society Supplementary Volume* 32:123–42.

Lewy, Casimir. 1976. *Meaning and Modality*. Cambridge: Cambridge University Press.

Makinson, David. 1965. "The paradox of the preface." *Analysis* 25:205–7.

Makinson, David. 2005. *Bridges from Classical to Nonmonotonic Logic*. London: King's College Publications.

Martin, Erroll and Meyer, Robert. 1982. "S (for syllogism)." Unpublished manuscript.

McGee, Vann. 1992. "Two problems with Tarski's theory of consequence." *Proceedings of the Aristotelian Society* 92:273–92.

Moruzzi, Sebastiano and Zardini, Elia. 2007. "Conseguenza logica." In Annalisa Coliva (ed.), *Filosofia analitica*, 157–94. Roma: Carocci.

Paoli, Francesco. 2002. *Substructural Logics: A Primer*. Dordrecht: Kluwer.

Parikh, Rohit. 1983. "The problem of vague predicates." In Robert Cohen and Marx Wartofsky (eds.), *Language, Logic, and Method*, volume 31 of *Boston Studies in the Philosophy of Science*, 241–61. Dordrecht: Reidel.

Popper, Karl. 1955. "Two autonomous axiom systems for the calculus of probabilities." *British Journal for the Philosophy of Science* 6:51–7.

Priest, Graham. 2006. *In Contradiction*. Oxford: Oxford University Press, 2nd edition.

Quine, Willard. 1960. *Word and Object*. Cambridge, MA: MIT Press.

Quine, Willard. 1986. *Philosophy of Logic*. Cambridge, MA: Harvard University Press, 2nd edition.

Read, Stephen. 1981. "Validity and the intensional sense of 'and'." *Australasian Journal of Philosophy* 59:301–7.

Read, Stephen. 1988. *Relevant Logic*. Oxford: Blackwell.

Read, Stephen. 2003. "Logical consequence as truth-preservation." *Logique et Analyse* 183:479–93.

Rényi, Alfréd. 1955. "On a new axiomatic theory of probability." *Acta Mathematica Academiae Scientiarum Hungaricae* 6:286–335.

Restall, Greg. 2000. *An Introduction to Substructural Logics*. London: Routledge.

Routley, Richard, Plumwood, Val, Meyer, Robert, and Brady, Ross. 1982. *Relevant Logics and Their Rivals*, volume I. Atascadero, CA: Ridgeview.

Sainsbury, Mark. 2001. *Logical Forms*. Oxford: Blackwell.

Salmon, Wesley. 1967. *The Foundations of Scientific Inference*. Pittsburgh: University of Pittsburgh Press.

Scott, Dana. 1974. "Completeness and axiomatizability in many-valued logic." In Leon Henkin (ed.), *Proceedings of the Tarski Symposium*, 411–35. Providence: American Mathematical Society.

Smiley, Timothy. 1959. "Entailment and deducibility." *Proceedings of the Aristotelian Society* 59:233–54.

Smith, Nicholas. 2004. "Vagueness and blurry sets." *Journal of Philosophical Logic* 33:165–235.

Sylvan, Richard. 2000. *Sociative Logics and Their Applications*. Aldershot: Ashgate.

Tarski, Alfred. 1930. "Fundamentale Begriffe der Methodologie der deduktiven Wissenschaften. I." *Monatshefte für Mathematik und Physik* 37:361–404.

Tarski, Alfred. 1936. "O pojęciu wynikania logicznego." *Przegląd filozoficzny* 39:58–68.

Tennant, Neil. 1987. *Anti-Realism and Logic*. Oxford: Clarendon Press.

Varzi, Achille. 2002. "On logical relativity." *Philosophical Issues* 10:197–219.

von Wright, Georg. 1957. "The concept of entailment." In *Logical Studies*, 166–91. London: Routledge.

Walton, Douglas. 1979. "Philosophical basis of relatedness logic." *Philosophical Studies* 36:115–36.

Weir, Alan. 1998. "Naive set theory is innocent!" *Mind* 107:763–98.

Wójcicki, Ryszard. 1988. *Theory of Logical Calculi: Basic Theory of Consequence Operations*. Dordrecht: Kluwer.

Zardini, Elia. 2008a. "Living on the slippery slope: the nature, sources and logic of vagueness." Ph.D. thesis, Department of Logic and Metaphysics, University of St Andrews.

Zardini, Elia. 2008b. "A model of tolerance." *Studia Logica* 90:337–68.

Zardini, Elia. 2009. "Towards first-order tolerant logics." In Oleg Prozorov (ed.), *Philosophy, Mathematics, Linguistics: Aspects of Interaction*, 35–8. Moscow: Russian Academy of Sciences Press.

Zardini, Elia. 2011. "Truth without contra(di)ction." *Review of Symbolic Logic* 4:498–535.

Zardini, Elia. 2012a. "It is not the case that [*P* and 'It is not the case that *P*' is true] nor is it the case that [*P* and '*P*' is not true]." *Thought* 1:309–19.

Zardini, Elia. 2012b. "Truth preservation in context and in its place." In Catarina Dutilh-Novaes and Ole Hjortland (eds.), *Insolubles and Consequences*. London: College Publications.

Zardini, Elia. 2013a. "The bearers of logical consequence." Unpublished manuscript.

Zardini, Elia. 2013b. "Borderline cases." Unpublished manuscript.

Zardini, Elia. 2013c. "Closed without boundaries." Unpublished manuscript.

Zardini, Elia. 2013d. "Context and consequence: an intercontextual substructural logic." Unpublished manuscript.

Zardini, Elia. 2013e. "Higher-order sorites paradox." *Journal of Philosophical Logic* 42:25–48.

Zardini, Elia. 2013f. "Naive *modus ponens*." *Journal of Philosophical Logic* 42:575–93.

Zardini, Elia. 2013g. "Open knowledge of one's inexact knowledge." Unpublished manuscript.

Zardini, Elia. 2013h. "Sequent calculi for tolerant logics." Unpublished manuscript.

Zardini, Elia. 2014a. "Evans tolerated." In Kensuke Akiba and Ali Abasnezhad (eds.), *Vague Objects and Vague Identity*, 327–52. New York: Springer.

Zardini, Elia. 2014b. "Naive truth and naive logical properties." *Review of Symbolic Logic* 7:351–84.

Zardini, Elia. forthcoming a. "First-order tolerant logics." *Review of Symbolic Logic.*

Zardini, Elia. forthcoming b. "Getting one for two, or the contractors' bad deal: towards a unified solution to the semantic paradoxes." In Theodora Achourioti, Kentaro Fujimoto, Henri Galinon, and José Martínez (eds.), *Unifying the Philosophy of Truth*. New York: Springer.

9

Non-detachable Validity and Deflationism

Jc Beall

9.1 Introduction: History and Setup

This chapter began as a paper in St Andrews on validity and truth preservation, focusing on a point that I (and others) had observed: namely, that validity is not truth preserving in any *detachable* sense (to be explained in the chapter). The paper was later expanded for a conference in Princeton on the philosophy and logic of truth (and their interplay): one's views on validity can often be constrained by one's philosophy of truth (or allied notions). The chapter before you, which is a lightly modified version of the later conference presentation, focuses on one instance of such interplay: deflationism about truth and the issue of (non-) 'detachable validity'. My chief aim in the chapter—as in the talks that occasioned it—is simply to raise the issues rather than decisively answer them. With this aim in mind, I have attempted to leave this contribution in its 'talk form', highlighting only the essential points of the discussion, expanding only where clarity demands it, and often using bullets instead of paragraph form.

9.1.1 What is deflationism about truth?

Too many things. Deflationism about truth is not one but many views, united only by the thought that truth plays no explanatory role: truth (or 'true') serves as a vehicle for explanations (or, generally, generalizations) of the world, not as explaining anything in the world. Instead of trying to untangle all such views, I shall focus on the 'transparency' version of deflationism about truth, a strand of Quinean 'disquotationalism' whereby 'true' is a *see-through* device brought into the language (for the language) of practical necessity—the in-practice necessity of expressing long generalizations (e.g. 'everything in such-n-so theory is true', etc.).

This sort of 'transparency' view goes back to Quine (1970), was clarified by Leeds (1978), and in turn was widely advanced by Field (2001) as a 'pure' version of disquotationalism. My focus on this version of deflationism is not to suggest that others aren't important or that this one has the best chance of being true. I focus on it because it has some common intuitive appeal, and, besides, I've thought more about it than others, and advanced a version (Beall 2009).

9.1.2 Main issue of the chapter

The issue concerns validity and arises out of (not unfamiliar) truth theoretic paradoxes—particularly, curry paradox.[1] In what follows, I review a result to the effect that validity is not truth-preserving in any 'detachable' sense, spelling this out in terms of what this means for one's validity predicate or corresponding validity connective. In short: for curry-paradoxical reasons, there's no valid argument from the validity of an argument and the truth of its premises to the truth of its conclusion. Saving some sense in which validity detaches is a goal that motivates a stratified or hierarchical approach to validity (Myhill 1975; Whittle 2004). But, details of the stratified approach aside, a question concerning deflationists about truth immediately emerges: can *deflationists* about truth go stratified about validity? I briefly discuss this question and, relying on a proposal by Lionel Shapiro (2011), briefly suggest an affirmative answer.

9.2 Background Conception of Truth

The aim of this section is simply to set some terminology. The background 'transparency' conception of truth is along the lines mentioned above: our truth predicate is an expressive device and nothing more; it was not introduced to name an important or otherwise explanatory property in the world; it was brought into the language to serve as a vehicle for explanations or, more generally, the expression of generalizations.

9.2.1 Transparent or see-through predicate

- Let $\ulcorner\ \urcorner$ be some naming device over language L, some function that assigns each sentence A an appropriate name $\ulcorner A \urcorner$. (This may be via a suitable quotation convention, as in many natural languages, or may be something fancier, such as Gödel coding.)

[1] I use 'curry' in 'curry paradox' as a predicate that classifies various paradoxes that, while not exactly like Curry's original paradox (arising from combinatorial logic), are clearly of the type to which Curry originally pointed (Curry 1942).

- Let $\varphi(x)$ be a unary predicate in L.
- Let $\widehat{A}$ be the result of substituting $\varphi(\ulcorner B \urcorner)$ for all (non-opaque) occurrences of B in A.
- $\varphi(x)$ is said to be a *transparent* or *see-through* predicate for L just if A and $\widehat{A}$ are equivalent for all A in L.

Example: assuming that negation is non-opaque, $\varphi(x)$ is transparent for L only if $\varphi(\ulcorner \neg A \urcorner)$ and $\neg\varphi(\ulcorner A \urcorner)$ are equivalent, and similarly only if $A \odot B$ and $\varphi(\ulcorner A \urcorner) \odot B$ are equivalent for any (non-opaque) binary connective $\odot$, and so on. Throughout, 'equivalent', at the very least, involves *logical equivalence*, so that, for any A in the language, A and $\widehat{A}$ are (at least) logically equivalent if $\varphi(x)$ is transparent for L.[2]

9.2.2 Transparency conception

With the notion of a transparent or see-through predicate (or 'device') in hand, the transparency conception (or view) of truth may be characterized as follows.

- The transparency view has it that truth is a (logical) property expressed by a see-through device in (and for) our language.
- *Truth* is in-principle-dispensable but in-practice-*in*dispensable: God could fully specify our world without using the truth predicate; but our finitude requires that we use it—for familiar purposes of generalization.
- Truth, on this conception, is not at all an explanatory notion; it is involved in explanations in the way that our 'voice box' is involved: we use it to express the explanation, but the explanation doesn't itself invoke it. (All of this is standard 'deflationary' story.)

The crucial negative point is that truth is not an important explanatory notion—not explanatory at all. And this view spills over into other common notions: satisfaction and denotation, as is well known, must be treated along similar deflationary lines.

What about validity? *Must* it too receive a 'deflationary' philosophy? Against some (Shapiro 2011), I think not; but I will not engage on this issue here. The question I shall briefly address is twofold:

[2] By *logical equivalence* is meant *whatever, in the end, the given logic—validity relation—counts as equivalent*. If $\vdash$ is the logic, then A and B are logically equivalent, in the target sense, just when $A \dashv\vdash B$. (NB: what's important is that a logic—and, in particular, the notion of logical equivalence—be understood broadly enough to allow for the target notion of transparent predicates. The resulting logic may—and, in truth theory, generally will—involve rules governing special predicates. Think, for example, of identity or, more on topic, truth. My interest is not in whether these are 'really' logical expressions.)

- How, if at all, might deflationists about truth maintain that *validity* is 'detachable' (in a sense to be explained below)?
- How might deflationists about truth be similarly deflationary about (detachable) validity?

By way of answers, I shall suggest—though only suggest—a marriage of ideas already available: one from John Myhill (1975, 1984) and the other, more recent, from Lionel Shapiro (2011). But first I rehearse a (perhaps now-familiar) point about validity and truth preservation, and I make explicit a corollary concerning the (non-) 'detachability of validity'.

9.3 Predicates and Connectives

I assume, throughout, that we can take any binary (indeed, n-ary) sentential predicate—that is, a predicate defined over all sentences—and get an equivalent corresponding sentential operator:

$$A \odot B := \Pi(\ulcorner A \urcorner, \ulcorner B \urcorner)$$

Given a (transparent) truth predicate, one can go the other way too; but the predicate-to-operator direction is the important one for current purposes.[3]

9.4 Validity, Truth Preservation, and Detachment

It's known that, for *curry-paradoxical* reasons, transparent truth theorists need to reject that validity is 'truth preserving' in any *detachable* sense (Beall 2006, 2009; Field 2008; Priest 2006); they need to reject that there's a valid argument from the validity of arbitrary argument $\langle A, B \rangle$ and the truth of A to the truth of B (Beall 2009, 35ff). Let me make this plain.[4]

- Curry sentences arise in various ways, commonly via straightforward 'curry identities' such as $c = \ulcorner Tr(c) \odot \bot \urcorner$. (Example: in English, one might have it that 'Bob' denotes 'If Bob is true then everything is true', thereby grounding a suitably necessary link between 'Bob is true' and the given conditional.)

[3] The other direction relies explicitly on the truth predicate $Tr(x)$. In particular, where $\odot$ is a binary sentential operator, one defines the corresponding predicate Π by setting $\Pi(Tr(x), Tr(y))$ to be true just when $Tr(x) \odot Tr(y)$ is true.

[4] While I wave at examples, I assume familiarity with curry paradox and recent debate on it. Also, $\bot$ is any *explosive* sentence—a sentence implying everything.

- *Detachable.* A binary connective $\odot$ is *detachable* iff the argument from $A \wedge (A \odot B)$ to B is *valid*, iff the argument from $\{A, A \odot B\}$ to B is valid.[5]
- For curry-paradoxical reasons, transparent truth theorists need to reject the validity (or even unrestricted truth) of 'pseudo modus ponens' or PMP (Beall 2009; Restall 1993; Shapiro 2011) for *any detachable* connective $\odot$.[6]

$$\text{PMP.} \quad A \wedge (A \odot B) \odot B$$

1. $c = \ulcorner Tr(c) \odot \bot \urcorner$. [Empirical fact (let us say)]
2. $Tr(c) \wedge (Tr(c) \odot \bot) \odot \bot$. [PMP]
3. $Tr(c) \wedge Tr(\ulcorner Tr(c) \odot \bot \urcorner) \odot \bot$. [2; Transparency]
4. $Tr(c) \wedge Tr(c) \odot \bot$. [1,3; Identities]
5. $Tr(c) \odot \bot$. [4; Substitution of A for $A \wedge A$]
6. $Tr(\ulcorner Tr(c) \odot \bot \urcorner)$. [5; Capture/Transparency]
7. $Tr(c)$. [1,6; Identities]
8. $\bot$. [5,7; MP—i.e., $\odot$-detachment]

Question: how does the rejection of PMP for any detachable connective amount to the rejection of the detachability of validity? As follows:

- Suppose, now, that in addition to a truth predicate, we have an adequate *validity predicate* $Val(x, y)$ in our language. Then to say that $\odot$ is detachable is to say that $Val(\ulcorner A \wedge (A \odot B) \urcorner, \ulcorner B \urcorner)$ is true.
- Define a corresponding validity *operator*: $A \Rightarrow B := Val(\ulcorner A \urcorner, \ulcorner B \urcorner)$. Then an operator $\odot$ is detachable just if $A \wedge (A \odot B) \Rightarrow B$ is true.
- But now the point is plain: validity $\Rightarrow$ is detachable iff $A \wedge (A \Rightarrow B) \Rightarrow B$ is true iff PMP holds for the validity operator.
- Hence, validity itself is not detachable.[7]

Parenthetical note. I am embarrassed to say that I didn't sufficiently spell out this point in my *Spandrels of Truth* discussion (Beall 2009, ch. 3), though did spell it out enough for it to be an implication: 'I reject that valid arguments are ttruth-preserving in anything beyond the hook sense' (i.e. in any detachable sense) (Beall 2009, §2.5, 37), and 'one can know that an argument is valid and know that its premises are all ttrue, but nonetheless remain without a valid argument that takes one from such information [i.e. truth of premise and validity of argument] and

[5] I assume throughout that conjunction $\wedge$ is normal. Giving a non-standard account of $\wedge$ affords options—but I won't here look into that. (Similarly with respect to standard structural rules: I assume them, and do not here discuss giving them up.)

[6] For readability's sake, let $\wedge$ bind more tightly than $\odot$, so that $A \wedge B \odot C$ is $(A \wedge B) \odot C$.

[7] To use Restall's lingo (Restall 1993) now common in this area: transparent truth theorists need to be 'really contraction free' to avoid curry paradox—and this, I'm here noting, applies to *all* binary connectives (including the validity one). See Shapiro (2011) and Beall and Murzi (2013).

the [non-detachable sense of validity truth-preservation] to the given conclusion' (Beall 2009, 36). Only recently, after having returned to some of these issues in a paper with Julien Murzi (Beall and Murzi 2013), did I see things perfectly simply and clearly in the way I've laid it out above. Looking at the literature, it is clear to me that Lionel Shapiro (2011) was the first to make explicit what was nearly—but only nearly—explicit in my claim above, and I probably owe my appreciation of the point to him. (I briefly discuss some of his key work below.) But other work, cited in Beall and Murzi (2013), is also in the area—perhaps most explicitly John Myhill's (1975) and Bruno Whittle's (2004).

9.5 Validity: Detachable via Stratification?

Some might think that detachable truth preservation or, as I'll just say here, the detachment of validity is essential to our notion of validity. Suppose that that's right. How, then, are we to keep detachment without falling prey to the perils of PMP?[8]

The most natural thought points to a stratified or hierarchical notion of validity. For precisely the sort of reasons above (though put in different ways), John Myhill (1975) proposed that validity be understood along a stratified front, as did Bruno Whittle (2004) more recently. The idea, in a nutshell, is that we have no cover-all validity relation but many limited relations—or, if you want, we have one big stratified relation, with each stratum itself a validity relation. This way, we can have that each validity relation is truth preserving (and, so, detachable); we can truly say that validity$_i$ is 'detachable' by using some 'higher' (or extended, or etc.) relation:

$$A \wedge (A \Rightarrow_i B) \Rightarrow_{i+1} B$$

But a question arises: namely, whether any such stratified approach to validity is philosophically compatible with our target sort of deflationism about truth.

9.6 Compatible with Deflationism?

Is the stratified approach available to *deflationists* about truth deflationism—particularly, the sort of 'merely expressive device' ones at issue? I don't see why not. Moreover, I think that there's a clear path towards taking an expressive-device deflationary view of validity—a path cut recently by Lionel Shapiro (2011). Shapiro,

[8] I should note that, in recent work (Beall 2012), I have come to think that our language is entirely detachment-free—containing no detachable connectives (hence, no detachable validity connectives). I cannot go into these ideas here, and suppress them throughout. I think that the issues raised in the current paper are still very much worth putting forward for exploration and debate—my main aim in this chapter.

I should make plain, agrees that validity is non-detachable in the given sense.[9] While he does *not* consider stratified validity, Shapiro's idea for a way to see validity as 'deflationary' applies just as well in the stratified case. Let me present the basic idea, and then summarize its relevance here.

Shapiro's paper is rich with ideas, but I shall focus on only one thing. For present purposes, what Shapiro gives us is a sense in which the validity *predicate*—versus operator—may be seen as an expressive device, a generalizing device along the lines of truth. Importantly, Shapiro's picture is one in which we already have validity operators in the language, and we introduce a validity *predicate* to generalize over them. And this, on the Shapiro picture, is precisely what is going on with other expressive devices like the truth predicate and similarly falsity predicate. How does this go?

Invoking an analogy from Anderson and Belnap (1975), Shapiro's idea is strikingly simple. For convenience, let me set some terminology:

- A *negation* is a sentence whose main connective is negation.
- A *nullation* is a sentence whose main connective is the null operator. (Every sentence is a nullation.)

In turn, we are to see 'true' as generalizing over nullations in the same way that, for example, 'false' generalizes over negations: on a transparency conception, $\mathtt{Tr}(\ulcorner A \urcorner)$ and A are intersubstitutable in the way that $\mathtt{False}(\ulcorner A \urcorner)$ and $\neg A$ are. (On a transparency conception, falsity is generally the transparent truth of negation: $\mathtt{False}(\ulcorner A \urcorner)$, by definition, is $\mathtt{Tr}(\ulcorner \neg A \urcorner)$, equivalently $\neg \mathtt{Tr}(\ulcorner A \urcorner)$.)

Suppose, now, that we have an entailment or validity connective $\Rightarrow$ in the language, and let an *implication* be a sentence with $\Rightarrow$ as its main connective. Shapiro argues that a validity predicate generalizes over implications in exactly the way that the truth and falsity predicates generalize over nullations and negations, respectively. That's the basic idea. In a picture:

- 'true' generalizes over nullations;
- 'false' generalizes over negations; and
- 'valid' generalizes over implications.

In our stratified setting, we simply broaden the point about validity:

- 'valid$_i$' generalizes over implications$_i$ (i.e, $\Rightarrow_i$ claims).

[9] He explains this by going 'really contraction-free'—a logic that, unlike leading transparency theories, gives up *substructural* contraction, and thereby the PMP form of contraction. I will slide over these details for present purposes.

An example: *the argument from* $A \wedge (A \Rightarrow_i B)$ *to* B *is valid*$_{i+1}$. More ordinary examples, using 'consequence' instead of 'validity' (as the former has a more ordinary ring), are claims such as *Axiom 1 of So-n-so's theory is a consequence of something the Pope said*. Here, the validity (or consequence) predicate is generalizing over implications in a familiar way: either the Pope said x and *that* x *is true entails that Axiom 1 is true* or the Pope said y and *that* y *is true entails that Axiom 1 is true* or . . . so on. (Here, I use 'true' in its usual see-through role, just for convenience. This can be dropped.) Along the same lines: *everything in theory* T *is a consequence of something in theory* T′. And so on.

The examples themselves may be less important than the main point here: namely, that this provides at least one clear sense in which validity predicates—even if stratified (as we're assuming)—can be seen as expressive devices (generalizing devices) along the same front as 'true'. The sole role of an expressive device is to generalize over some fragment (possibly improper fragment) of the language; and the device achieves its role in virtue of simple rules (e.g. 'capture and release rules' or 'intro and elim rules', etc.)—and we needn't read into the basic rules any 'metaphysical baggage', but instead can see such devices as merely logical. Validity predicates can be seen as such—stratified or not.

9.7 Questions and Replies

But let me quickly answer a few questions, before summarizing and closing.

* *Question.* But this approach to 'deflationism about validity' only works if we already have validity operators in the language. How is 'validity' (the predicate) then seen as on par with our so-called *merely see-through device* 'true'?

* *Answer.* There are differences: the truth one is in-practice indispensable, while the others aren't (ignoring propositional quantification). For example, we can generalize over negations and implications using only 'true' in its standard role. If we get rid of 'true', we'd be stuck again—regardless of whether 'valid' can do its generalizing role. But all that this shows is that truth is indispensable in a way that, so long as we have truth, 'valid' and 'false' aren't; it doesn't undermine—as Shapiro himself notes—that the sole role of the given *predicates* is the given generalizing work (even if that work can be done by other devices in the language).

In short, we can clearly acknowledge, along the Shapiro picture, that all of 'true', 'false', and 'valid' (the predicates) are *expressive devices*, generalizing devices that do their job via their basic rules (e.g. Release and Capture, or Intersubstitutability, or some such). But we can also distinguish between *in-practice-indispensable* ones and *in-practice-useful* (or the like) ones. Truth, on the transparency view, is in the former category, and the others—stratified or not—along the latter.[10]

* *Question.* If you go stratified for 'validity', why not also for 'true'? There may not be an *incompatibility* between transparency about truth and stratified validity, but if you go stratified for one, why not for both?!

* *Answer.* The expressive role of 'true' requires more than what that of 'valid' may require. You can't generalize over all sentences (all nullations) in your own language with stratified truth. (Indeed, if we focus on the Shapiro picture of *expressive device*, it's plain that *validity* is a notion for which stratification makes sense, much like negation itself. But the *null operator* can't be stratified: this marks again the special status that truth enjoys.)

9.8 Summary

There's a wide variety of deflationary views, perhaps each with its own peculiarities and problems and virtues. I've focused on the *transparency view*, one to which 'true' is nothing more than an expressive device—a full *see-through* device over one's entire language. One issue topic of fundamental concern is *validity* in a transparent-truth setting. I've argued that such theorists need to reject that validity is detachable. A natural way towards *some* sense of 'detachable validity' is via stratification. I've suggested that stratification is *not* philosophically incompatible with an appropriate device-deflationary view of 'valid', and waved at Shapiro's approach as one way (probably among others) to see stratified validity predicates as 'expressive devices'. There are no doubt lots more issues worth thinking about with respect to both validity and truth, but my aim has been to highlight at least one—and I hope I've done that.[11]

[10] Accordingly, I think that, contrary to Shapiro (2011), transparent-truth theorists needn't be similarly deflationary with respect to 'validity'. But I do think that they can be, and indeed can maintain that validity—qua stratified notion—is both detachable *and* 'deflationary' in the sense discussed.

[11] I am grateful to the organizers of the FLC workshops and conference(s) in St Andrews that occasioned many of the ideas in this chapter. I'm very grateful to many people for discussion, in one form or another, on many different occasions, including (but probably not limited to) Alexis Burgess, John Burgess, Andrea Cantini, Colin Caret, Roy Cook, Aaron Cotnoir, Hartry Field, Michael Glanzberg, Ole Hjortland, Hannes Leitgeb, Vann McGee, Julien Murzi, Charles Parsons, Graham Priest, Agustin Ráyo, Stephen Read, Greg Restall, David Ripley, Marcus Rossberg, Josh Schechter, Lionel Shapiro, Bruno Whittle, and many participants at the 'Pillars of Truth' conference in Princeton.

References

Anderson, Alan Ross and Belnap, Nuel D. 1975. *Entailment: The Logic of Relevance and Necessity*, volume I. Princeton, NJ: Princeton University Press.

Beall, Jc. 2006. "Truth and paradox: a philosophical sketch." In Dale Jacquette (ed.), *Philosophy of Logic*, volume X of *Handbook of the Philosophy of Science*, 187–272. Dordrecht: Elsevier.

Beall, Jc. 2009. *Spandrels of Truth*. Oxford: Oxford University Press.

Beall, Jc. 2012. "Non-detachable dialetheism." This was a five-lecture series of talks delivered at the Arché Research Centre, University of St Andrews; it is the basis of a book project. Some of the material was also presented at the University of Otago and Auckland University in 2011.

Beall, Jc and Murzi, Julien. 2013. "Two flavors of curry paradox." *Journal of Philosophy* 110:143–65.

Curry, Haskell B. 1942. "The inconsistency of certain formal logics." *Journal of Symbolic Logic* 7:115–17.

Field, Hartry. 2001. *Truth and the Absence of Fact*. Oxford: Oxford University Press.

Field, Hartry. 2008. *Saving Truth from Paradox*. Oxford: Oxford University Press.

Leeds, Stephen. 1978. "Theories of reference and truth." *Erkenntnis* 1:111–29. Reprinted in *Deflationary Truth*, B. Armour-Garb and Jc Beall (eds)., Chicago, IL: Open Court Press, 2005.

Myhill, John. 1975. "Levels of implication." In A.R. Anderson, R.B. Marcus, and R.M. Martin (eds.), *The Logical Enterprise*, 179–85. New Haren, CT: Yale University Press.

Myhill, John. 1984. "Paradoxes." *Synthese* 60:129–43.

Priest, Graham. 2006. *In Contradiction*. Oxford: Oxford University Press, 2nd edition.

Quine, W.V. 1970. *Philosophy of Logic*. Englewood Cliffs, NJ: Prentice-Hall.

Restall, Greg. 1993. "How to be *really* contraction free." *Studia Logica* 52:381–91.

Shapiro, Lionel. 2011. "Deflating logical consequence." *Philosophical Quarterly* 61:320–42.

Whittle, Bruno. 2004. "Dialetheism, logical consequence and hierachry." *Analysis* 64:318–26.

PART IV

Applications of Logical Consequence

10

Embedding Denial

David Ripley

10.1 Introduction

10.1.1 A puzzle about disagreement

Suppose Alice asserts p, and the Caterpillar wants to disagree. If the Caterpillar accepts classical logic, he has an easy way to indicate this disagreement: he can simply assert $\neg p$. Sometimes, though, things are not so easy. For example, suppose the Cheshire Cat is a paracompletist who thinks that $p \vee \neg p$ fails (in familiar (if possibly misleading) language, the Cheshire Cat thinks p is a *gap*). Then he surely disagrees with Alice's assertion of p, but should himself be unwilling to assert $\neg p$. So he cannot simply use the classical solution. Dually, suppose the Mad Hatter is a dialetheist who thinks that $p \wedge \neg p$ holds (that is, he thinks p is a *glut*).[1] Then he may assert $\neg p$, but it should not be taken to indicate that he disagrees with Alice; he doesn't. So he too can't use the classical solution.

The Cheshire Cat and the Mad Hatter, then, have a common problem, and philosophers with opinions like theirs have adopted a common solution to this problem: appeal to denial.[2] Denial, these philosophers suppose, is a speech act like assertion, but it is not to be understood as in any way reducing to assertion. Importantly, denial is something different from the assertion of a negation; this

[1] There's a pesky feature in terminology here. A 'paracompletist' is typically taken to be someone who thinks that some instances of excluded middle fail (see e.g. Field 2008), but a 'paraconsistentist' need not think that some contradictions are true. That's a 'dialetheist'. So parallel terms don't pick out parallel points in philosophical space (and the position dual to paraconsistentism doesn't even have a standard name). Nevertheless, here I follow standard, if suboptimal, terminology.

[2] For details on specific ways of doing this, see for example Priest (2006a), Field (2008), Beall (2009), and Parsons (1984). The description I offer here of denial is common to all these authors. Note that some of these authors prefer to speak of *rejection* (an attitude) rather than *denial* (a speech act); I don't think the difference matters for my purposes here (simply swap in 'acceptance' where I say 'assertion'), and I will henceforth ignore it.

is what allows it to work even in cases where assertion of negation does not. Just as importantly, denial must express disagreement, since this is the job it's being enlisted to do.

Let's consider in this light the case of the Cheshire Cat. He disagrees with Alice's assertion of p, but is unwilling to assert $\neg$p. He can indicate his disagreement simply by denying p. A paracompletist who denies p is not committed to asserting $\neg$p; in fact, the Cheshire Cat would deny $\neg$p as well. Dually, a dialetheist who asserts $\neg$p is not committed to denying p; the Mad Hatter can assert $\neg$p, and we need not take that as indicating his disagreement with Alice. If he were to go on to *deny* p, however, we'd understand him as disagreeing.

Nothing about this story, let me hasten to note, undermines the Caterpillar's strategy for expressing disagreement. Since the Caterpillar accepts classical logic, we can take his assertion of $\neg$p as committing him to a denial of p. Thus, his assertion of $\neg$p is still successful in expressing disagreement with Alice, so long as he is committed to classical logic. This theory about denial, then, has the virtue of allowing us to make sense of a range of disagreements, even when those disagreements occur between characters with different logical commitments.[3]

10.1.2 *The denier paradox*

In order to work as I've described it, denial must satisfy both of the following principles:

> D-exclusivity: Asserting p and denying p are *incompatible* speech acts; they rule each other out (otherwise we could have denial without disagreement)
>
> D-exhaustivity: Having a settled opinion about p requires being either willing to assert p or willing to deny p (otherwise we could have disagreement without denial)[4]

These are parallel to the classical principles of $\neg$-exclusivity (that p and $\neg$p can't both be true) and $\neg$-exhaustivity (that at least one of p, $\neg$p must be true). Since both the paracompletist and the dialetheist give up the conjunction of these negation principles to avoid paradox, one might naturally worry that invoking a notion like denial, one that satisfies the analogues of the negation principles, might lead right back to paradox.

[3] For more on this particular virtue, see Restall (2013).

[4] In ordinary life, one might have a settled opinion about p without being willing to say much at all about it—maybe p is a secret, or too long or boring to bother with. Here, when I talk about willingness to assert or deny, I mean *in a complete statement of one's theory*. (This is one reason why the difference between assertion/denial and acceptance/rejection can be ignored here.)

Here's the threat: consider a sentence (or a content) δ (we'll call it the *denier*) such that asserting δ is equivalent to denying δ. That is, asserting it commits us to denying it, and denying it commits us to asserting it. If we assert it, we're committed to denying it, but that's unacceptable (by D-exclusivity). If we deny it, we're committed to asserting it, but that's unacceptable (by D-exclusivity). So we'd better neither assert it nor deny it. But since this is (must be!) our settled opinion, this is unacceptable too (by D-exhaustivity). So we're in trouble no matter what we do.

This is only a threat, however, and not genuine trouble, so long as there is no such δ. Paracompletists and dialetheists who appeal to denial must take care that their theories do not countenance such a δ, and indeed they do take such care.

Avoiding such a δ requires avoiding certain operations on content. Suppose we were to consider a unary content operator D such that to assert DA, for any content A, is equivalent to denying A. Then there would be a δ as specified above: simply let $\delta = DT\langle\delta\rangle$, where $T\langle\rangle$ is an intersubstitutable truth predicate, augmented with the usual sort of self-reference-allowing naming device.[5] So it's crucial to the tenability of our denial story that there be no such operator D.

In section 10.2, however, I'll argue that we need an operator like D to make full sense of denial. Once we have D (and therefore δ), the denier paradox pushes us to give up either D-exclusivity or D-exhaustivity; but then the initial puzzle about disagreement cannot be solved in the way I've outlined. I consider other possible ways to solve the puzzle about disagreement in section 10.3 and section 10.4.

10.1.3 Does denial exist already?

In Parsons (1984), denial is assimilated to 'metalinguistic negation' in the sense of Horn (1989). Metalinguistic negation is the sort of negation occurring in, for example, 'Bryce doesn't have THREE helicopters; he has FOUR'. It seems distinct from an ordinary negation because the utterer of such a sentence doesn't mean to say that Bryce doesn't have three helicopters. In fact, for the sentence to be true, Bryce *has* to have three helicopters. He just also needs to have a fourth one. The utterer of such a sentence is rejecting the *appropriateness of an assertion of* 'Bryce has three helicopters', not because the sentence is false, but instead because it would be misleading.

[5] Both Field (2008) and Beall (2009) advocate for such a truth predicate in their respective frameworks. While Priest (2006a) and Priest (2006b) argue against truth's being intersubstitutable, there's no reason not to define an intersubstitutable predicate in Priest's framework. He might insist it's not truth, but for its role in constructing δ it doesn't matter whether or not it's truth; it only matters that the predicate be intersubstitutable.

Parsons focuses on this use of metalinguistic negation—to object to a sentence without necessarily committing oneself to an assertion of the sentence's negation—and says that this is our real-world denial. According to Parsons, we use the word 'not' in two different ways (Horn (1985) calls this a 'pragmatic ambiguity'): sometimes it modifies the content of our sentence, and sometimes it leaves the content alone, but indicates that the speaker rejects the sentence on some grounds or other. (In the above example, the grounds are grounds of misleadingness.) When that rejection is based on the sentence's content, we have a denial, according to Parsons's theory.

There's a serious problem with understanding denial in terms of metalinguistic negation, however. Denial is supposed to be a certain type of speech act, and speech acts are something we do with a content. We can't build contents with speech acts as parts. However, as Geurts (1998) points out, metalinguistic negation has no difficulty embedding into larger contents; the following sentences are perfectly fine:

(1) If Bryce doesn't have THREE helicopters but FOUR, then he has one more helicopter than I thought.

(2) Mary thinks Bryce doesn't have THREE helicopters but FOUR.

Geurts points this out as part of an attack on Horn's 'pragmatic ambiguity' thesis, but it tells just as strongly against Parsons's analysis of denial as metalinguistic negation.

In fact, I don't know of any phenomenon studied outside the realm of philosophical logic that could fill the theoretical role occupied by denial in our philosopher's theories.[6] I'm not sure, though, that that in itself is a problem for these theories. After all, denial in the present sense only needs to be invoked (that is, it only differs from asserting a negation) in the presence of gaps or gluts. If most speakers, most of the time, are not worried about gaps or gluts (as seems plausible), then we shouldn't necessarily expect there to be an obvious ordinary-language correlate of denial. Those of us who are worried about gaps and gluts may simply have to *introduce* a new sort of speech act to make plain our meanings, and I don't see that there's anything to stop us from doing this.

Indeed, Field (2008, 96) seems to treat denial roughly along these lines, proposing that saying something 'while holding one's nose' or writing something in a certain ugly font could serve, if we so chose, as marks of denial. There's obviously

[6] Geurts (1998) mentions another phenomenon traveling under the name of 'denial', but it is clearly unrelated. I'll mention it again in n. 12.

something arbitrary about these choices, but arbitrariness should be no barrier here.

Of course, if we are *introducing* denial rather than *discovering* it, it's important to be clear about just what we're introducing. Which features does this speech act have? Which does it lack? I think when we try to get clear about these questions, we will see that a speech act that behaves like denial does ought to have a corresponding operation (like D) on content. And, as we've seen, that leads to revenge paradox. Section 10.2 spells this out.

10.2 Denial in the Image of Assertion

Denial, for these theorists, is not any kind of assertion, but it's at least something *like* assertion.[7] In fact, it's more like assertion than either of them is like questioning or ordering. Both assertion and denial are *informative*; they attempt to tell us something about the way things are. This suggests that, to learn about denial, we should look at theories about assertion, and make the appropriate modifications.

In this section, I'll do just that. First, I'll look at norms assertion is sometimes alleged to fall under, and point out some trouble for stating the appropriate parallel norms on denial. Second, I'll look at uses of truth together with assertion to express agreement, and point out some trouble for using truth to express agreement in the presence of denial. Third, I'll point to a certain sort of question about priority that is a sensible question to ask about assertion, and point out some trouble for stating a parallel question about denial. All three of these troubles can be solved if we have a D operator in the language. (Of course, as we've already seen, D brings its own troubles. Nothing in this section will alleviate any of those. The purpose of this section is rather to make clear the troubles that arise from a *lack* of D.) Fourth, I'll show how to generalize a Stalnakerian theory of assertion to encompass denial, and point out that this allows us simply to define D. All these considerations, taken together, provide a compelling argument for having D around. I close this section with some reason to think having D around isn't totally hopeless, despite the risk posed by the denier paradox.

10.2.1 Norms governing denial

Assertion is often taken to be subject to certain norms. Just what these norms are is a matter of some dispute; here I'll consider two options:

[7] For example, Field (2008, 74) offers: '[W]e should regard acceptance and rejection as dual notions. And how exactly one thinks of rejection will depend on how one thinks of the dual notion of acceptance.'

(Assert-T) Assert A only if A is true

(Assert-K) Assert A only if you know A

10.2.1.1 (ASSERT-T)

Let's start with (Assert-T). Suppose this is a norm that governs assertion. The natural question is: what is the corresponding norm governing denial? Here are two possibilities:

(Deny-F) Deny A only if A is false

(Deny-NT) Deny A only if A is not true

Neither of these, though, should be acceptable to the proponent of denial. Let's consider (Deny-F) first. If (as is standard) we take falsity to be truth of negation, then (Deny-F) tells us to deny something only when its negation is true. But the Cheshire Cat couldn't then use denial to indicate his gappy take on p, not without (by his own lights, anyway) flouting (Deny-F). On the other side of the coin, the Mad Hatter ought to find (Deny-F) unacceptably weak; he thinks there are plenty of false things (Alice's p among them) that ought not be denied. What's more, it's not that they shouldn't be denied for reasons of politeness or some such; it's that one would be *mistaken* to deny them, in the same way one is mistaken in asserting something that isn't true. Such cases ought to be covered by a norm governing denial parallel to (Assert-T), but (Deny-F) fails to do the work.

If we turn to (Deny-NT), things are little better. In fact, if we have an intersubstitutable truth predicate, things are no better at all, since there is no difference between A's being false (that is, $T\langle\neg A\rangle$) and A's not being true (that is, $\neg T\langle A\rangle$), and so no difference between (Deny-F) and (Deny-NT). This already rules out (Deny-NT) for Field-style paracompletists and Beall-style dialetheists.

As above, however, not every theorist who appeals to denial accepts that truth is intersubstitutable. In particular, Priest-style dialetheists and dual sorts of paracompletist do not. (There are probably no actual instances of Priest-duals, but there certainly could be.) For example, Priest takes the simple liar—$\lambda' = T\langle\neg\lambda'\rangle$—to be false (he asserts $T\langle\neg\lambda'\rangle$), but he doesn't take it to be untrue (he denies $\neg T\langle\lambda'\rangle$). His dual would do just the reverse: assert $\neg T\langle\lambda'\rangle$, but deny $T\langle\neg\lambda'\rangle$. Indeed, when untruth and falsity come apart in either of these ways, (Deny-NT) outperforms (Deny-F): Priest won't deny λ', and his dual will. This is in accord with (Deny-NT), but not (Deny-F).

Unfortunately, this works as nicely as it does only because the simple liar is a special case. When we turn to the *strengthened* liar—$\lambda = \neg T\langle\lambda\rangle$—things don't work as cleanly. This is because Priest and his dual agree that falsity and untruth *can't* come apart for λ. Suppose λ is false. Then we can argue that it is untrue as follows:

1. $T\langle\neg\lambda\rangle$ Assumption
2. $\neg\lambda$ 1., Release[8]
3. $\neg\neg T\langle\lambda\rangle$ 2., Substitution
4. $T\langle\lambda\rangle$ 3., Double Negation
5. λ 4., Release
6. $\neg T\langle\lambda\rangle$ 5., Substitution

On the other hand, suppose λ is untrue. Then we can argue that it is false by running the above argument backwards (each step is valid in the other direction as well). Thus, when it comes to λ, we lose the distinction between (Deny-F) and (Deny-NT). This is to (Deny-NT)'s detriment; Priest and his dual should have the same complaints about (Deny-NT) applied to λ as the Mad Hatter and Cheshire Cat had about (Deny-F) in general: it is unacceptably weak.

10.2.1.2 PRIEST AND (DENY-NT)

This is enough to indicate the problems that the paracompletist and dialetheist should have with (Deny-F) and (Deny-NT). Neither can work as a parallel to (Assert-T). However, there is a potential objection to this conclusion, suggested by remarks in (Priest 2006a, section 6.5). (I owe this observation to Priest (pc).)

Priest considers endorsing something like (Deny-NT) as a principle governing rejection. His principle: 'One ought to reject something if there is good evidence for its untruth.' Ignoring the difference between rejection and denial, this is an evidentially flavored version of (Deny-NT). Rather than consider this principle directly, I'll consider how Priest's defense of it would apply to (Deny-NT). (This affects nothing of substance.)

Recall that according to Priest the strengthened liar λ is both true and untrue. As such, (Deny-NT) comes into conflict with (Assert-T) in the case of λ; according to these principles we should both assert it and deny it. Priest's suggestion is that this may simply be a rational dilemma; that in virtue of (Deny-NT) and (Assert-T) both holding, we are under obligations that we cannot fulfill.

From one point of view, this is very close to the approach I'll recommend in section 10.4.2. However, it is more radical than Priest acknowledges. For example, consider a discussion between Priest and the (paracompletist) Cheshire Cat about

[8] I assume that $T\langle\rangle$ validates two inferences: Capture (from A to $T\langle A\rangle$) and Release (from $T\langle A\rangle$ to A). This is weaker than the assumption of intersubstitutivity, and is acceptable to Priest and his dual.

the strengthened liar. The Cheshire Cat denies it, while Priest asserts it. But Priest can have no leverage with which to criticize the Cheshire Cat; by (Deny-NT) the Cheshire Cat is behaving as it ought to. Of course, the Cheshire Cat is also behaving as it oughtn't (by (Assert-T)); but so too is Priest, by (Deny-NT). On the rational-dilemma approach, Priest has no firmer grounds for criticism of the Cat than he has for self-criticism.

Thus, (Deny-F) and (Deny-NT) should leave the orthodox dialetheist and paracompletist alike unsatisfied as parallels to (Assert-T). But (Assert-T) is not the only norm of assertion on offer. Perhaps we can do better finding a parallel to (Assert-K)?

10.2.1.3 (ASSERT-K)

Here are two that won't work:

(Deny-KF) Deny A only if you know A is false

(Deny-KNT) Deny A only if you know A is not true

These fail for the very same reasons as (Deny-F) and (Deny-NT), respectively.

There is another strategy we have available here, though. Parallel to the distinction between assertion and denial (speech acts), dialetheists and paracompletists often draw a distinction between acceptance and rejection (attitudes). By asserting, one indicates acceptance, and by denying, one indicates rejection. Rejection, they say, cannot be understood as acceptance of negation, for the same reasons that denial can't be understood as assertion of negation.

This suggests a more general strategy of bifurcation. To construct a norm on denial parallel to (Assert-K), maybe we shouldn't try to build up anything workable out of knowledge and negation. Instead, maybe we should postulate a new attitude, *knowledge*$_D$, parallel to knowledge in the same way rejection is parallel to acceptance, and denial to assertion. You might worry that this bifurcation strategy is a bit clumsy. But it turns out to face more trouble than that.

If we understand knowledge as justified true belief, we should probably understand knowledge$_D$ as justified D-true rejection (where A is D-true iff $T\langle DA\rangle$). This, of course, requires us to have the paradoxical D operator in our language, and so is not acceptable. Fortunately, there's no need here to accept the JTB account of knowledge; we can simply take both knowledge and knowledge$_D$ as primitives. Then the proposed norm on denial is:

(Deny-K$_D$) Deny A only if you know$_D$ A

Unfortunately, knowledge$_D$, even without the D operator on its own, is problematically paradoxical. Consider the know$_D$er paradox—$\kappa_D = \langle\kappa_D\rangle$ is know$_D$n.

Suppose someone know$_D$s κ_D. Then κ_D is true; but true things can't be know$_D$n.[9] We can't assert our supposition, so, by D-exhaustivity we can deny it; in fact, we now know$_D$ our supposition. But our supposition just was κ_D, so we know$_D$ κ_D. We must both assert and deny our supposition, and this is no good, by D-exclusivity.

So while (Deny-K$_D$) forms a fine parallel to (Assert-K), it mires us in paradox. It does not seem that either (Assert-T) or (Assert-K) has a parallel that can apply to denial without trouble: the proposed parallels either fail to be properly parallel, are unstatable without D, or mire us in D-style paradox on their own.

10.2.1.4 FINDING THE RIGHT NORMS

Now, suppose we have our denial-embedding content operation D around. Then we can formulate two new principles:

(Deny-D) Deny A only if DA is true

(Deny-KD) Deny A only if you know DA

These are not only parallel to (Assert-T) and (Assert-K); they are instances! Remember, denying A is equivalent to asserting DA. So (Deny-D) amounts to: Assert DA only if DA is true, an instance of (Assert-T). Similarly, (Deny-KD) amounts to: Assert DA only if you know DA, an instance of (Assert-K). So the presence of D allows us to state norms on denial in a simple and straightforward way. These norms are, as we want, parallel to our norms on assertion.

Of course, just as with knowledge$_D$, D causes problematic paradox. But it allows us to give a straightforward theory of norms governing denial, which we do not seem to be able to do without paradox. What's more, it allows us to avoid the clumsy bifurcation strategy in general. Knowledge$_D$ is just ordinary knowledge of a D-content; rejection is just acceptance of a D-content; etc.

10.2.2 Agreement and generalizations

One use we have for the truth predicate is to make certain generalizations:

(3) Everything Alice says is true

(4) If everything Alice says is true, I'll eat my hat

The motivations have been stressed by Field (2008) and Beall (2009), who use them to argue for an intersubstitutable truth predicate. However, although their

[9] On a dialetheist line, one can know the negation of a true thing (if that negation is also true), but knowledge$_D$ and truth remain incompatible—that's the point of distinguishing knowledge$_D$ from knowledge of negation.

theories provide fully intersubstitutable truth predicates, those predicates don't quite work to generalize in the way they seem to hope, because of denial.

If Alice only asserts, and never denies, then (3) and (4) serve their purpose well. But suppose that Alice has asserted some things and denied some others, and that Humpty Dumpty is in full agreement with Alice. That is, Humpty Dumpty is willing to assert everything Alice asserted, and willing to deny everything Alice denied. How can he indicate this?[10] He might try something like (5):

(5) Everything Alice said is true

There are two readings we can give to (5), depending on how we interpret 'says'. Unfortunately, neither reading gives us what we want; on neither reading does (5) express agreement with Alice.

Suppose that Alice said something iff she either asserted it or denied it. Then (5) clearly does not convey agreement; if Humpty Dumpty thinks that Alice denied some true things, then he *disagrees* with her. So suppose Alice said something iff she asserted it; she does not count as having said the things she denied. Then (5) still doesn't convey agreement; Humpty Dumpty might agree with all of Alice's assertions, but think that some of her denials were mistaken, and still be willing to assert (5).

We might try some simple modifications:

(6) Everything Alice asserted is true, and everything she denied is false

(7) Everything Alice asserted is true, and everything she denied is not true

But these modifications get us nowhere. For the same reasons that 'false' and 'not true' would not serve to solve the initial disagreement problem—remember, only denial would do—they cannot solve the agreement problem.

Suppose Alice denies q. If the Cheshire Cat is a paracompletist who thinks that $q \vee \neg q$ fails, then he agrees with her. But he will not assert either (6) or (7). If the Mad Hatter is a dialetheist who thinks that $q \wedge \neg q$ holds, then he disagrees with her. But he will still be willing to assert both (6) and (7).[11] Thus, there seems to be no way to express agreement with those who deny things as well as asserting.

Now, suppose we have D in the language, and consider (8):

(8) Everything Alice asserted is true, and everything she denied is D-true

[10] He can say 'I agree totally with Alice', but this will not behave properly in embedded contexts; try it in (4) to see the trouble.

[11] As we saw in section 10.2.1, the difference between falsity and not-truth depends on a non-intersubstitutable truth predicate. If there is such a difference, there might be some cases where (7) could work; but it still cannot work in full generality. For example, if q in the above example is the strengthened liar, it cannot work.

(For A to be D-true, recall, is for DA to be true; D-truth is thus to D as falsity is to $\neg$.) If truth is intersubstitutable, this expresses agreement with Alice; to assert (8) is to commit to asserting everything Alice asserted and denying everything she denied. (If truth is not intersubstitutable, this might not quite be the case, but the desire to use truth to express agreement motivates an intersubstitutable conception; it shouldn't surprise us that that remains the case here. And, as above, we can add an intersubstitutable predicate for these purposes, even if we don't think that predicate is a *truth* predicate.)

10.2.3 Questions of priority

(9) and (10) are truisms about conjunction:

(9) If you're committed to asserting both conjuncts, you're committed to asserting their conjunction

(10) If both conjuncts are true, their conjunction is true

One might explain (9) in terms of (10), explain (10) in terms of (9), or take both to stand on their own. Whatever route you're tempted by, there's clearly at least a question to be answered here. Are constraints on assertion like (9) prior to, posterior to, or unrelated to, constraints on truth like (10)?

Since this makes sense as a question about assertion, it ought to make sense as a question about denial. There's trouble, though. (9) has a parallel involving denial:

(11) If you're committed to denying at least one conjunct, you're committed to denying the conjunction

Unfortunately, there is no corresponding parallel to (10). (Again, we can try to build such a parallel using falsity or not-truth, and again, such efforts will fail, for the same reasons as above.) So what is a perfectly sensible question about assertion cannot be stated as a question about denial.

That is, of course, unless we have D around. Then, parallel to (10), we get (12):

(12) If one conjunct is D-true, the conjunction is D-true

We can sensibly ask the same questions about the relation between (11) and (12) as we asked about (9) and (10). Once again, it is only in the presence of D that assertion and denial are really parallel.

We've thus seen that three theories of assertion—to do with norms, interaction with truth and agreement, and questions of priority—*have no parallel* relating to denial, unless D is in our language. For all I've said here, we might give a theory of denial in a different way, a way that doesn't hold it to parallel our theories of assertion. But that is not how denial is understood in Priest (2006a), Field (2008), or Beall (2009). These authors take denial to be parallel to assertion; and this simply cannot be so unless D is around.

10.2.4 *Stalnakerian denial*

There's a quite natural way to understand denial on a Stalnakerian picture of pragmatics. But once denial is understood that way, we can simply *define* D. Here, I explain.

In the framework of Stalnaker (1978), each stage of a conversation is associated with a set of possible worlds—the *context set* for that stage. Roughly, we can understand the context set as the set of possible worlds that are still live possibilities for the conversation. (Note that even if some (or even all) of the participants in a conversation don't personally consider a possibility live, it can still be live for the conversation, and vice versa.) The context set at any given stage constrains which conversational moves are acceptable, and how they will be interpreted. In turn, conversational moves change the context set in certain predictable ways.

In particular, assertion works by ruling out certain possibilities from the context set. When someone asserts A, the context set shrinks; any world at which A fails to hold is removed.[12] If there were no worlds at which A fails to hold left in the context set, then the assertion is infelicitous. This may trigger reinterpretation or censure of various sorts.

This can straightforwardly be extended to denial: when someone denies A (and the denial goes unchallenged), the context set shrinks; any world at which A holds is removed. This not only gives us a clear picture of denial as parallel to assertion, but it also allows us to understand what it is that assertion and denial have in common as opposed to, say, questioning. Both assertion and denial shrink the context set; this is how they are *informative*. They help us narrow down possibilities for how things might be.

The Stalnakerian picture is attractive on its own, and its extension to denial has some nice features. But it leads directly to embeddable denial. Call a move in a conversation *appropriate* iff it does not rule the world in which it is made out of the context set. Then we can build an embeddable denial: just let DA be: 'It would now be appropriate to deny A'. Asserting DA would thus rule out a world w iff a denial of A fails to be appropriate in w—that is, iff a denial of A would rule out w. Thus, asserting DA amounts to denying A; this was our condition on embedding denial.

[12] At least if the assertion goes unchallenged. We can ignore this caveat here, but it's worth noting that some authors (e.g. Geurts 1998) use 'denial' to pick out these assertion-challenges. That's, of course, not the kind of denial I'm focusing on.

10.2.5 Reason for hope

It does not seem, then, that we can have a theory of denial parallel to our theory of assertion unless we include the D operator in our language. And, as we've seen, D, in the presence of D-exhaustivity and D-exclusivity, causes problematic paradox. But all is not lost—we have reason to question D-exclusivity and D-exhaustivity (or at least their conjunction) anyway. This reason is provided in Restall (2013): we have problematic paradox from these two principles alone, even without D!

Restall presents constraints on assertion and denial in a sequent calculus with a particular interpretation (motivated in Restall (2005)). Where Γ and Δ are sets of sentences, Restall reads $\Gamma \vdash \Delta$ as: it is incoherent to assert everything in Γ and deny everything in Δ.

Given this reading of $\vdash$, D-exclusivity is easy to express:

$$\text{(Id)} \quad \Gamma, A \vdash A, \Delta$$

That is, no matter what else we assert (Γ) or deny (Δ), it's incoherent to both assert and deny A. D-exhaustivity is a bit trickier; however, it justifies:

$$\text{(Cut)} \quad \frac{\Gamma, A \vdash \Delta \quad \Gamma \vdash A, \Delta}{\Gamma \vdash \Delta}$$

That is, if asserting A is incoherent given your other commitments, and denying A is incoherent given your other commitments, then your other commitments are already incoherent on their own.

From (Id), (Cut), and a number of principles of naive set theory (or naive truth), Restall shows how to derive $\vdash p$. That is, for any p, it's incoherent to deny p. We could question the principles of naive set theory or naive truth, of course, but then we wouldn't be playing the dialetheist/paracompletist game we set out to play. So it looks like either D-exclusivity or D-exhaustivity has to go, even with no D in the language.

This is *good* news. After all, it was only in the presence of D-exclusivity and D-exhaustivity that D caused any trouble. Perhaps, then, we can add D to our language while avoiding the trouble, if we figure out just how to weaken these principles.

10.3 Negation and Denial

10.3.1 $\neg$ and D side by side

In this section, I briefly consider ways to add a unary operator D to the language to express denial. So far, I've considered both dialetheist and paracomplete theories of negation. For concreteness here, I'll assume negation works as dialetheists suppose (that is, that it doesn't satisfy $A \wedge \neg A \vdash \bot$). If we were to instead look at paracomplete theories of negation (where $\top \vdash A \vee \neg A$ fails), this would all play the same, *mutatis mutandis*.

We can't have both (D-EFQ) $A \wedge DA \vdash \bot$ and (D-LEM) $\top \vdash A \vee DA$. Remember the denier—$\delta = DT\langle\delta\rangle$, where $T\langle\rangle$ is intersubstitutable, and consider the following argument:

1	$\top$	
2	$\delta \vee D\delta$	1, D-LEM
3	δ	
4	$T\langle\delta\rangle$	3, Capture
5	$DT\langle\delta\rangle$	3, Substitution
6	$T\langle\delta\rangle \wedge DT\langle\delta\rangle$	4, 5, $\wedge$-I
7	$\bot$	6, D-ECQ
8	$D\delta$	
9	$DT\langle\delta\rangle$	8, Intersubstitutivity
10	$DDT\langle\delta\rangle$	8, Substitution
11	$DT\langle\delta\rangle \wedge DDT\langle\delta\rangle$	9, 10, $\wedge$-I
12	$\bot$	11, D-ECQ
13	$\bot$	3–7, 8–12, $\vee$-E

10.3.2 Gappy D

Here, I consider relaxing D-LEM, while retaining D-ECQ. Can this get us a denial fit to express disagreement? We've got half of what we want: if I assert DA, I'd better disagree with A (on pain of explosion). Unfortunately, there will be sentences I disagree with, but will not assert the D-sentence of.[13]

For example, consider the denier δ. As the above argument shows, $\delta \vee D\delta$ must fail, and so I should disagree with anyone who asserts δ, and disagree with anyone who asserts $D\delta$. But how? If I disagree with a δ-asserter by asserting $D\delta$, I'm in

[13] The situation for D here is exactly the situation for the so-called 'arrow-falsum' connective used by Priest and Beall to force denial in some cases.

trouble. I want to disagree with the $D\delta$-asserter too, but now I *am* a $D\delta$-asserter. So something's gone wrong.

Negation is no more help here than it ever was. I might assert $\neg\delta$ and $\neg D\delta$ (in fact, I have to, if $\top \vdash A \vee \neg A$ holds), but this isn't enough to express disagreement—I assert $\neg\lambda$ and $\neg D\lambda$ as well, but I don't disagree with a λ-asserter. (Remember, I'm here assuming the dialetheist is right about negation, and in particular about the liar λ.)

If I need to invoke a new speech act, say of *shmenial*, to express my disagreement with the denier, then something's gone very wrong. So it doesn't look like I have the resources to express disagreement with a gappy D.

10.3.3 Glutty D

On the other hand, if D is glutty (that is, if we relax D-ECQ instead of D-LEM), then it's not clear how D helps to express disagreement any more than we already could with negation alone. We should assert $D\delta$, but we don't disagree with δ. The situation is just the same as with negation and the liar.

We might try taking both $\neg$ and D to be glutty, require both LEM and D-LEM, and add a new requirement: $A \wedge \neg A \wedge DA \vdash \bot$. In this system, negation can't express disagreement on its own, and neither can denial. But together, they suffice to force disagreement. Is this any better?

Actually, it's worse; the system is trivial. Let $\iota = \neg\iota \wedge D\iota$. (The proof relies on distribution to show $\top \vdash A \vee (\neg A \wedge DA)$. From there, it's familiar.)

10.4 Making Do

To sum up so far: classicalists can use negation to express disagreement, but para-completists and dialetheists have trouble following suit. To express disagreement, they appeal to denial as a separate sort of speech act, one that satisfies D-exclusivity and D-exhaustivity, and so can be used to express disagreement where negation fails to work. The arguments in section 10.2 have tried to show that giving a complete theory of denial requires us to have some connective D in the language that embeds denial, in the sense that asserting DA is equivalent to denying A, for every A. I've argued that one of D-exclusivity and D-exhaustivity must fail, if we are to have D without triviality. In section 10.3, we saw that, without D-exclusivity and D-exhaustivity, this D is going to have the same trouble expressing disagreement that negation originally had. We seem to have gotten nowhere.[14]

[14] To summarize the summary: we want a connective D to embed denial, but any such connective will fail to express disagreement for the same reasons negation failed originally.

This suggests that the strategy of appealing to denial as a separate speech act, and distinguishing D from negation, was pointless in the first place. The para-completist and dialetheist alike should accept, with the classicalist, that negation embeds denial,[15] and that asserting the negation of A is equivalent to denying A. There is thus no need to distinguish denial from assertion of negation, and our theory of denial can become part of our theory of assertion. (Alternately, we could maintain the distinction between denial and assertion of negation, and simply use negation as section 10.2 suggests we use D to construct theories of denial from our theories of assertion.)

We have one remaining choice to make: we can continue to accept the arguments that negation can't express disagreement, and so accept that denial doesn't express disagreement, or we can find a flaw in those original arguments, and keep negation, denial, and disagreement all tied together. I'll explore one variety of each option in turn, although there may well be others.

10.4.1 Exclusion

Suppose we accept the arguments that negation can't express disagreement. If we suppose that negation embeds denial, then denial doesn't express disagreement either. How can we understand disagreement on such a view?

One possibility is to invoke a binary relation on the propositions themselves, and say that two people disagree when there is a pair x, y of propositions such that one asserts x, the other asserts y, and x and y are related. Following Marques (2014), I'll call this relation *exclusion*. For our purposes here, we can take exclusion to be primitive: some things simply exclude other things. For example, we can assume that that the Cheshire Cat is on fire excludes that the Cheshire Cat is in a lake. Then if Alice asserts that the Cheshire Cat is on fire, Humpty Dumpty can indicate his disagreement by asserting that the Cheshire Cat is in a lake. No appeal to denial or negation is necessary to disagree on this account.[16]

Something like this might work, but there is familiar trouble lurking not far away. Consider the *excluder*: η = 'η is excluded by some true content'. Well, if η

[15] That negation embeds denial is argued for, on very different grounds, in Price (1990).

[16] Marques's discussion takes careful note of possible shifts in context and circumstance of utterance, which I am ignoring here. In some ways, this notion of exclusion is quite like the relation of *incompatibility* drawn on in discussions of negation by Dunn (1993); Restall (1999). But that relation isn't between contents or propositions. The discussion of negation in Brady (2006, 20–1) is also related, as is Millikan (1984, 224–9).

All of these discussions tie exclusion and its relatives very closely to negation, which is no good for the purposes I'm exploring here. I'm assuming that the tie can be severed without too much loss. But since I'm about to argue that this strategy won't work anyway, that's not really a worrisome assumption.

is true, then it's excluded by some true content. But it at least seems that that last sentence expresses a true content that excludes η. So η is excluded by some true content, and so it's true. We should assert η and a content that excludes η. So it looks like exclusion isn't sufficient for disagreement.

Perhaps there's a way to give an account of exclusion, or something like it, that gets around this. But I think there is a more natural account available anyway, and it's to that that I now turn.

10.4.2 Paracoherentism

The more natural account takes negation, denial, and disagreement all to be tied together, just as the classicalist thinks. That is, on this account, negation embeds denial, and denial expresses disagreement. Thus, the dialetheist, in asserting $\lambda \wedge \neg\lambda$, disagrees with herself. This amounts to a different response to the original problem of disagreement. On this approach, the dialetheist picture is *incoherent*: it asserts and denies the same thing, and it takes assertion and denial to express disagreement. The question for the dialetheist thus becomes: how bad is incoherence?[17]

If incoherence is to be anything less than crippling, it had better be possible to be incoherent in a limited way. That is, we must be able to have an incoherent take on a content A without necessarily having an incoherent take on any old content B. If we replace 'incoherent' here with 'inconsistent', though, this is a familiar problem, and it is to be solved by adopting an appropriate paraconsistent logic, in which one can be inconsistent about A without having to be inconsistent about B. In the present setting inconsistency amounts to incoherence, and thus a paraconsistent approach amounts to a *paracoherent* approach, in which one can be locally incoherent without global incoherence. Since we know there are appropriate paraconsistent logics, there should be no trouble here. Let's christen this sort of view—dialetheism plus the view that negation embeds denial and denial expresses disagreement—*paracoherentism.*

Taking the paracoherentist option I'm suggesting here amounts to choosing to maintain a certain amount of *reflective tension.* This is not an equilibrium position; someone who adopts it disagrees with themselves, after all. On this view, that is the only way to believe truly. Although the truth cannot be coherently stated or believed, it still can be both stated and believed. It simply requires cultivating the right sort of non-equilibrium state.

[17] I focus on dialetheism in this section; as usual, the paracomplete approach plays largely the same. For example, on this picture, the paracompletist, in refusing to assert $\lambda \vee \neg\lambda$, disagrees with nobody, not even the dialetheist. The parallel question is: how bad is it to fail to disagree with one's opponents? How close is it to not having a view at all?

10.4.2.1 ADVANTAGES TO PARACOHERENTISM

Paracoherentism has several nice features. For one thing, it allows us to answer the demands of section 10.2 by providing an operation on content that embeds denial. In particular, we can 1) state norms on denial parallel to our norms on assertion—(Deny-F) and (Deny-KF) will do as parallels to (Assert-T) and (Assert-K), respectively; 2) express agreement using truth; and 3) explore the relation between denial and falsity as parallel to the relation between assertion and truth. We can define D as suggested from the Stalnakerian framework, and we see that D just is $\neg$; our theory already includes it.

What's more, paracoherentism allows us to use denial to express disagreement; we don't need a separate theory of disagreement, as we would on the approach explored in §10.4.1. And we solve the initial disagreement puzzle.

10.4.2.2 CHALLENGES FOR PARACOHERENTISM

Many of the natural objections to paracoherentism have natural analogs as objections to dialetheism. For example, Slater (1995) objects to dialetheism on the grounds that if $A \wedge \neg A$ can be true, $\neg$ must not be a real negation. One could similarly object to paracoherentism by claiming that if one can rationally disagree with oneself, 'disagree' must not pick out real disagreement. Priest (2006a) responds to Slater by pointing out how many of negation's features his $\neg$ has. Not least among these is the preservation of $\neg(A \wedge \neg A)$ as a theorem-scheme. Similarly, we can respond to the corresponding objection by pointing out how many of disagreement's features disagreement retains on this theory. Not least among these is Dis-exclusivity:

Dis-exclusivity: Agreement and disagreement are *incompatible* states

We can (and should) hold to this. Agreement and disagreement really are incompatible; it's incoherent to do both.

Now, it's possible to find this reply unconvincing. But then, I suggest, one ought to find Priest's original reply to Slater unconvincing as well. *For the dialetheist*, paracoherentism is no extra cost on these grounds. The argument purporting to show that incoherence is bad mirrors the argument purporting to show that inconsistency is bad. If one is unconvinced by the latter, one ought to be unconvinced by the former as well.

A novel problem for paracoherentism (that is, a challenge faced by paracoherentists but not by dialetheists) is in giving an account of logical consequence. For example, suppose we adopt the suggestion in Restall (2005) of taking $\Gamma \vdash \Delta$ to indicate that it's incoherent to accept all of Γ and reject all of Δ. Then, on the present view, $A \wedge \neg A \vdash B$; since it's incoherent to accept $A \wedge \neg A$, it's incoherent to accept $A \wedge \neg A$ and reject B. Thus, a paracoherentist view can't be closed under $\vdash$, on this understanding of $\vdash$; this is not a paracoherent relation.

Alternately, we might follow Beall and Restall (2006) in taking $\Gamma \vdash \Delta$ to indicate that there is no case in which everything in Γ is true and everything in Δ is not true. This would require us to give some theory of truth-in-a-case. In particular, we would have to give a theory about when a negation is true-in-a-case. And we should be careful. If there is no case where $A \wedge \neg A$ holds, then $A \wedge \neg A \vdash B$ will hold. A paracoherentist view can't be closed under this reading of $\vdash$ either. This is so even if there are some cases where $A \wedge \neg A$ holds and B fails to hold; in this case, $A \wedge \neg A \vdash B$ would fail as well. Priest (2006a) suggests something like this, but phrases the definition slightly differently (using a restricted universal quantification: every case at which everything in Γ is true is also such that something in Δ is true). He does not get the bad result, but it is avoided only by appealing to a non-classical metalanguage that, as yet, awaits full development. The paracoherentist might pursue this general approach, however; she just needs to be a bit careful.

A final possibility for understanding consequence follows Brady (2006) in taking $\Gamma \vdash \Delta$ to express a *containment* between the contents of Γ and Δ. Brady takes this quite literally, using sets of sentences as contents and taking containment to be ordinary set-theoretic containment. On his view, $\Gamma \vdash \Delta$ whenever the union of the contents of the Γs, closed in a certain way, contains the intersection of the contents of the Δs. As far as I yet see, this route holds no pitfalls for the paracoherentist, but further exploration will have to wait for another day.

10.5 Conclusion

The initial disagreement problem was supposed to show that negation can't express disagreement for a dialetheist or paracompletist. By and large, dialetheists and paracompletists have accepted this argument and appealed to a speech act of denial, separate from assertion and negation, to express disagreement. They've taken denial to be parallel to assertion. However, when we follow that through, we see that there is real trouble in taking the parallel seriously unless there is an operation D on content such that an assertion of DA is equivalent to a denial of A. What's more, some ways of understanding denial give us the resources to define such a D. This seems problematic, since we can form a new paradox (the denier) with D that dialetheism and paracompletism alone don't address.

In trying to address this paradox, we've seen that we end up saying the same things about D that the dialetheist and paracompletist already said about negation.

Thus, we end up facing the same puzzle about using D to express disagreement as they already faced about using negation to express disagreement. This suggests that distinguishing $\neg$ from D in the first place was a mistake. Negation embeds denial. This, of course, leaves us with our initial puzzle about disagreement intact.

We can try to solve the puzzle by appealing to something other than negation/denial to express disagreement, or we can attempt to keep the three closely linked. I recommend the latter course. The dialetheist approach, seen in this light, is incoherent, but only locally so. It is possible to be sensibly incoherent. The approach has several advantages over orthodox dialetheism, and seems to face few new troubles not also faced by dialetheism. One new trouble—over understanding logical consequence—may well be solvable.[18]

References

Beall, Jc. 2009. *Spandrels of Truth*. Oxford: Oxford University Press.

Beall, Jc and Restall, Greg. 2006. *Logical Pluralism*. Oxford: Oxford University Press.

Brady, Ross. 2006. *Universal Logic*. Stanford, CA: CSLI Publications.

Dunn, J. Michael. 1993. "Star and perp: two treatments of negation." *Philosophical Perspectives* 7:331–57.

Field, Hartry. 2008. *Saving Truth from Paradox*. Oxford: Oxford University Press.

Geurts, Bart. 1998. "The mechanisms of denial." *Language* 74:274–307.

Horn, Laurence R. 1985. "Metalinguistic negation and pragmatic ambiguity." *Language* 61:121–74.

Horn, Laurence R. 1989. *A Natural History of Negation*. Chicago, IL: University of Chicago Press.

Marques, Teresa. 2014. "Doxastic disagreement." *Erkenntnis* 79:121–42.

Millikan, Ruth Garrett. 1984. *Language, Thought, and Other Biological Categories*. Cambridge, MA: MIT Press.

Parsons, Terence. 1984. "Assertion, denial, and the liar paradox." *Journal of Philosophical Logic* 13:137–52.

Price, Huw. 1990. "Why 'not'?" *Mind* 99:221–38.

Priest, Graham. 2006a. *Doubt Truth to be a Liar*. Oxford: Oxford University Press.

Priest, Graham. 2006b. *In Contradiction*. Oxford: Oxford University Press, 2nd edition.

[18] For discussion regarding earlier versions of this chapter, thanks to audiences at the University of Melbourne, the Arché Logic of Denial workshop, the Propositional Content and Proposition-Related Acts workshop, and PALMYR IX: Logic and the Use of Language. Comments from Aaron Cotnoir, Catarina Dutilh Novaes, Graham Priest, Greg Restall, and Stewart Shapiro proved especially helpful at these events. This research was partially supported by the Agence Nationale de la Recherche, program 'Cognitive Origins of Vagueness', grant ANR-07-JCJC-0070, by the project 'Borderlineness and Tolerance' (Ministerio de Ciencia e Innovación, Government of Spain, FFI2010-16984), and by the Australian Research Council Discovery Project 'Paraconsistent Foundations of Mathematics'; I'm grateful to all of them.

Restall, Greg. 1999. "Negation in relevant logics: how I stopped worrying and learned to love the Routley star." In D. Gabbay and H. Wansing (eds.), *What is Negation?* 53–76. Dordrecht: Kluwer.

Restall, Greg. 2005. "Multiple conclusions." In P. Hajek, L. Valdez-Villanueva, and D. Westerståhl (eds.), *Proceedings of the Twelfth International Congress on Logic, Methodology and Philosophy of Science*, 189–205. London: King's College Publications.

Restall, Greg. 2013. "Assertion, denial, and non-classical theories." In Koji Tanaka, Francesco Berto, Edwin Mares, and Francesco Paoli (eds.), *Paraconsistency: Logic and Applications*, 81–100. Dordrecht: Springer.

Slater, Hartley. 1995. "Paraconsistent logics?" *Journal of Philosophical Logic* 24:451–4.

Stalnaker, Robert. 1978. "Assertion." *Syntax and Semantics* 9:315–32.

11

Assertion, Denial, Accepting, Rejecting, Symmetry, and Paradox

Greg Restall

11.1 Introduction

Here are my assumptions about assertion, denial, accepting, rejecting, consequence, and negation. I will not take the time to defend these assumptions here: I discuss them elsewhere (Restall 2005).

Assertion and denial are speech acts, related to the cognitive states of accepting and rejecting. Assertion typically expresses acceptance, and denial typically expresses rejection. If I assert that the cat is on the mat, then this typically expresses my accepting that the cat is on the mat. If I deny that the cat is on the mat is the case, then this typically expresses my rejecting that the cat is on the mat. These speech acts and related cognitive states are equally rationally constrained by a relation of logical consequence. If the claim that the cat is on the mat (abbreviate this claim as "A") entails the claim that some feline is on the mat (abbreviate this claim as "B", so the claim of entailment is written "$A \vdash B$") then accepting that the cat is on the mat and rejecting that some feline is on the mat is logically incoherent. Similarly, asserting that the cat is on the mat and denying that some feline is on the mat is likewise logically incoherent. We do not need to make any assumption about the relationship between these properties of speech acts and cognitive states: perhaps the coherence or otherwise of collections of speech acts arises out of the coherence or otherwise of the cognitive states they express. Perhaps, on the other hand, the coherence or otherwise of the cognitive states depends on the commitments incurred by engaging in the speech acts which express them. Perhaps instead, the notion of coherence, between claims, states, and perhaps other items too, is fundamental and it is only through identifying relations of coherence or incoherence that we identify what is said or what is believed. Such niceties are important for the general project of understanding logic, language, belief, and meaning, but we do not need to settle those issues here.

More important for us are the properties of coherence already inherent in any picture like this. The standard account of logical consequence takes the relation $\vdash$ to be reflexive and transitive. This tells us immediately that it is never coherent to simultaneously assert and deny the same claim A; that it is never coherent to both accept and reject A. This is the reflexivity of consequence: $A \vdash A$. Similarly, the transitivity of consequence tells us that if $A \vdash B$ and $B \vdash C$ then $A \vdash C$. Contraposing, it follows that if A / ∈ (that is, if it is coherent to assert A and deny C) then it is coherent to assert A and deny B, or coherent to assert B and deny C. This means, after a little fiddling, that if it is coherent to assert A and deny C then either it is also coherent to assert B, along with asserting A and denying C, or it is also coherent to deny B, along with asserting A and denying C.[1] The arbitrary B is either coherently assertible, though not necessarily assertible with any *warrant* or *evidence*, or it is coherently deniable. There will never be a situation which, on the basis of logic alone, rules out asserting B and also rules out denying it. One way to see this is that if logic alone rules out asserting B, then B has been ruled out, and we have every reason to deny it. This gives us very mild versions of the law of "non-contradiction"—it is incoherent both to assert and to deny A, or $A \vdash A$—and of the law of "excluded middle"—it is coherent either to assert A or to deny it. I say that these are very mild versions of these laws, because they are acceptable to the gap and glut theorists, who reject the law of the excluded middle and the principle of non-contradiction respectively. Some gap theorists take $A \vee \neg A$ to fail for some instances of A, and hence, allow for the possibility that both A and $\neg A$ be rejected. That is, the friend of gaps sometimes rejects A without thereby accepting $\neg A$. Symmetrically, the friend of gluts takes $A \wedge \neg A$ to succeed for some instances of A, and hence, allows for the possibility that both A and $\neg A$ be accepted. That is, the friend of gluts sometimes accepts $\neg A$ without thereby rejecting A.

Noticing this symmetry between gap and glut treatments and their connection to assertion, denial, and negation is not original with me (Parsons 1984; 1990). The original contribution in this chapter is the treatment of this phenomenon in the context of multiple premise, multiple conclusion consequence, and it's to this that we turn.

The vocabulary of assertion and denial gives us a way to explain consequence not only as applying to arguments with a sole premise or a sole conclusion, but as relating premise*s* to conclusion*s*. If $A \vdash B$ tells us that asserting A and denying

[1] The fiddling involves using the "weakening" structural rule, and transitivity in its full generality. If $X \vdash B, Y$ and $X, B \vdash Y$ then $X \vdash Y$: the so-called "*cut* rule."

B is to make an error of logic, then if X and Y are collections of statements, we can think of $X \vdash Y$ as telling us that asserting each element of X and denying each element of Y is to make an error of logic.

This generalization to multiple premise and multiple conclusion logic is not startling or original. It has been known to proof theorists at least since Gentzen's groundbreaking work in the 1930s (Gentzen 1935). The philosophical significance of multiple conclusion consequence is, however, much disputed (Shoesmith and Smiley 1978). Instead of developing the argument here, I will take it as read that multiple conclusion consequence makes sense and leave the defence of the notion to another place (Restall 2005). One way to make sense of the notion of multiple conclusion consequence is to work with it and consider its properties, and to this we will briefly turn.

Allowing sets of statements as premises and conclusions opens the way for the *empty* set of premises. To say that $\{\} \vdash Y$ (which we write simply as '$\vdash Y$') means that the combination of asserting every member of the empty set while denying every member of Y is ruled out. The empty set has no members, so we need do nothing to assert every member of the empty set. So this means simply that denying every member of Y is ruled out. In the case where Y is a single element set, $\{A\}$, then $\vdash \{A\}$ (which we write as '$\vdash A$') tells us that A is undeniable—it is true on the basis of logic alone, a tautology. This, of course, trades on the move from undeniability to truth. It is important to recall here that we are not talking of what evidence or warrant may be available for a claim, but simply what position (*pro* or *con*) may be coherently taken up concerning it. If we wish to come to a position concerning A—if the question of whether or not A has arisen, and we wish to resolve it—then if A is undeniable, the only resolution possible is to assert it.

Similarly, we may replace "assert" by "deny" and consider an empty conclusion set of the sequent $X \vdash Y$. In that case we have $X \vdash$, which tells us that asserting every member of X while denying every member of the empty set is ruled out. Just as with asserting, denying every member of the empty set is trivial, as there's nothing to do: we never fail to deny each member of the empty set. So, $X \vdash$ if and only if asserting each member of X is ruled out. Logic tells us that at least one member of X is false. In the case where X is a single element set $\{A\}$, then $A \vdash$ tells us that A is false on the basis of logic alone: it is inconsistent. An empty conclusion set gives us a guide to what to *deny* on the basis of logic alone, given that we wish to resolve the issue of whether to assert or to deny A.

Finally, consider the transitivity of consequence. We would like to generalize the straightforward claim that if $A \vdash B$ and $B \vdash C$ then $A \vdash C$. In the case of more than one premise, we have the following deduction:

$$\frac{X \vdash B \quad X', B \vdash C}{X, X' \vdash C}$$

since the extra premises X and X′ simply pile up. If X suffices to deduce B, and X′ and B suffice for C, then simply plug X in for B in the proof of C. Similarly, if we have multiple conclusions (but single premises) we get

$$\frac{A \vdash B, Y' \quad B \vdash Y}{A \vdash Y, Y'}$$

as we plug in the proof from B to Y in the conclusion of the proof from A to either B or Y′, to conclude from A the multiple conclusions Y or Y′. Now, in the case of multiple premises and multiple conclusions, we combine the insights:

$$\frac{X \vdash B, Y' \quad X', B \vdash Y}{X, X' \vdash Y, Y'}$$

If we can infer from X to either B or Y′, then plug in the proof from X′ and B to Y here. To do this we gain some extra premises X′, and we leave the alternate conclusions Y′ to one side. If B applies, though, we can (with X′) deduce Y. So, if we have X and X′ we may deduce either Y (via X′ and B) or Y′ (via X), as desired. This is the *Cut* rule, and it will be important in what follows. Let's return, now, to the consideration of the paradoxes. Priest takes symmetry between gap and glut accounts to be broken by the presence of the strengthened liar paradox.

⟨SL⟩ This sentence is not true.

Priest reasons as follows: the gap theorist is committed to ⟨SL⟩ not being true, since it is a truth-value gap. But ⟨SL⟩ itself *says* that it's not true, so the gap theorist contradicts herself by both saying that ⟨SL⟩ is not true (by virtue of the analysis admitting that it has a truth-value gap) and thereby asserting ⟨SL⟩.

But we have already seen that the gap theorist need not follow Priest in this reasoning. The gap theorist can avail herself of the distinction between asserting a negation and denying. The gap theorist can deny that ⟨SL⟩ is true without thereby committing herself to the negation of ⟨SL⟩. The friend of gluts cannot rule out such a move by the defender of gaps, since friends of gluts *also* split assertion of negations and denials, albeit in the opposite direction. If the gap account involves *denying* ⟨SL⟩ without asserting its negation (or rejecting ⟨SL⟩ without accepting its negation) the strengthened liar paradox is circumvented. This circumvention comes at a price. We have a denial (the denial of ⟨SL⟩ itself) which cannot be expressed as an assertion of a negation. This denial must be expressed in some other way. Is this a *cost* for the friend of gaps? Perhaps it is: after all, it seems that we generally express denials when we assert negations, so breaking this nexus is a price to pay.

THE PRICE OF GAPS: It's not generally true that a denial of a statement is expressible by asserting the negation of that statement. We must deny ⟨SL⟩ without asserting its negation.

However, we shall see that it is exactly the same kind of price which must be borne by the friend of gluts in response to exactly the same phenomenon. As a result, this cost *cannot*, on its own, be seen as a reason for favouring gluts over gaps.

To see this, consider the glut theorist's account of ⟨SL⟩. The glut theorist's account is, on the face of it, straightforward. The sentence ⟨SL⟩ is true, and also, it is not true. According to the friend of gluts, the strengthened liar paradox is a canonical case of a true contradiction. However, it does not follow that in the case of ⟨SL⟩ (in the mouth of the friend of gluts, at least) the assertion "this sentence is not true" expresses a denial. The dialetheist does not both assert and deny ⟨SL⟩. He asserts ⟨SL⟩ and its negation. Just as we saw with the gap theorist, assertions of negations on the one hand, and denials on the other, diverge. Here, we have an assertion of a negation which cannot express a denial.

THE PRICE OF GLUTS: It's not generally true that a negation of a statement expresses the denial of that statement. We must assert the negation of ⟨SL⟩ without denying it.

Symmetry is preserved: the friend of gluts, just like the friend of gaps, pays a price. It follows that if the strengthened liar paradox is a problem for the *gap* theorist and not for the glut theorist, then some symmetry-breaking principle must be applied. The onus is on partisans—both partisans of gaps, and partisans of gluts—to elucidate such a principle. But what kind of principle could that be? Let's first consider some arguments due to Priest, in *In Contradiction* (Priest 2006, section 4.5–4.7), then an elaboration of the argument, and lastly an argument from a defender of gaps.

11.2 The Aim of Assertion

Priest seeks symmetry breaking principles in an account of the significance of assertion. He follows Dummett in taking assertion to aim at truth. An assertion succeeds when (and only when) it is true.[2]

[2] This is, no doubt, correct under *some* understanding of what it is for an assertion to succeed, but it is by no means an unproblematic principle.

> To speak truly is to succeed in a certain activity. And in the context of asserting, anything less than success is failure. There is no question of falling into some limbo between the two. (Priest 2006, section 4.7)

This is fine, as far as it goes, but it does not go far enough to establish Priest's desired conclusion (as Parsons 1990 notes). The friend of truth-value gaps can agree that anything less than success (in this case, truth) is failure. Of course, anything that is not true is not true (that is, it fails to be true). The friend of gaps does not deny *this*. Rather, the friend of gaps *denies* that the liar sentence is true, and she denies that it is not true. Denial outstrips assertion of negation. Some untruths are denied without the friend of gaps asserting their negations.

However, I want to do more than repeat Parsons' straightforward explanation of why Priest's argument fails. I wish to add a reason of my own. If Priest's argument against the gap theorist *works*, then so does a parallel argument against the glut theorist.

> To correctly deny is to succeed in a certain activity. And in the context of denial, only that which isn't a failure is success. There is no question of lying in some overlap between the two.

(It is a little hard to get the parallel right, because Priest conflates "speaking truly" with assertion.) Regardless, the argument goes as follows. The aim of denial is untruth. Anything not true is fit for denial, and *only* untruths are fit for denial. But pick a sentence, like ⟨SL⟩, that the glut theorist takes to be both true and untrue. If the aim of denial is untruth, then this is to be denied. But wait a minute! According to the friend of gluts, ⟨SL⟩ is true, and to be accepted. But then its denial *fails* because only untruths are to be denied, not truths. So, ⟨SL⟩ falls into the overlap between what is to be denied and what is to be asserted.

Priest would agree that this is a *bad* argument against dialetheism. Not all negations express denials for Priest. The assertion of the negation of ⟨SL⟩ is not (in the dialetheist's mouth) a denial. But how is it any worse than Priest's own argument against the gap theorist? Priest presupposes that the denial of ⟨SL⟩ in the gap theorist's mouth is a negation. (Or at least, he has to, if he wants to move from the "failure" of the assertion (the denial that it is true) to the truth of its negation.)

Priest's second argument on the nature of assertion is similar to the first. He approvingly quotes Dummett:

> A statement, so long as it is not ambiguous or vague, divides all states of affairs into just *two* classes. For a given state of affairs, either the statement is used in such a way that a man who asserted it but envisaged that state of affairs as a possibility would be held to have spoken misleadingly, or the assertion of the statement would not be taken

as expressing the speaker's exclusion of that possibility. If a state of affairs of the first kind obtains, the statement is false; if all actual states of affairs are of the second kind, it is true.

Parsons notes four different problems with this argument. (1) Not all friends of gaps accept Dummett's premise. (2) The last sentence (concerning falsity) does not follow from what is said previously, since what is said up to this point does not concern falsity. (3) Priest's sense of "false" and Dummett's differ. (4) The desired conclusion (that the statement is either true or false) depends crucially on a close relative of the law of the excluded middle (that either a state of affairs of the first kind obtains or all actual states of affairs are of the second kind) that the friend of gaps is free to deny (Parsons 1990, n. 7). It seems to me that all four suggestions are well put. However, it is conceivable that Priest will find some way to repair the argument on these measures. Nonetheless, if Priest is able to present a principle to ensure that the final excluded middle succeeds (so, some grounds upon which to assert that either a state of affairs of the first kind obtains or all actual states of affairs are of the second kind) then the foe of gluts may argue for a similar principle. After all, Dummett's considerations rule against *gluts* just as well as they rule against gaps. The foe of gluts need simply argue for the denial of any conjunction of the form (a state of affairs of the first kind obtains, and all actual states of affairs are of the second kind). If that charge sticks, then this undercuts Priest's grounds for taking a statement to be both true and false. Priest's use of Dummett's condition against gaps relies on an inexplicit endorsement of an excluded middle. If that is legitimate, then we need some reason to see why an inexplicit rejection of contradiction is not also acceptable. In the absence of a symmetry breaker, we may conclude that these considerations relating assertion to truth and falsity lead us no further in choosing between gaps and gluts.

11.3 LEM and LNC

In a later paper, "What not? A defence of dialethic theory of negation" (Priest 1999), Priest attempts a different argument to the conclusion that there are truth-value gluts but no gaps. The crux of the matter is, perhaps surprisingly, the assumption that negation is a contradictory forming operator. That is, the negation $\neg A$ of a statement A is such that the following two theses, the *law of the excluded middle* (henceforth LEM) and the *law of non-contradiction* (henceforth LNC), both hold:

$$A \lor \neg A \qquad\qquad \neg(A \land \neg A)$$

Priest argues that nothing that fails to satisfy either LEM or LNC counts as a *negation*.[3] Such an operator might be contrary forming, or some other negation-like connective, but it does not count as negation. This argument, if it succeeds, is surely a symmetry breaker. Fortunately or unfortunately, the argument fails.

The first thing to note is that satisfying LNC and LEM does not suffice to isolate a single negation operator. After all, if $\neg$ satisfies LNC and LEM, then if C is some true statement so does $\neg^C$, where we set

$$\neg^C A := \neg A \wedge C$$

For it is surely the case that $A \vee (\neg A \wedge C)$ (at least, if $A \vee \neg A$, which we granted), and it is the case that $\neg(A \wedge (\neg A \wedge C)) \wedge C$ (since $\neg(A \wedge \neg A)$, which we granted), given the truth of C. Similarly, if C is false, then $\neg_C$, defined by setting

$$\neg_C A := \neg A \vee C$$

also satisfies LNC and LEM, given LNC and LEM for the original connective $\neg$. But these are *not* negation connectives. If C is merely contingently true, then while $A \vee \neg^C A$ happens to be true, it could very easily be false (if A and C both fail). Merely *contingently* satisfying LNC and LEM is enough to count as being a negation connective.[4]

What more could we ask for? Well, a natural requirement for Priest to demand is that the statements $A \vee \neg A$ and $\neg(A \wedge \neg A)$ not be just *contingently* true, but *logically* true. That is, we require:

$$\vdash A \vee \neg A \qquad\qquad \vdash \neg(A \wedge \neg A)$$

That would rule out bizarre "connectives" such as $\neg^C$ and $\neg_C$. But now you can see the symmetry objection. *Why* privilege one side of the turnstile over the other? Why is the way barred from the friend of gaps to say that we require LNC and LEM in the form:

$$A \wedge \neg A \vdash \qquad\qquad \neg(A \vee \neg A) \vdash$$

That is, $A \wedge \neg A$ is to always be rejected—and this is an *extremely* natural way to read the law of *non*-contradiction—and $\neg(A \vee \neg A)$ is always to be rejected too. This is, to be sure, not a prominent formulation of the law of the excluded middle, but it has something to be said for it. After all, on Priest's dialethic account of negation, no statement truly *excludes* any other, so one might think that a reading

[3] And it certainly *looks* in this paper that there is nothing else to being a contradictory forming operator than satisfying LNC and LEM. Priest does not give any other account of what is required.

[4] Note that these connectives $\neg^C$ and $\neg_C$ suffice to show that satisfying LNC and LEM does not suffice for proving that A is equivalent to its double negation, for $\neg^C\neg^C A$ is $\neg(\neg A \wedge C) \wedge C$, which is certainly not equivalent (in general) to A.

of a thesis that is intended to exclude a possibility must be phrased in terms that make use of denial. What is excluded by the law of the excluded middle? The possibility that A is in the "middle"—that is, that A is neither true nor false. But $\neg(A \wedge \neg A)$ is one way to say that A is neither true nor false. I conclude that, at the very least, this treatment of the law of the excluded middle is at least as plausible as Priest's rendering of $\vdash \neg(A \wedge \neg A)$ as a law of non-contradiction, and hence, the idea that negation is governed by the laws of excluded middle and non-contradiction is as good a justification for a glut theory of negation as it is for a gap theory of negation. We find no symmetry-breaking principles here, either.

11.4 The "Reason Why" Objection

One objection to gaps which does not appear to apply to gluts goes like this: ask the friend of gaps why she denies ⟨SL⟩. The only answer she could give is something like "it's not true." But that is the one kind of answer not available to her. Reasons for assertions or for denials are propositions, and the friend of gaps does not have a reason for denying ⟨SL⟩ the canonical reason for denying something: "it's just *not* true" is not available to her. This objection does not apply to the friend of gluts, because he does not deny ⟨SL⟩.

What can the friend of gaps say? She can start by noting that reasons for denials are many different things, just lite reasons for assertions. Suppose I already have reason to accept X and to reject Y. Then, a proof that $X, A \vdash Y$ gives us reason to reject A, since A, in the context of X, which we accept, entails Y, which we reject. This is exactly what the friend of gaps thinks of ⟨SL⟩. So, let λ be a strengthened liar sentence, and let $X : Y$ be a theory of truth sufficient to ensure the problematic fixed-point properties of λ. That is, the theory consists of statements X to be accepted and Y to be rejected. Perhaps X is empty. Perhaps Y is empty. So, we have

$$X, \lambda \vdash \neg\lambda, Y \qquad X, \neg\lambda \vdash \lambda, Y$$

Then, applying the gap theorist's assumption $A, \neg A \vdash$ that there are no gluts, we can reason as follows:

$$\frac{\lambda, \neg\lambda \vdash \qquad X, \lambda \vdash \neg\lambda, Y}{X, \lambda \vdash Y}\,[\text{CUT}]$$

This proof is *exactly* symmetric to the glut theorist's proof that ⟨SL⟩ is true. As before, $X : Y$ is whatever is needed to prove that the strengthened liar is a fixed point, and this time we apply the glut theorist's assumption $\vdash A, \neg A$ that there are no gaps.

$$\frac{\vdash \lambda, \neg\lambda \qquad X, \lambda \vdash \neg\lambda, Y}{X \vdash \lambda, Y}\ [\text{CUT}]$$

The gap theorist's reason to deny has a parallel structure to the glut theorist's reason to assert. We merely exchange premises for conclusions, and no-gap principles with no-glut principles. There is no symmetry breaking here. The one and the same kind of *proof* (to the conclusion that $X, \lambda \vdash Y$ in the gap theorist's case, or $X \vdash \lambda, Y$, in the glut theorist's case) counts as the reason to reject (for the friend of gaps) and a reason to accept (for the friend of gluts).

The potential rejoinder, that a proof (by *reductio*) that $X, A \vdash Y$ *should* entail $X \vdash \neg A, Y$ (and hence, that we should go on to accept the negation of A), will cut no more ice with the gap theorist than the parallel rejoinder to be made to a glut theorist that the proof of $X \vdash A, Y$ should entail $X, \neg A \vdash Y$, and hence, to the rejection of $\neg A$. Symmetry is preserved here too.

11.5 Sentences and Propositions

Here is a reason to favour gaps over gluts. The liar sentence is not true because it does not express a proposition. Its negation is also not true, because that sentence does not express a proposition either. So neither ⟨SL⟩ nor its negation is true (we deny both). This reason seems to favour gaps over gluts. Doesn't this break symmetry?

Certainly, symmetry is broken if we leave things at this point. (This is not to say that the foe of gaps and friend of gluts would be *moved* by this reasoning. The argument to the conclusion that ⟨SL⟩ does not express a proposition will use principles rejected by the friend of gluts: the reasoning crucially passes to the conclusion that if ⟨SL⟩ *were* to express a proposition, it and its negation would be true, but this is rejected ...) However, once we take it that seemingly syntactically well-formed sentences (such as ⟨SL⟩) do not express propositions, then presumably, a parallel case is to be found with sentences that don't fail in virtue of expressing too few propositions, but rather, succeed all too well in expressing too many. If we have a sentence that expresses *two* propositions, then when is an assertion of that sentence acceptable? There seem to be two different choices. Either it is true when *either* of the propositions expressed is true (the *existential* reasoning), or it is true when *both* of the propositions expressed are true (the *universal* reading). In the existential case, then both the assertion of that sentence and the assertion of its negation may be acceptable. In the universal, we may have a gap. What reason might we have to choose between these two options? Do we wish to encourage assertion by allowing it to be acceptable more often, or do we wish to restrict it by imposing the tougher condition that both propositions expressed be true? The first option brings gluts and the second brings gaps. The mere fact of multiple expression gives us no reason to choose between them.

But if this is the case, it seems to be an option *even* in the case where a proposition has not been expressed. The friend of gluts could say that the assertion of a sentence expressing no proposition could, after all, be acceptable, by analogy with the *universal* reading in the case of multiple propositions. After all, in the empty case, *every* proposition expressed by the sentence (which expresses no proposition!) is true. An assertion of this sentence fails, to be sure: it fails to rule anything out. Therefore (on the universal reading, at least) the assertion is acceptable. It seems that even here, the friend of gluts has some room to manoeuvre.

My tentative conclusion, to the effect that gap and glut theories are on a par, is not popular. Almost all defenders of dialethic accounts of the paradoxes take their account to be superior to gap accounts because of ⟨SL⟩. Almost everyone else takes truth-value gluts to be crazy, and truth-value gaps to be sane. Symmetry is a minority position.[5] Furthermore, a symmetricalist can take one of two forms. Gap and glut theories could be equally *good* (so, we could endorse a form of Parson's *agnostaletheism*, an agnosticism between gaps and gluts, or perhaps a more liberal semantic theory which allows for both gaps and gluts) or they are equally *bad*.

References

Brandom, Robert B. 2000. *Articulating Reasons: An Introduction to Inferentialism*. Cambridge, MA: Harvard University Press.

Dunn, J. Michael. 2000. "Partiality and its dual." *Studia Logica* 65:5–40.

Dunn, J. Michael and Hardegree, Gary M. 2001. *Algebraic Methods in Philosophical Logic*. Oxford: Oxford University Press.

Gentzen, Gerhard. 1935. "Untersuchungen über das logische Schliessen." *Mathematische Zeitschrift* 39:175–210, 405–31. English translation 'Investigations concerning logical deduction' in Szabo (1969, 68–131).

Parsons, Terence. 1984. "Assertion, denial, and the liar paradox." *Journal of Philosophical Logic* 13:137–52.

Parsons, Terence. 1990. "True contradictions." *Canadian Journal of Philosophy* 20:335–54.

Priest, Graham. 1999. "What not? A defence of dialetheic theory of negation." In D. Gabbay and H. Wansing (eds.), *What is Negation?* 101–20. Dordrecht: Kluwer.

Priest, Graham. 2006. *In Contradiction*. Oxford: Oxford University Press, 2nd edition.

[5] But friends of symmetry are in very good company. See, for example, Dunn's work on negation (Dunn 2000), and (with Hardegree) on multiple-premise, multiple-conclusion consequence as an appropriate general setting to characterize properties of different 'logics' (Dunn and Hardegree 2001, especially section 6.8).

Restall, Greg. 2005. "Multiple conclusions." In P. Hajek, L. Valdez-Villanueva, and D. Westerstahl (eds.), *Proceedings of the Twelfth International Congress on Logic, Methodology, and Philosophy of Science*, 189–205. London: King's College Publications.

Shoesmith, D.J. and Smiley, T.J. 1978. *Multiple-Conclusion Logic*. Cambridge: Cambridge University Press.

Szabo, M.E. (ed.). 1969. *The Collected Papers of Gerhard Gentzen*. Amsterdam and London: North-Holland.

12

Knowability Remixed

Heinrich Wansing

12.1 Introduction

The idea of truth preservation from the premises to the conclusions of an inference (or falsity preservation in the opposite direction) is usually seen to be fundamental to the notion of logical consequence. An inference from a set of premises Δ to a set of conclusions Γ is said to be valid if and only if necessarily the truth of all formulas from Δ implies the truth of a least one formula from Γ. The controversy about realistic versus anti-realistic notions of truth therefore is of obvious relevance to the conception of logical consequence. The predominant, though not the unique, anti-realistic conception of truth is epistemic. It identifies the truth of a statement with its knowability and gives rise to an epistemic notion of logical consequence. This conception is threatened by the Knowability Paradox.

Intuitively, it is *knowable* that p if and only if it is possible to know that p. A straightforward way of formalizing the phrase 'it is possible to know that p' is to translate it into the language of modal epistemic logic as $\Diamond Kp$. If p is instantiated by a so-called Fitch-conjunction $q \wedge \neg Kq$, the Fitch–Church Paradox of Knowability (see Brogaard and Salerno 2009; Kvanvig 2006; Salerno 2009) is a serious challenge to anti-realists who claim that every true proposition is knowable: $\forall p(p \rightarrow \Diamond Kp)$. The Fitch–Church Paradox can be interpreted quite differently and, depending on how it is analyzed, different strategies to resolve the paradox can be pursued. If the problem is seen in the derivation of $\forall p(p \rightarrow Kp)$ from $\forall p(p \rightarrow \Diamond Kp)$, then the derivation itself is the target in a resolution of the paradox, (see, for example, Wansing 2002).[1] If the Fitch-Church Paradox is first of all seen as calling into doubt

[1] Johan van Benthem (2004, 95) describes this strategy as follows:

> Some weaken the logic in the argument still further. This is like turning down the volume on your radio so as not to hear the bad news. You will not hear much good news either.

the plausibility of the anti-realist thesis $\forall p(p \rightarrow \Diamond Kp)$, then a revision of the anti-realist thesis is in focus. Note that these two strategies of dealing with the Fitch–Church Paradox are independent of each other. In view of the paradox, an anti-realist might want to block the derivation of $\forall p(p \rightarrow Kp)$ from $\forall p(p \rightarrow \Diamond Kp)$ *and* restrict the anti-realist thesis or replace it by another principle. Also, while revision may be seen as changing the problem, solving the original problem might be regarded as uninteresting from the metaphysical, epistemological, or some other point of view. The most pertinent restriction of the anti-realist thesis has been suggested by Neil Tennant. This restriction to sentences expressing Cartesian propositions, however, has been criticized for being ad hoc. A sentence p is Cartesian iff Kp is not provably inconsistent, and being non-Cartesian is the crucial property of Fitch-conjunctions in the Fitch–Church Paradox. In this chapter I will present another revision of the anti-realist thesis.

12.2 Knowability Rearranged

Interestingly, the neat way of analyzing 'it is knowable that p' as $\Diamond Kp$ does not fully reveal the standard reading of 'it is knowable that p' that can be found in the literature, which seems to be the following: *it is possible that it is known by someone at some time that* p (cf. Brogaard and Salerno 2009). With this reading, the knowledge operator occurs within the scope of an existential quantification over agents and moments of time and has narrow scope with respect to p. The standard demonstration showing that a Fitch-conjunction is non-Cartesian makes use of the fact that the K-operator is factive and distributes over conjunction: $K(q \wedge \neg Kq)$ logically implies Kq and $K\neg Kq$, and $K\neg Kq$ logically implies $\neg Kq$. Usually, it is assumed that there exists at least one true Fitch-conjunction. If the anti-realist thesis is true and is instantiated by a true Fitch-conjunction $q \wedge \neg Kq$, one obtains $\Diamond K(q \wedge \neg Kq)$, which is, however, unsatisfiable.

If it is the anti-realist (or verificationist) thesis one is concerned about in the first place, the Fitch–Church-style derivation of a contradiction can be avoided if one unburdens the K-operator from part of its intuitive reading and internalizes the implicit quantification over moments of time into the object language. Quantification over epistemic subjects may still be suppressed in the surface grammar and be relegated to the semantics. Let $\blacklozenge p$ express that p is true at some moment of time. The standard reading of 'it is knowable that p' can then be translated into the

I do not quite see the analogy. Whereas suitably weakening (or changing) the logic *resolves* the Knowability Paradox, turning down the volume on the radio only *blinds outs* the (bad and other) news.

reading	variant of the Fitch–Church Paradox	epistemic accessibility
$\Diamond\blacklozenge Kp$	✓	strong sense
$\blacklozenge\Diamond Kp$	✓	strong sense
$\Diamond K\blacklozenge p$	–	weak sense
$\blacklozenge K\Diamond p$	–	weak sense

Figure 12.1 Four readings of 'it is knowable that p'

language of modal epistemic temporal logic as $\Diamond\blacklozenge Kp$. My proposal is to rearrange the formal explication of 'it is knowable that p' as $\Diamond K\blacklozenge p$. In other words, it is suggested to read 'it is knowable that p' as *it is possible that it is known by someone that at some time* p. The formula $\Diamond K\blacklozenge(q \wedge \neg Kq)$, the result of substituting the Fitch-conjunction $q \wedge \neg Kq$ for p in $\Diamond K\blacklozenge p$, is satisfiable.

Note that I do not suggest to understand 'it is known that p' as 'it is known that at some time p'. What I *do* claim, however, is that it is neither unreasonable nor implausible to understand 'every truth is knowable' as $\forall p(p \rightarrow \Diamond K\blacklozenge p)$. If this claim is correct, the anti-realist could adopt $\forall p(p \rightarrow \Diamond K\blacklozenge p)$ as the formal expression of verificationism. In the presence of both $\Diamond$ and $\blacklozenge$, another paradox-free rearrangement of the standard reading of 'it is knowable that p' is, of course, available: $\blacklozenge K\Diamond p$. As a result, we obtain the four readings of 'it is knowable that p' listed in Figure 12.1.

Epistemic anti-realism is sometimes presented as the claim that reality is epistemically accessible (to us or to at least one epistemic subject) (see, for example, Jenkins 2007).[2] An anti-realist who in order to escape the Fitch-Church Paradox holds that the formula $\forall p(p \rightarrow \Diamond K\blacklozenge p)$ (or $\forall p(p \rightarrow \blacklozenge K\Diamond p)$) is a formal expression of verificationism might maintain that states of affairs which render the proposition expressed by p true are epistemically accessible *in a weak sense* if $\Diamond K\blacklozenge p$ is true (or $\blacklozenge K\Diamond p$ is true).

12.3 Knowability in Branching Time

In this section I will present an interpretation of $\Diamond K\blacklozenge p$ and $\Diamond\blacklozenge Kp$ in branching-time models, (see, for example, Belnap et al. 2001; Semmling and Wansing 2008). The purpose of this exercise is to argue that $\Diamond K\blacklozenge p$ is an at least as natural formalization of 'it is knowable that p' as $\Diamond\blacklozenge Kp$. Given the focus on these two readings, let us call the former formalization *the diamond-*K*-diamond analysis* of knowability and the latter one *the diamond-diamond-*K *analysis.*

[2] Note that epistemic accessibility in this sense is not a relation between possible worlds or situations.

A branching-time frame (or just frame) is a structure of the form $\mathcal{F} = (T, \leqslant)$, where T is a non-empty set understood to be a set of moments of time, and $\leqslant$ is a partial order on T satisfying the following property: if $n \leqslant m$ and $l \leqslant m$, then $n \leqslant l$ or $l \leqslant n$, for all $m, n, l \in T$ (no backward branching). The partial order $\leqslant$ thus is a tree order on T. A maximal linearly ordered subset $h \subseteq T$ is said to be a *history* in T. The set of all histories of a given frame $\mathcal{F}$ is denoted by H, and the set $H_m = \{h | m \in h\}$ is said to contain all histories which pass through moment m. A moment/history pair (m, h) with $m \in h$ is called a situation in the given frame, and (m, h) is said to be a situation featuring m. The set of all situations in the frame $\mathcal{F} = (T, \leqslant)$ is the set of all moment/history pairs (m, h) such that $m \in T$ and $h \in H_m$; it is denoted by $S_{\mathcal{F}}$. If $\mathcal{F} = (T, \leqslant)$ is a frame and v is a function mapping every atomic formula to a set of situations in the frame, then $\mathcal{M} = (T, \leqslant, v)$ is called a branching-time model (or just model), and $S_{\mathcal{M}}$ is defined as $S_{\mathcal{F}}$. If, moreover, *Agent* is a non-empty finite set of epistemic subjects and $\mathcal{R}$ a set containing for every $\alpha \in Agent$ a binary relation R_α on $S_{(T,\leqslant)}$, then $(T, \leqslant, Agent, \mathcal{R}, v)$ is said to be an epistemic model.[3] We do not assume that the set of agents may vary from situation to situation because for our resolution of the Fitch–Church Paradox it does not matter whether we work with constant or varying domains. Usually, the relations in $\mathcal{R}$ are taken to be equivalence relations. We may then denote by $|\ (h, m)\ |_\alpha$ the equivalence class of R_α containing situation (h, m). Formulas are evaluated as true or false at situations in epistemic models. The notion $\mathcal{M}, (m, h) \models A$ ("A is true at situation (m, h) in model $\mathcal{M}$") is inductively defined as follows:

$\mathcal{M}, (m, h) \models p$	iff	$(m, h) \in v(\phi)$, if p is an atomic formula.
$\mathcal{M}, (m, h) \models \neg A$	iff	not $\mathcal{M}, (m, h) \models A$.
$\mathcal{M}, (m, h) \models (A \wedge B)$	iff	$\mathcal{M}, (m, h) \models A$ and $\mathcal{M}, (m, h) \models B$.
$\mathcal{M}, (m, h) \models \Diamond A$	iff	$\mathcal{M}, (m, h') \models A$ for some $h' \in H_m$.
$\mathcal{M}, (m, h) \models \blacklozenge A$	iff	$\mathcal{M}, (m', h) \models A$ for some $m' \in h$.
$\mathcal{M}, (m, h) \models KA$	iff	$(\exists \alpha \in Agent)\ \mathcal{M}, s \models A$ for all $s \in S_{\mathcal{M}}$ such that $(m, h) R_\alpha s$.

With this definition, $\Diamond$ is interpreted as "historical possibility" (see Belnap et al. 2001). If $\Diamond$ is historical possibility, then $\blacklozenge$ may be called "momentary possibility". Using terminology introduced by Jaakko Hintikka (1962), if $sR_\alpha s'$, then s' is a situation compatible with what agent α knows at s, and the set $\{s' \mid sR_\alpha s'\}$ then

[3] The relations in $\mathcal{R}$ are often said to be epistemic accessibility relations. This notion of epistemic accessibility is not to be confused with the notion of epistemic accessibility mentioned in the previous section.

is the set of agent α's epistemic alternatives to s. The histories passing through m may be regarded as possible developments of the world at moment m.

From this perspective, one might expect that in evaluating knowability statements at a situation (m, h), the relevant epistemic alternatives are epistemic alternatives to a situation featuring the very moment m, for some agent. If, according to the diamond-K-diamond analysis, 'it is knowable that p' is read as $\Diamond K \blacklozenge p$, and $\Diamond K \blacklozenge p$ is evaluated at a situation (m, h), then $\blacklozenge p$ is indeed evaluated only at epistemic alternatives to a situation featuring m. This conception results in a certain notion of *actual knowability* with respect to the moment from the situation at which a knowability statement is evaluated. According to the diamond-diamond-K analysis, however, the relevant epistemic alternatives in evaluating 'it is knowable that p' at a situation (m, h) need not feature m. The formula $\Diamond \blacklozenge Kp$ may be true at (m, h) if p is true at some agent α's epistemic alternatives, although none of these situations features m. Both formulas $\Diamond \blacklozenge Kp$ and $\Diamond K \blacklozenge p$ may be seen to represent an understanding of 'it is knowable that p'. Whereas the strong notion of epistemic accessibility assumed in the diamond-diamond-K analysis runs into a variant of the Fitch–Church Paradox,[4] the weak notion of epistemic accessibility employed in the diamond-K-diamond analysis does not.

Figure 12.2 presents part of an epistemic model with $Agent = \{\alpha\}$, satisfying $\Diamond \blacklozenge Kp$ at the displayed situation (m, h_0). In this model, there is a moment distinct from m, namely m', and an equivalence class of R_α such that (i) for every situation s from this class, p is true at s and (ii) no situation from this class features m.

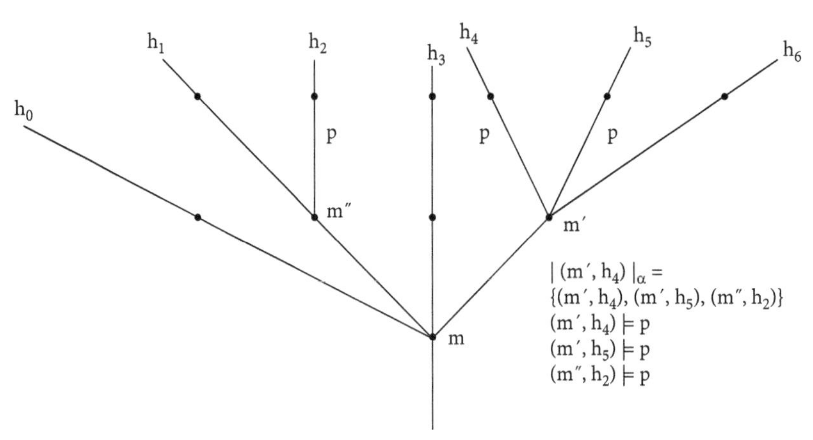

Figure 12.2 $(m, h_0) \models \Diamond \blacklozenge Kp$

[4] Suppose that $(q \land \neg Kq)$ and $\forall p(p \rightarrow \Diamond \blacklozenge Kp)$ are true. It follows that $\Diamond \blacklozenge K(q \land \neg Kq)$ is true, which, however, is unsatisfiable. Therefore, $\forall p(p \rightarrow \Diamond \blacklozenge Kp)$ logically implies $\neg \exists p(p \land \neg Kp)$, which is classically equivalent to $\forall p(p \rightarrow Kp)$.

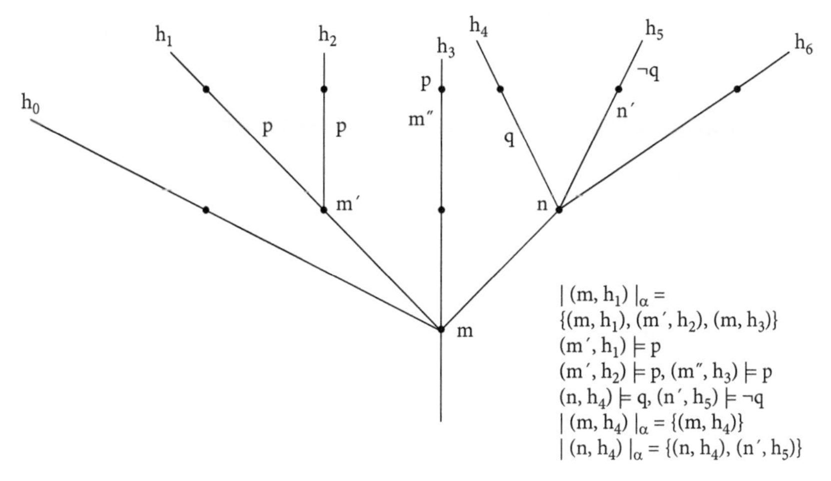

Figure 12.3 $(m, h_0) \models \Diamond K \blacklozenge p$

Figure 12.3 depicts part of an epistemic model with $Agent = \{\alpha\}$, satisfying $\Diamond K \blacklozenge p$ at the displayed situation (m, h_0). In this model, there is an equivalence class of R_α such that (i) for every situation s from this class, $\blacklozenge p$ is true at s and (ii) each element of this class is an epistemic alternative to a situation featuring m, namely (m, h_1). Also, by reflexivity of R_α, there exists a situation featuring m in this class. Moreover, $\Diamond K \blacklozenge (q \wedge \neg Kq)$ is true at (m, h_0).

12.4 A Semantical Framework for Reasoning about Knowability

In the previous section, I have presented four different notions of knowability that permit an interpretation in epistemic models. Branching-time frames provide a basis for an even richer semantical framework for reasoning about knowability. In suitable extensions of epistemic models, one may define a notion of "metaphysical possibility" $\Diamond^*$. Let $\mathcal{M} = (T, \leqslant, Agent, \mathcal{R}, v)$ be an epistemic model and let R be an equivalence relation on $S_\mathcal{M}$. Then $\mathcal{M} = (T, \leqslant, Agent, \mathcal{R}, \mathsf{R}, v)$ is an extended epistemic model. The notion $\mathcal{M}, (m, h) \models A$ ("A is true at situation (m, h) in model $\mathcal{M}$") is inductively defined as before, except that we have the following additional clause:

$$\mathcal{M}, (m, h) \models \Diamond^* A \quad \text{iff} \quad \mathcal{M}, s \models A \text{ for some } s \in S_\mathcal{M} \text{ with } (m, h)\mathsf{R}s.$$

In Figure 12.4 four additional readings of 'it is knowable that p' are listed.

One may also define a notion of "historical knowledge" K^* by introducing a set R containing for every $\alpha \in Agent$, an equivalence relation $\mathsf{R}_{\alpha,m}$ on H_m and

reading	variant of the Fitch–Church Paradox	epistemic accessibility
$\Diamond^{*}\blacklozenge Kp$	✓	strong sense
$\blacklozenge\Diamond^{*}Kp$	✓	strong sense
$\Diamond^{*}K\blacklozenge p$	–	weak sense
$\blacklozenge K\Diamond^{*}p$	–	weak sense

Figure 12.4 Four additional readings of 'it is knowable that p'

stipulating that $\mathcal{M}, (m, h) \models K^{*}A$ iff $(\exists\alpha \in \textit{Agent})\ \mathcal{M}, (m, h') \models A$ for all $h' \in H_m$ such that $hR_{\alpha,m}h'$. Let $\Box A$ abbreviate $\neg\Diamond\neg A$. I do not consider here versions of the verificationist thesis involving K^{*} because the following interaction principle is valid: $\Box A \to K^{*}A$ (historical necessity implies historical knowledge).[5]

References

Belnap, Nuel D., Perloff, Michael, and Xu, Ming (eds.). 2001. *Facing the Future: Agents and Choices in our Indeterminist World.* Oxford: Oxford University Press.

Brogaard, Berit and Salerno, Joe. 2009. "Fitch's Paradox of Knowability." In Edward N. Zalta (ed.), *The Stanford Encyclopedia of Philosophy.* CSLI, fall 2009 edition. <http://plato.stanford.edu/archives/fall2009/entries/fitch-paradox/>.

Hintikka, Jaakko. 1962. *Knowledge and Belief.* Ithaca, NY: Cornell University Press.

Jenkins, C.S. 2007. "Anti-realism and epistemic accessibility." *Philosophical Studies* 132:525–51.

Kvanvig, Jonathan. 2006. *The Knowability Paradox.* Oxford: Oxford University Press.

Salerno, Joe (ed.). 2009. *New Essays on the Knowability Paradox.* Oxford: Oxford University Press.

Semmling, Caroline and Wansing, Heinrich. 2008. "From BDI and stit to bdi-stit logic." *Logic and Logical Philosophy* 17:185–207.

van Bentham, Johan. 2004. "What one may come to know." *Analysis* 64:95–105.

Wansing, Heinrich. 2002. "Diamonds are a philosopher's best friend: the knowability paradox and modal epistemic relevance logic." *Journal of Philosophical Logic* 31:591–612.

[5] I would like to thank Andrea Kruse and Caroline Willkommen for comments on an earlier version of this chapter.

13

Accuracy, Logic, and Degree of Belief

J. Robert G. Williams

13.1 Introduction

Why care about being logical? Why criticize people for inconsistency? Must we simply take the normative significance of logic as brute, or can we explain it in terms of goals on which we have an independent grip: the merits of true (or knowledgeable) belief, for example? This chapter explores Jim Joyce's argument for probabilism in the light of these questions—arguing that it provides a plausible route for *explaining* the value of consistency.

13.2 Logical Norms, Probability, and Accuracy

It seems good to be consistent; and bad to be inconsistent. If you believe B, and also believe ¬B, then *something has gone wrong*. Something has gone wrong, too, if B obviously follows from what you already believe, and you don't believe that B when the question is raised. These sorts of principles apparently report a *normative* role for logic. Logic is a source of principles about how we ought or ought not to conduct ourselves.

Some prefer to talk in terms of partial beliefs (whether as a replacement for, the explanatory basis of, or a supplementation to, talk of all-or-nothing belief). Here too we find similar theses about what belief states should look like. Most familiar is *probabilism*: the doctrine that our partial beliefs *should* be representable as (or extendable to) a probability.

Probabilism, like logical norms, recommends or condemns certain *patterns of attitudes*. But the connection is even tighter: arguably, probabilism *embeds* logical norms. Probabilities can be "locally" characterized in terms of logical relations among partial beliefs—thus, each of the following are necessary for P to be a probability.

- If B follows from A, then $P(A) \leqslant P(B)$.
- If A and B partition C, then $P(C) = P(A) + P(B)$.[1]
- If A is a logical truth, then $P(A) = 1$.

If one's partial beliefs violate any one of the above, then they are not representable via a probability. So probabilism might be redefined "locally" as the view that partial beliefs ought to meet the particular constraints just given. My concern in this chapter is with logical norms on partial belief, as expressed by this form of probabilism.

(Some caveats: it's one thing to say that logic norms partial belief. It's another to say that probabilism describes these norms. Other formal treatments of belief functions might equally make play with logical notions, and can be taken as rival characterizations of the logical norms in play. Within the generally probabilist camp, one might wish to hedge and qualify matters in various ways—saying that we ought to avoid *obvious* violations of the probabilistic constraints, for example, rather than violations *simpliciter*. In order to have a clean model to work with, I'll set aside these complications for now.[2])

For the sake of argument, suppose we agree that probabilism sounds initially attractive. Then we face the question raised at the beginning: are we prepared to take these principles about how our beliefs ought or ought not to be as explanatorily basic, or do we seek an illuminating explanation?

What's the motivation to take the latter view? Niko Kolodny believes that brute logical norms would seem like a mere fetish for a certain patterning of mental states:

> Simply put, it seems outlandish that the kind of psychic tidiness that ... any ... requirement of formal coherence, enjoins should be set alongside such final ends as pleasure, friendship, and knowledge.[3]

How much more attractive if we could reveal a commitment to psychic tidiness as *implicit in* or *following from* a respect for the kind of basic values Kolodny lists.

[1] We say that A and B "partition" C, if we have a logical guarantee that C is true iff exactly one of A/B is true. For example, in a classical setting, when A, B are inconsistent, A and B will partition $A \vee B$, and hence the above constraint gives us a familiar additivity principle: $P(A) + P(B) = P(A \vee B)$. I like this formulation rather than the additivity proper because it doesn't bake in assumptions about how particular connectives behave. For discussion, see Williams (forthcoming).

[2] For alternative models of partial belief, see Halpern (1995). For the formulation of the norms, see Field (2009).

[3] Kolodny (2007).

Kolodny's discussion is set within a wider discussion of the normativity of *rationality*, whose subject-matter includes not only logic but also coherence between beliefs, desires, and intentions. We may assume for the sake of argument that we have identified a number of constraints (such as those axioms of probability theory) that must be met if a subject is to count as "perfectly rational". Call these *rational requirements*. Crucially, we treat it as an open question whether someone *should* be rational—whether rational requirements are normative requirements. Perhaps "rational" seems too normatively loaded to allow one to hear the latter question as substantive. If so, let's simply take our candidate cluster of principles, and stipulate that they are to be treated as conditions of being "R"—with the substantive question being whether we *ought* to be R. If the answer is negative—if it is *not* the case we ought, even *pro tanto*, to be rational—then we end up with a view whereby conforming to rational requirements is like membership of a kind of club. To be in that club, you have to do certain things or be a certain way (roll up your trousers in appropriate circumstances); but there's no obligation to join.[4]

The negative answer in the case of logic is perfectly possible. Harman and Maudlin think of logic as something like "the science of guaranteed truth-preservation"—consisting of facts about what follows from what, which are unrelated to questions about what patterns of attitudes one *ought* or *ought not* to have.[5] Clearly such answers avoid the challenge to explain the basis of logical norms, by denying their existence (though, to be fair, Maudlin at least allows a surrogate notion which *does* seem to have a normative role—and we can fairly raise the question about the source of those norms).

Positive answers to the question "Why be logical?" come in various grades. Here are three. First, the grade one position: one says of each individual constraint, that there's a norm against violating it. This is a version that I read Hartry Field as subscribing to in the paper cited—though his main focus is on exactly what constraints it is that have this status. A grade two norm is to say *generically* that we ought not to violate logical constraints. Compare: generically you ought not to jump off cliffs—even though on a particular occasion (when escaping from a bear) jumping off a cliff may be the thing you ought to do. This is John Broome's favoured view. Finally, there is the lowest grade, debunking position, which claims that there are no logical norms as *such*. Rather, any violation of the constraints laid down will *guarantee* that some *other* norm is violated. Kolodny argues, for example, that believing an inconsistency guarantees that your beliefs are out of line with the evidence—hence failing to be logical is an invariable sign of failing to meet evidential norms on belief. The final position (by design) does not recognize a

[4] See Kolodny (2005); Broome (2005).

[5] Harman (1986); Maudlin (2004).

normative difference between the case of the perfectly rational agent who (perhaps because of odd priors) is out of line with the objective evidence, and one who has grossly inconsistent beliefs. That's why it debunks, rather than vindicates, logical norms.

As well as asking questions about the grades of normativity attaching to the logical norms, we can equally ask about the type of normativity in question—and in particular, the extent to which an agent's subjective take on what the relevant facts or values are, should factor into an evaluation. A hardliner might say it is *the facts* about whether attitudes or actions fit with what are *in fact* the proper values, that are relevant to normativity. If pleasure, friendship, and knowledge are the aim, then actions not conducive to those ends are bad, independently of whether an agent has the mistaken view of the consequences of their actions (thinking that showing off will win them friends) or a mistaken take on what the proper ends are (thinking that power rather than pleasure is the goal, perhaps). Our subjective take needs to be accounted for somewhere, but we can argue about how to frame it. One view—advocated by Williamson in his account of the aim of belief[6]—is that the "fundamental" norm that one should believe p only if that would count as knowledge, gives rise to a "derivative" norm that one should believe p only if one believes it would count as knowledge. And I take it that part of the idea here is that we can explain our concern for derivative norms in terms of underlying concern for the fundamental norm (so at least in these cases they don't appear as a fetish for mental tidiness).

Kolodny recommends we describe this sort of case differently. Rather than admit there's anything "normative" to the rule that you should believe p only if you believe that this would count as knowledge, Kolodny highlights the fact that violating this rule means that *from the subjects' point of view* they will have violated the fundamental norm. One does not have to posit a "respect" for the derivative norm as such, to explain why violations of this rule are relevant—by calling the subjects' attention to violations, their respect for the fundamental norm kicks in.

There are really two issues here. First is whether we want to call the "rules" that arise from subjectivizing a fundamental norm themselves "normative". The second is exactly how to formulate the derivative rule. Notice, for example, that Williamson's formulation of a "derivative" norm talks about what a subject believes they are in a position to know. So the rule is sensitive to subjects' limited information about what they know. But the rule is not sensitive to subjects' limited information about what the aim of belief really is. A subject who confidently

[6] See Williamson (2000, ch. 11) for a discussion of fundamental and derivative norms of assertion; and (2000, 48) and *passim* for the claim that belief aims at knowledge.

believed (perhaps on good evidence) that the aim of belief was to cohere with the opinions of one's peers, would violate the derivative norm *even if they exactly satisfied, and knew they satisfied* what they take the fundamental norm to be. Kolodny's formulation is more thoroughgoingly subjective—it will seem to the subject just mentioned that they meet their obligations.

The contrast can be illustrated in the case of consequentialist evaluations of action. Take the fundamental norm on action to be that one should do what maximizes good consequences. A partial subjectivization is given by the rule to do what maximizes the expected goodness of the consequences. A full subjectivization is given by the rule to do what maximizes the expected *believed* goodness of the consequences.

Part of Kolodny's case that logic has only the third, debunking grade of normativity is that he doesn't think that the rules arising from subjectivizing fundamental norms in general have normative punch. His overall project is to display "rational requirements" as arising from factoring subjectivity into the fundamental norms—roughly, even though there's strictly speaking no reason to be rational, it will always *appear to the subject* that there is such a reason.

The project to be pursued here cuts across this debate. What I intend to do is to look at prospects for explaining the apparent normativity of logic in terms of subjectivizations of fundamental norms on belief. The project is substantive and of interest whether or not we think that success would count as vindicating (derivative) normativity of logical constraints, or merely explaining away the appearances of normativity, Kolodny-style. It is within this context that I wish to examine Joyce's work on accuracy and probabilism.[7]

13.3 Accuracy Norms

Joyce's starting point is a norm for belief that is not itself a norm of coherence—rather, it is a norm of *truth*. For all-or-nothing beliefs, the norm might simply have been that one ought to believe the true and disbelieve the false; Joyce generalizes this to partial belief, holding that a belief is *better* the *more accurate* it is—where the accuracy of a belief is a measure of how close it is to the truth value. An immediate issue is what this talk of "closeness to the truth value" amounts too; and this has been the focus of much discussion. I want to concentrate on the downstream issues, so to fix ideas I will simply stipulate a particular accuracy measure: the inaccuracy of a degree of belief $\mathrm{b(p)}$ in the proposition p is given by

[7] Koldony himself discusses Joyce—albeit very briefly—in Koldony (2007, 256).

$|t(p) - b(p)|^2$—and the inaccuracy of an overall belief state is simply the average inaccuracy (this is the "Brier score").

How much is built into the postulation of a norm of accuracy or truth? Must we, for example, buy into an ambitious teleological account of epistemic normativity, on which every epistemic norm on belief must ultimately be defended in terms of its producing the "final value" of truth? It's clear, I hope, that nothing so ambitious can be read off the mere postulation of a truth or accuracy norm on belief (even at a "rock bottom" level). For one thing, the existence of such a norm doesn't exclude the existence of other dimensions along which to assess belief; for another, we haven't made any commitment here to the general project of reducing justification (say) to the norm of truth. The assumption that we've made is comparable, in the ethical case, to claiming that the goodness of a state of affairs is measured by how much welfare it produces; that alone doesn't suffice to commit one to a teleological account of the rightness of action in terms of the production of welfare. Even if we did think that all epistemic norms are to be understood in relation to a norm of truth (compare Wedgwood on the aim of belief)[8] we're not thereby committed to explicating that relationship in aggregative or maximizing/satisficing terms.

However, there is something explicitly aggregational about the Brier score as a norm of accuracy (and also with all ways of measuring inaccuracy that meet Joyce's (1998) axioms). After all, the inaccuracy of a total belief state is a straight average of the inaccuracy of individual beliefs. Whether this form of aggregation is problematic should be measured by its fruits—but notice how limited a thesis it is. An aggregational theory of a *single person's* welfare *at a time*, doesn't commit one to aggregational theory of a single person's welfare *across time*, or to aggregational theories of overall welfare across persons. Likewise, we shouldn't assume that an aggregational theory of the goodness of a belief state at a time commits one to aggregating the value of truth over time or over whole communities.[9]

With these observations in mind, let's stick with the Brier score as a measure of overall epistemic goodness of belief states. Joyce then proves a theorem:

- **Accuracy domination.** If a set of partial beliefs B is not a probability, then there's a probability, C, such that C is more accurate (so, better) than B no matter which world is actual.

We say in such a case that C "accuracy dominates B". One can also show:

- **Probabilistic anti-domination.** No probability function C is accuracy dominated by any belief state.

[8] Wedgwood (2002).

[9] Thanks to Selim Berker for discussion here. For discussion, see Berker (2013).

It's easy to show that every probability function "minimizes expected inaccuracy" by its own lights (the Brier score is "proper", in the jargon). It's also easy to see that when x accuracy-dominates y, the expected inaccuracy of x is strictly less than the expected inaccuracy of y. Anti-domination follows.

Joyce uses accuracy domination to argue for probabilism: and it's certainly suggestive—violating probabilistic norms means that in a certain sense you are being *needlessly inaccurate*. Joyce is content to take as a further premise that this accuracy domination "gives us a strong, purely epistemic reason to prefer the [dominating credence] over the [dominated credence]".[10] Perhaps—but it would be nice to have the case for this spelled out. What sort of epistemic reason is it, and how do such reasons arise out of the underlying concern for accuracy? Our question here is how the (alleged) vindication of probabilism ensues in detail.

[*Aside*: part of my interest in investigating Joyce's argument is that it generalizes very nicely. Joyce proves accuracy domination for *classical* probabilities assuming *classical* truth-value assignments. But his argument generalizes to show accuracy domination for specific sorts of *generalized* probabilities assuming specific sorts of *non-classical* truth-value assignments. Moreover, one can use non-classical logic to spell out "localizations" of the generalized probabilities, just we familiarly use classical logic to localize classical probability.[11] So a route from accuracy domination to the corresponding probabilism promises a *very general* bundling together of truth values, logic, and norms on belief.]

13.4 Accuracy and Logical Norms as Such

Can we turn accuracy domination into a defence of grade-one probabilistic norms? And if not, how far can we get along this route? In this section, I argue in the light of Joyce's result that:

(i) Probabilism is true for agents who are *a priori omniscient* (APO agents), so long as the strategy for "deriving" subsidiary norms from fundamental norms given earlier is ok.

(ii) There's no such direct case for probabilism for ordinary agents (you and me). But there's a specific sense in which according to the logical constraints is a *virtue* for such an agent—which can form the basis for a defence of a Broome-style norm of *generic* logicality.

[10] Joyce (2009, 285).

[11] See Williams (forthcoming) for an overview; Williams (2012a) for a discussion of the relevant generalized Brier-score domination results (originally due to de Finetti); and Williams (2012b) for a discussion of Joyce's own argument in the generalized setting.

Recall that Joyce's underlying norm was one of accuracy: of getting one's degrees of belief as close as possible to the truth values. There's thus an immediate, debunking, but rather uninteresting sense in which we ought to be probabilists. For the *best* belief state to adopt, by the Joyce norm, is one that exactly matches the truth-value distribution. And such credence functions are (limiting cases) of probabilities—*extremal* cases where every probability is either 1 or 0. The most fundamental point here is that this rationale is so demanding as to be uninteresting. For *any* responsible (non-godlike) believer presumably will have intermediate credences in some things. And thus we *all* violate the norm of "matching the truth values". So there's no basis here for discriminating criticism—we're all condemned.

Rather than appeal to the accuracy norm itself, we'll appeal to the rule that says that when you know that C is more accurate than B, you ought to have belief state C rather than B. This is a derivative norm in the Williamsonian sense given earlier—and while Kolodny may deny it is *normative*, if we can explain the (apparent) normativity of probabilistic constraints in terms of the (apparent) normativity of this rule, that can feed into Kolodny's eliminative project.

It's pretty clear that APO agents will violate the derivative norm, if their belief state B is not a probability, by the following argument:

Core argument:

1. B is not a probability. (premise)
2. B is accuracy dominated. (by accuracy domination theorem and 1)
3. An APO agent will be able to calculate a specific x (call it C) that she can tell is more accurate than B. (from 2)
4. If one knows of C that it is more accurate than B, then one ought not to have belief state B. (premise)
5. An a priori-omniscient agent ought not to have belief state B.

In sum: the APO agent will always know of a better belief state than any candidate improbabilistic state she might consider adopting.[12]

[12] Two notes on this. Though we can argue for the contrastive claim that one ought to have C rather than B in these circumstances, and it follows that one ought not have belief state B, it doesn't follow that one ought to adopt belief state C. After all, there might be an even better belief state C*. Hájek notes, for example, that midway between the probability C and the probability B there will be many improbabilities that accuracy-dominate B and are accuracy dominated by C. Plausibly, we ought to adopt C* rather than B, but in fact we ought to have neither of these. Hájek notes that if probabilities were accuracy-dominated in turn (perhaps with an unending-chain of accuracy domination) we might never reach a belief state that it's permissible to adopt. Fortunately, for the Brier score (and proper scoring rules in general) we can show that no probability will be accuracy-dominated.

For APO agents, violations of probabilistic constraints guarantee violation of the derivative accuracy norm. But furthermore, we can argue that the derivative norm is violated only when the probabilistic norms are. Probabilistic anti-domination doesn't give us this—for that tells us only that there's no state which is *necessarily* more accurate than one that meets probabilistic constriants, and for all that, the agent with belief state B could know of another that it is more accurate than C. However, because the accuracy measure is "proper", we can show that the *expected inaccuracy* of B is less than that of any other belief state—by the lights of B. So if the agent was to know—therefore believe—of C that it is more accurate than B, this belief must be combined with C having *no higher* expected accuracy than B. The relation between flat-out belief and partial belief is vexed, but I think we're entitled to assume that this situation won't arise. So—for APO agents—we have a grade-one vindication of probabilistic norms so long as we suppose there is a *derivative* accuracy norm.

The interest of such results is limited, given the restriction to APO agents. What we'd really like to know is whether logical norms constrain *us*, not some hypothetical idealization. The trouble is that to use the above argument for ordinary agents, we'd need to weaken (3)—perhaps to say that it is "in principle possible" to calculate an accuracy-dominating credence; and correspondingly weaken the antecedent of (4). But then (4) will no longer be supported by our derivative norm, or by the spirit of "subjectivizing" what we fundamentally care about. Suppose that the oracle tells me that some miners are in a specific mine shaft iff the answer to some horrendous mathematical puzzle is 42. I have to choose almost instantly whether to take action that will save all of them if they're in that shaft, or kill them all if they're not—or I can opt out of the decision, letting a small fraction die but saving the majority. If they are in fact in the shaft, I should "objectively" take the action, but "subjectively" I ought to opt out since it would be grossly irresponsible for me to take the risk of them all dying. It's the actual information state, not the "in principle available" one, that is relevant to the subjective norms.

13.5 Persistent Partial Beliefs

In this section, I'm going to add an extra element to the setup. Ultimately, this allows us to say something about the way accuracy arguments make logic

Even if our "target" probability accuracy-dominates and is not itself accuracy-dominated, it still doesn't follow that we ought to adopt it. After all, for all we've said, many probabilities may accuracy-dominate B—with the APO agent aware of each. It seems bad to stick with B, but *ceteris paribus* permissible to shift to any one of the dominating probabilities. So the most we can hope for, I suspect, is the permissibility of shifting to C, alongside the impermissibility of sticking with B. (Thanks to Al Hájek, Samir Okasha, and Richard Pettigrew for discussion of these issues.)

constrain ordinary, non-ideal agents. But along the way we get other interesting results.

The overarching idea is that in adopting doxastic attitudes to a proposition, we incur commitment to *persist* in those attitudes if no new evidence is forthcoming (where persistence is understood as *not changing one's mind*—that is, not adopting a different attitude to the same proposition. I discuss cases of simply ignoring the proposition below). In the limiting case, consider a situation where one simply moves from one moment to the next, with no new input or reflection. It would be bizarre to change one's (non-indexical) beliefs in such circumstances. Insofar as action, over time, is based on one's beliefs, it would mean that a course of action started at one time might be abandoned (since it no longer maximizes expected utility) without any prompting from reflection or experience.

Persistence might be construed as a (widescope) diachronic *norm* on belief. Alternatively, a disposition to retain an attitude to the proposition over time might be *constitutive* of belief. If what makes something count as a belief is its functional role, then the reflections on extended action above motivate this kind of claim.

In what range of circumstances must one's beliefs be persistent? It's clear that when new empirical evidence comes in, persistence is off the table—when the evidence changes, your mind changes. Less clear is a case where one learns something new, by "pure reflection". On the one hand, it is clear that when we figure out how to prove p, it is perfectly proper to shift from agnosticism over whether p, to endorsing p. That might seem to be a violation of persistence. I'll argue in a little while that this appearance is misleading, but for now let's set this aside, and consider an intermediate case.

Between the clear case where there's no change at all, and the questionable cases where the only relevant changes are as a result of pure reflection, there are other cases in which (I contend) we are committed to persist in attitudes. In particular, there are cases where the only changes are shifts in attention: where one simply stops taking an attitude to a proposition, because it is no longer relevant to one's purposes (one may, of course, still be disposed to adopt that same attitude, should the question rearise). Suppose an agent believes that she'll win the lottery to degree 0.1; that she'll have sweet potato for lunch to degree 0.9; and that she'll be given a job offer to degree 0.5. Now, without new evidence coming in, if she starts ignoring the proposition about the lottery, it would be completely bizarre if she thereupon lowered her credence in sweet potato for lunch, or raised her credence in the obtaining of a job offer. As before, it's questionable whether an agent who is disposed to shift their attitudes to p in these circumstances really had a degree of *belief* in the first place.

I'll assume persistence in this still very restricted sense. Then we get the following principle: if an agent has belief state B, over propositions P, then if P* is a subset of P, they are committed to adopting the belief state B*, where B* is simply the restriction of B to P*, in the event that their attention shifts so that P* are now the only propositions they have attitudes towards. The sense of "commitment" I'll need is at least this: if it's bad to have belief state B* over P*, it's also bad to be committed to belief state B* over P*. This turns out to be a powerful assumption in this context. Let's start to put it to work.

To begin with, there's an assumption in Joyce's formal arguments that may seem worrying. His argument for accuracy domination works *on the assumption that the belief states in question are defined over a finite algebra of propositions.*[13] But one might think this is a serious restriction. On many conceptions of belief, I presently believe not only that the number of my hands is two, but also that for each integer n other than two, the number of my hands is not n. The trouble is that we have no argument for accuracy domination for infinite belief states of this kind.

In comes persistence. If in having a specific partial belief in each of infinitely many propositions, I am committed to the finite subsets of those beliefs, then we can meaningfully appeal to accuracy domination. Suppose that some finite subset of my infinite beliefs violates one of the probabilistic constraints. One cannot apply Joyce's theorem to my actual belief state as a whole. But we can consider the restricted belief state to which I am committed, concerning only the finite subset in question. And this belief state is accuracy-dominated, hence bad. By the assumption that it's bad to be committed to a bad belief state, any locally improbabilistic infinite belief state is a bad thing to adopt.

A second application is the following preface-paradox-like situation. Modest agents should perhaps concede that they violate probabilistic constraints. So suppose we now detect some specific local inconsistency, and see how to tweak our beliefs to regain local consistency. Does the above give us motivation for doing so? One would like to argue that the original beliefs have the vice of being accuracy-dominated; and we can remove this flaw by the tweak. But if we're pretty confident that the local inconsistency in question isn't the *only* such inconsistency in our overall belief state, we'll believe the resulting belief state is accuracy-dominated, even after tweaking. So what exactly is gained?

Again, persistence can help. If we see that our attitudes to P, Q, R violate probabilistic constraints, then tweaking them so they no longer do so at least does this: it means we can be confident that the *restricted* belief state involving only

[13] For cousins of Joyce's arguments in the infinite setting—though not of the dominance form I have been discussing—see Easwaran (2013).

P, Q, R is not accuracy-dominated, whereas the original belief state we knew was. So even if we haven't improved matters for the belief state as a whole, we've at least removed one identifiable respect in which our commitments are open to criticism.

A final application of persistence involves the more *general* project of defending accuracy scores. We've been assuming that the Brier score measures inaccuracy. But Joyce (2009) proves a more general theorem, which one might wish to have as a backstop if one is worried about the specific Brier-score proposal. Joyce shows that one can prove accuracy domination for any accuracy score which at least makes belief states meeting probabilistic norms *admissible* (i.e. which gives us probabilistic anti-domination, as described earlier) and which satisfies a certain "truth-directedness" principle (in that, if two belief states agree on every proposition but P, and B is linearly closer to the truth value on P than is C, then B is overall more accurate than C). And those conditions are really quite weak constraints to impose on an accuracy score. However, there's a crucial qualification: Joyce proves this theorem only for belief states defined over propositions that form a *partition* (are pairwise inconsistent and mutually exhaustive—he shows that accuracy domination follows if the sum of one's credences across the partition is other than 1). But there's an obvious worry: what is the relevance of the result to belief states like ours, where we take attitudes not only to *grass being green* and *grass not being green*, but also to conjunctions and disjunctions thereof?

It turns out that any improbabilistic belief state will be such that, on some partition, credences in propositions in that partition do not sum to 1. We can focus on the restricted belief state where we adopt attitudes only to propositions in that partition. Given persistence, we are committed to a belief state over that partition alone, which sums to other than 1. Joyce's (2009) theorem can then kick in—*this restricted state* is accuracy-dominated. And since it is bad in this sense, the overall belief state believing which committed you to it is derivatively bad.

Persistence in the extreme limiting case is extremely plausible. The slight extension to cases of ignoring seems intuitively well motivated; and adding it to our account increases the explanatory power of the setup considerably. This itself, I think, gives us reason to think it is on the right track.

But what of the more contentious case mentioned earlier: persistence under pure reflection? As noted above, this seems a different case at first glance. Whatever one might say about ideal agents, for real agents like you and me, pure reflection can produce what seems like new information. If we're faced with sealed boxes, and told that there is a bar of gold inside the box that is labelled with a root of polynomial P; then the pure reflection required to solve the polynomial will change me from a state where (intuitively) I should divide credences evenly as to which box contains the gold; to a state where I'm relatively certain which one contains it. There seems little wrong with that movement in thought; it certainly doesn't seem constitutive that we avoid it.

There is another way to look at it, however. To be sure, the change is a positive one, given the information made available by reflection; in the situation in question, once we've reflected, we should change our belief state to incorporate the now-manifest information. But that isn't inconsistent with the persistence principle applied to this case. For that principle should be understood in a wide-scope way: one shouldn't both have credence d in p at t; and have credence other than d in p at t*, given that only pure reflection occurs in the meantime. It is quite consistent with the truth of this normative claim that one should have credence other than d in p at t*: one obvious way to make this the case arises when one was wrong to have credence d in p at the earlier time. So, consistently with the pure reflection persistence principle, we can see the change in view from 50-50 credence in the proposition that n is a root of the polynomial, to credence 1 in that proposition, as something to be recommended but only necessary because of a flawed initial state.

Underlying this is the recognition that, in cases of pure reflection like the solving of a polynomial, the information in question was *at the earlier time* within one's epistemic reach. One has to *wait* on empirical information—until the sense data impact, the empirical information is in a strong sense *unavailable*. But information reached by pure reflection was *already* available, at least in principle. (Notice that the case of the polynomial is special in that there's an algorithm one can use to get the answer in a short period of time. Cases where there's no decision procedure like this might well be treated differently.)

I think the *obvious* criticism of a norm of persistence under pure reflection is misguided, therefore. But is there anything to the idea? The cases where one culpably violates this norm will be those where one adopts a set of attitudes, being fully aware that pure reflection would lead one to change them—that is, one knowingly adopts unstable beliefs (an example is the gold bar/polynomial case above). Such a course is often excusable: there are competing demands on our energy and resources, and practical agents can't spend all day in a priori reflection. But there's a tradition in philosophy of considering impractical agents—Descartes' pure enquirer, for example. As Bernard Williams describes this thinker, she is one whose sole project is to gain beliefs; whose sole goal is truth; and who has created a space (practically) free of distractions or limitations of time and resources.[14] The pure enquirer commits herself to not appealing to the sort of excuses just

[14] Williams (1978).

mentioned. Such an agent, I think, should have stable belief states under pure reflection, and their beliefs should persist under such reflection.

The pure enquirer is not a priori omniscient—she may very well (as Descartes recommends) be initially agnostic about a whole range of questions, including those that can be settled by pure reflection. Nevertheless, once she adopts an attitude, she is committed to persisting with that state through any pure reflection (if she gives it up at any point, that will simply reflect badly on her initial adoption of the attitude). Despite not being a priori omniscient, we can use the accuracy domination results to argue that she should be a probabilist. Suppose she had improbabilistic belief state B. Pure reflection alone allows the agent to know, of C, that it accuracy-dominates B. And the result is that improbabilistic belief state B is unstable under pure reflection. Pure enquirers, therefore, shouldn't adopt B.

We are not pure enquirers. But the relation between us and pure enquirers is interestingly different between us and a priori omniscient agents. A priori omniscient agents differ in capacity from us—so we can't sensibly hold ourselves to their standards. On the other hand, pure enquirers have the same capacities as us—we could in principle become the pure enquirer—and so insofar as we share their goals, we can sensibly hold them up as a model of what our beliefs should look like, if only untainted by other interests. Dispositions possessed by the pure enquirer will be epistemic virtues in us. And since improbabilistic credences are bad for the pure enquirer, the proper conclusion is that probabilism is an epistemic virtue.

If this is all accepted, then we can use it and accuracy domination to argue for a norm binding on actual agents: one should meet logical constraints. For any belief state that violates such constraints will be accuracy-dominated, and so pure reflection would put us in the position of the APO agent discussed earlier. The improbabilistic APO agent directly violates a derivative accuracy norm, and so believes badly. The improbabilistic pure enquirer is committed to violating this norm, and so believes badly. The improbabilistic ordinary-joe fails to implement the rules that govern the pure enquirer and, to that extent, betrays an *epistemic vice*.

13.6 Probabilistic Evidence

Kolodny (2007, 256) discusses Joyce's results—though his main focus is on logical norms for all-or-nothing, rather than partial belief. The dialectic is somewhat involved by that stage of the paper, but he seems to be at least open to the idea that they can play a role in arguing that "the set of degrees of belief that epistemic reason requires are probabilistic". The focus here is not fundamental norms of accuracy, but norms of evidence—of proportioning one's credence properly to the evidence.

This section will explore how Joyce-style reasoning can assist in arguing for something like probabilism in a revised setting where matching evidence, rather than truth value, is taken as one's fundamental aim. So let us assume in a given context that each proposition has a specific *degree of evidential support*.[15] In the partial belief case, we'll assume an analogous form of evidentialism: that evidential support norms partial belief, in that one's degree of belief in p *should match* the corresponding degree of evidential support for p.

If we could assume that degrees of evidential support are in fact probabilistically structured, then we could immediately conclude that degrees of belief should be probabilistically structured. This would fit the Kolodny-esque debunking pattern. For the entire normative weight of this claim would hang on *matching one's evidence*. Someone who met all logical constraints, but had probabilities that are out of whack with the evidence, would not have anything more going for them than someone with improbabilistic beliefs. If there's no more to be said, then there would be no logical norms as such—though nevertheless anybody whose beliefs violates logical constraints would *ipso facto* not be believing as they should.

This is very similar to a picture we considered and dismissed as uninteresting in our initial discussion of Joyce's accuracy norm—based on the observation that truth-value distributions were probabilities, and that matching truth values minimizes inaccuracy. But in that context, we could say the same about violations of *extremality*, and since all reasonable agents will violate the latter, all reasonable agents will equally violate the norm—depriving it of any discriminating purpose.

But the evidential norm is much better placed. For a start, it's not a non-starter to think that some of us do properly align our credences to the evidence. Given this, one cannot straight away show it to be undiscriminating. At the very least, we don't have the "bad company" of the putative constraint "be extremal"—for there's no reason to think that violations of extremality violate the evidence norm. (Presumably, whether or not it's a practical option for agents to follow the evidence norm depends on how "externalistically" we construe evidence. A Williamson-like knowledge-based conception of evidential probability might make most if not all ordinary agents fail to meet the norm; on more internalistic conceptions perhaps it is easier to respect).

However, this whole case hangs on the assumption that *degrees of evidential* support are themselves structured probabilistically (as does the central argument

[15] The context will be relativized to the agent—partly because different agents will bring different background evidence to bear; and partly because there may be non-evidential facts about an agent (the "stance" they adopt) that factor into the ideal belief to have given the evidence available (which is how I'm thinking about "evidential support"). One's initial priors, attitudes to epistemic risk, and the like, are candidates for non-epistemic determinants of support.

of Kolodny (2007, 234; cf. n. 14). But that looks like it assumes the point to explained, in our context. Certainly one can imagine cases where there's prima facie strong evidential support for P, and strong evidential support for ¬P. Why couldn't it turn out that all-things-considered evidential support ranks assigns high degrees to each of P and ¬P—so the sums of their degrees is greater than 1? Such evidential support couldn't be represented probabilistically. The task is to explain why such cases don't arise.

This is where Joyce's accuracy domination theorem can kick in again. Here's an assumption that simply seems *right* to me:

> **Plausible Premise:** if an assignment of degrees Q is demonstrably closer to the truth than R (in any world), then Q beats R as a candidate to systematize "evidential support".

But Joyce's results show (relative to his way of accuracy or "closeness to the truth") that *only* probabilities are immune from such "trumping" (and, on a proper scoring function like the Brier score, all probabilities are so immune). Given the Plausible Premise, we can conclude that evidential support will therefore be probabilistically structured. And given an evidential norm on belief, we can get the Kolodny-esque debunking explanation going.[16]

Moreover, the availability of this debunking explanation needn't *exclude* the points made earlier. For just as we may generalize a truth norm to a gradational accuracy norm, we may generalize our evidential norm into a graded form. We would say that B is better than C to the extent that B is *closer to the degrees of evidence* than C. The Brier score mentioned earlier is straightforwardly generalizable—closeness being a matter of minimizing the average square difference between the degree of evidence and credence (with the minimum achieved when credence and evidential support match). An evidential analogue to the accuracy domination theorem can be proved.[17] And then our earlier discussion can be rerun. Overall, then, in this setting we may get the following three justifications of forms of probabilism:

[16] I give an alternative route to the conclusion that degrees of evidence are structured probabilistically in "A non-pragmatic dominance argument for conditionalization" (MS). However, the argument there *assumes* that the Joyce-style dominance situation is sufficient for a belief state being rationally flawed.

[17] The generalized forms of domination argument given in Williams (2012a) and Williams (2012b) can be interpreted to this effect—the focus there is on non-classical truth-value assignments, but the results are easily reinterpreted. The strategy of replacing truth with another "aim" in evaluating belief states is also explored in Pettigrew (2012), where the role played here by evidential support is played there by Chance.

(i) We ought to have credences that meet logical constraints on partial belief, in virtue of the fact that we should match our degrees of belief to the evidence.
(ii) Ideal agents are subject to logical norms as such (irrespective of whether they manage to match their credences to the evidence).
(iii) Ordinary agents' credences are epistemically virtuous if they meet the requirements; and epistemically vicious if they don't. Ordinary agents should be disposed to meet logical norms (irrespective of whether they manage to match their credences to the evidence).

13.7 The Accuracy Score

I've now finished discussing what we can extract from Joyce's accuracy domination result *if we spot him an appropriate accuracy score* (we have worked with the Brier score). But many have thought the discussion of the accuracy score itself the weak point of Joyce's whole discussion. I'm much more sanguine than others appear to be; and I close this chapter by explaining why.

One thing that we need to be clear on is the dialectical role we want the accuracy domination considerations to play. One conception of their role is as a bludgeon to use against theorists who endorse some rival to probabilistic norms—so conceived, Joyce would need to offer *suasive* considerations for each element of his setup, and would have to watch out for "begging the question" against his rivals. In particular, we certainly couldn't appeal to the probabilistic formulation of logical norms in support of a choice of accuracy score.

Here is a different project. Start by *assuming* certain that logic norms partial belief—at least for APO agents or pure enquirers. Advocates of *different* formulations of requirements that logic places on belief will thus already have got off the boat. Our task is not to *convince the unconvinced* that those norms are in force, but to *explain where they get their normative punch from*. It's then legitimate here to consider various candidate explanatory hypotheses. Accuracy domination suggests that the Brier score (for example) is in a position to explain the probabilistic norms in the relevant setting—as well as dispositional versions of that norm for ordinary agents. Of course, since we were initially unsure about the right formulation of the probabilistic norms and what strength to target, there would be some back and forth as the data are refined in the light of theoretical considerations. But the essential epistemic structure is that the accuracy score is not *independently* justified at all, nor does it need to be.

This may nevertheless have some teeth against rival conceptions of logical norms on credence. For we can legitimately ask fans of alternative models to provide an account of the source of the norms *they* think obtain, that achieves comparable success to our model here. Exploring whether (for example) advocates of Dempster Shafer theory can do this is a worthwhile project.

Patrick Maher objects that Joyce hasn't provided a sufficient justification for favouring scoring functions that will support his theorem, over the simple "linear score" that will not.[18] Advocates of Joyce need to fight Maher in the trenches if they do intend to convince the unconvinced to be probabilists. But challenge from the linear score, at least, has an easy answer within the explanatory project. The reason that we favour the Brier over the linear score is that *only the former is even a candidate explainer of the norms we're trying to explain.*

While Maher objected to Joyce on grounds of *bad company*—prima facie reasonable measures that don't deliver the needed results; Aaron Bronfman (2006) and others have objected on grounds of too much *good company*.[19] Bronfman's objection trades on an *overabundance* of candidate measures of accuracy—all of which, we may assume, allow us to derive accuracy domination (and probabilistic anti-domination). The trouble that Bronfman points to is that *these may disagree*. If one's credences B violate probabilistic constraints, then candidate accuracy measure 1 might tell you that C accuracy-dominates B; candidate accuracy measure 2 might tell you that D does so. It's an open possibility that the first accuracy measure says that D is in some worlds less good then B; and the second accuracy measure says the same about C. If you're getting *conflicting advice* from the various candidates, then, intuitively, your best bet might be to stick with B, rather than taking a risk on actually lowering your overall accuracy.

Exactly how we respond to the objection depends on how it is supposed to work. One reading (suggested by Joyce's characterization) is that the worry is that it turns out to be *indeterminate* which accuracy measure is the right one. The point then is that for all the candidate-by-candidate accuracy domination results tell us, there need be no probability P which is *determinately* more accurate than B. If that is the worry, then the position could be stabilized by *denying* that this kind of indeterminacy exists: this is in effect the proposal that Joyce (2009) and Huber (2007) advocate. (Joyce (1998) lays out a number of axioms for the accuracy score. Several different accuracy measures satisfy these axioms: the various quadratic loss scores, for example. If one thought of the axioms as *exhausting* the conventions governing the use of the term then there might be a prima facie case that there was "no fact of the matter" which satisfier deserves the name "accuracy". The alternative conception of the enterprise has it as purely

[18] Maher (2002). [19] Cf. Huber (2007); Joyce (2009).

epistemic: if Joyce's arguments work, we know that the One True accuracy measure satisfies the axioms he lays down, but that is all.)

I'll come back to the indeterminacy version of Bronfman's objection shortly. But I focus first on a variant that threatens even once we've made the "realism" assumptions that Joyce and Huber suggest—and which may seem particularly worrying if we want to give a Kolodny-style gloss on the enterprise. APO agents, we can agree, believe that C dominates B, where B is an improbabilistic belief state, and C is the probability that accuracy dominates according to the Brier score. It's true that, in virtue of their omniscience on matters a priori, APO agents will believe:

C dominates B *given that accuracy is articulated via the Brier Score.*

But APO agents (or pure enquirers) need the unconditional belief that C accuracy-dominates B (i.e. that C is epistemically better than C) if our argument is to work as stated. Can we make the required move? Even assuming it's true that accuracy is given by the Brier score, why do we assume that APO agents are aware of this? True, it's not as if *empirical* information has any obvious role to play in finding out what the accuracy norm is. But that simply raises the question of whether there's *any* way to find out what the accuracy norm is. We may, in the spirit of realism, postulate that epistemic value has a determinate shape; but realism alone doesn't entitle us to the assumption that what that shape is, is epistemically accessible.

How should we respond to this? Well, we might argue directly that the One True accuracy measure is a priori accessible—perhaps they will be able to *know* that it is the Brier Score (for example). I don't think this is out of the question, but I prefer a different approach.

Consider an analogy to *moral* oughts. Inflicting suffering is bad. In virtue of this, it is plausible one morally ought not take action A if one believes that A will inflict suffering. This remains so even if in fact no suffering would result from A (more carefully, I think there's at least *one reading* on which one morally ought not to A, even if there's a second, non-information-dependent, reading on which it's morally OK). By contrast, consider a case where someone performs act A, which they *know* will inflict suffering, while *believing* that what they do is OK because they have whacky beliefs about right and wrong (that causing suffering is a good thing, for example). There's no reading on which it's *morally permissible* for them to perform A, I claim. You might well think that they're not *irrational*—there's no internal tension in their attitudes, since they're doing what's good-by-their-lights—but that's quite a different claim. Conclusion: deontic modals aren't relative to beliefs about *value*, even if they get relativized to what factual information one has available.

A similar distinction can be made in our case. It's one thing to accuse an APO agent of irrationality—of having a suboptimal-by-their-own-lights belief state. It's another to accuse them of having a belief state that (relative to their information state) they ought not to have. For the latter, where V are in fact the relevant facts about epistemic value, it suffices that the APO agent knows that *if V are the value facts*, then belief state x is epistemically better than their own belief state B. And this a modification of the core argument can deliver:

Core argument (revised):

0. F is the accuracy measure. (premise)
1. B is not a probability. (premise)
2. B is F-accuracy dominated. (domination theorem, 1)
3. There is a specific probability (call it C), such that an APO agent will know of C that it is more F-accurate than B. (from 2)
4. If F is the accuracy measure, and one knows of C that it is more F-accurate than B, then one ought not to have belief state B. (premise)
5. An APO agent ought not to have belief state B. (0,3,4)

One consequence of these considerations is that the accuracy domination argument will *not* automatically reduce the "irrationality" of violating logical norms to a broader species of irrationality (of conflict between what one does, and what is best-by-one's-own lights)—that requires the *additional* premise about the a priori accessibility of the correct scoring rule. But even without that premise, our argument *does* make a strong case that one *ought not* to have improbabilistic credences. To bring this back to the original discussion of the grade and type of normativity of logic: recall that the Williamson-eseque structure of "derivative norms" springing from "fundamental norms" only partially subjectivized the norm—whether the agent believed they knew the proposition mattered, but whether they thought of knowledge rather than coherence-with-others as the fundamental norm was not factored in. This fits nicely with what I've just been urging. But Kolodny's formulations were to look *either* to the non-subjectivized "oughts" themselves, or to what the agent thinks "ought" to be the case. So epistemic limitations over the accuracy measure are more threatening for those who favour Kolodny's strategy for explaining away the appearance of normativity.

13.8 The Original Bronfman Objection

I earlier set aside the original version of Bronfman's objection, directed against those who think there's no fact of the matter about which scoring rule describes accuracy (albeit that accuracy domination is provable for each one). So our results above assume determinacy in the accuracy norm. Can we get anything similar if we loosen that assumption, and allow indeterminacy in which scoring rule gives the accuracy measure? The worry, recall, is that the precisifications of the score might give conflicting advice about which credence is the one that accuracy-dominates; each might condemn the others' recommendation.

Surprisingly (and some might think worryingly), if the argument above is valid, it looks like we can argue for probabilism even in this setting.[20] For simplicity, suppose that it is indeterminate whether F or G correctly describes the accuracy measure. For improbabilistic B, and accuracy measures F and G that allow an accuracy domination theorem, we can treat the earlier argument as a conditional proof of the conditionals:

If F is the accuracy measure,
then an APO agent ought not to have belief state B.
If G is the accuracy measure,
then an APO agent ought not to have belief state B.

On a supervaluation-style treatment of indeterminacy we have:[21]

Either F is the accuracy measure or G is the accuracy measure.

But then disjunction elimination gives us the unconditional:

An APO agent ought not to have belief state B.

So we have an argument for the epistemic badness of improbabilities, even when the relevant measures are indeterminate.

(I believe the arguments as presented are valid. Two concerns might be raised—one over the logic of the indicative conditionals especially with deontic modals in the consequent; the other over the logic of indeterminacy. In each case, the use of the metarules conditional proof and disjunction elimination have been questioned. However, I don't think we need worry about this. On the former front, the whole argument can be recast in terms of material conditionals rather than English indicatives, thus bypassing delicate issues about the interaction of deontic modality and natural language conditionals. On the latter, it's true that disjunction elimination and conditional proof are not *generally* valid in certain kinds of supervaluational logics. However, restricted versions of both *are* valid

[20] Thanks here to Hannes Leitgeb, who urged me to think about this kind of extension of the earlier argument.

[21] Of course, on a law-of-excluded-middle rejecting interpretation of indeterminacy (in an intuitionistic or Kleene logic, for example) the argument will be unsound at this point.

(essentially, those where no special "indeterminacy exploiting" move is made in the subproofs), and this is sufficient for our purposes).

If one thinks that the conclusion of this argument is implausibly strong, one might wonder whether this undermines the plausibility of one of our premises. But I suspect that this is the wrong diagnosis, for it seems to me that what we have here is a *general* puzzle about indeterminate value. Suppose we have acts A, B, C, and moral-theory-1 says that A is optimal, B OK, and C evil; while moral-theory-2 says that C is optimal, B OK, and A evil. If it is indeterminate which of the moral theories is true, by the disjunction-elimination pattern we can argue that B ought not to be done; even though there's no action that *determinately* is better than it. This has, for me, exactly the same strangeness that we feel in the Bronfman case—and if indeterminacy in value is possible at all, it's worth thinking through how to react.

In other work, I've explored how indeterminacy interacts with assignments of value. There are certainly conceptions of indeterminacy available on which the argument by cases above is valid, and the proper conclusion is indeed that agents ought not to have dominated belief states. I've explored one such framework elsewhere (and with quite independent motivations). The core idea is that when a subject is certain that p is indeterminate, then the subject is free to *groundlessly opt* either to judge that p, or to judge that ¬p.[22] In the case at hand, indeterminacy over which accuracy measure is the correct one makes it *permissible* to groundlessly opt for one candidate or another to guide one's epistemic evaluations. Since improbabilistic beliefs are bad no matter which one chooses, the defence of probabilism stands despite the indeterminacy.

There's no consensus in the literature on indeterminacy and vagueness about how to think of indeterminate belief, desire, and value, and so we can't appeal to an "off the shelf" model to resolve these sorts of question. I've pointed to one model where the Bronfman objection *wouldn't* be an obstacle to the accuracy domination argument for probabilism. It would be interesting to see presented a model of indeterminate value on which it *is* an obstacle. At that point, we could evaluate the success of the argument (on the assumption that accuracy measures are indeterminate) by tackling the broader question of which conception of indeterminate value is right.

[22] See, in particular, Williams (2014). The discussion there is framed in terms of desirability rather than objective moral or epistemic value, but the structure should be quite analogous. I think this model is the proper conception of indeterminacy for anybody who favours a non-epistemic but classical logic and semantics of the relevant indeterminacy. The treatment of vague desirability in Williams (forthcoming) provides further models for thinking about vague desirability.

13.9 Conclusion

Our starting point was the question: why be logical? The Joycean gradational results (based on a Brier-score articulation of accuracy) allow a substantive answer to this question. Agents like us should be probabilists because any failure to do so means we fail to match our beliefs to the evidence; but further, we should be *disposed* to meet logical constraints on partial beliefs since only belief states meeting this condition have the virtue of reflective stability. Reflective stability, I've argued, is a commitment of the pure enquirer; and departures from the rules binding on the pure enquirer count as epistemic vices.

The Joyce argument is at its strongest when we assume that there is some determinate scoring rule that describes accuracy, which leads to an accuracy-domination theorem. Perhaps a Joyce-style project of providing independent constraints on legitimate accuracy measures can help with this, but in principle our justification for the assumption need not be *independent* of a commitment to probabilism (indeed, one of the most persuasive arguments for me that there is some such accuracy measure is by inference to the best explanation from probabilism itself). I've argued that we do not need to assume that it's even knowable what accuracy measure is the right one, in order for this argument to go through. The assumption of determinacy may be unnecessary—I've pointed to one conception of indeterminacy where this is so.[23]

References

Berker, Selim. 2013. "Epistemic teleology and the separateness of propositions." *Philosophical Review* 122:337–93.
Bronfman, Aaron. 2006. "A gap in Joyce's proof of probabilism." Unpublished manuscript.
Broome, John. 2005. "Does rationality give us reasons?" *Philosophical Issues* 15:321–37.
Easwaran, Kenny. 2013. "Expected accuracy supports conditionalization—and conglomerability and reflection." *Philosophy of Science* 80:119–42.
Field, Hartry. 2009. "What is the normative role of logic?" *Aristotelian Society Supplementary Volume* 83:251–68.
Halpern, Joseph Y. 1995. *Reasoning under Uncertainty*. Cambridge, MA: MIT Press.
Harman, Gilbert. 1986. *Change in View*. Cambridge, MA: MIT Press.

[23] Versions of this paper were presented in Bristol and in St Andrews at the Foundations of Logical Consequence conference, and at the Centre for Ethics and Metaethics at Leeds. Thanks to all who've helped me with these issues; I'd like to mention in particular Selim Berker, Daniel Elstein, Al Hájek, Ulrike Heuer, Hannes Leitgeb, Andrew McGonigal, Richard Pettigrew, and Pekka Vayrynen. This work was carried out with the support of the British Academy grant "The cognitive role of indeterminacy", BARDA53286.

Huber, Franz. 2007. "The consistency argument for ranking functions." *Studia Logica* 86:299–329.

Joyce, James M. 1998. "A nonpragmatic vindication of probabilism." *Philosophy of Science* 65:575–603.

Joyce, James M. 2009. "Accuracy and coherence: prospects for an alethic epistemology of partial belief." In F. Huber and C. Schmidt-Petri (eds.), *Degrees of Belief*. 263–97. Springer.

Kolodny, Niko. 2005. "Why be rational?" *Mind* 114:509–63.

Kolodny, Niko. 2007. "How does coherence matter?" *Proceedings of the Aristotelian Society* 107:229–63.

Maher, Patrick. 2002. "Joyce's argument for probabilism." *Philosophy of Science* 69:73–81.

Maudlin, Tim. 2004. *Truth and Paradox*. Oxford: Oxford University Press.

Pettigrew, Richard. 2012. "Accuracy, chance, and the principal principle." *Philosophical Review* 121:241–75.

Wedgwood, Ralph. 2002. "The aim of belief." *Philosophical Perspectives* 16:267–97.

Williams, Bernard. 1978. *Descartes: The Project of Pure Enquiry*. Sussex: Harvester Press.

Williams, J. Robert G. 2012a. "Generalized probabilism: Dutch books and accuracy domination." *Journal of Philosophical Logic* 41:811–40.

Williams, J. Robert G. 2012b. "Gradational accuracy and non-classical semantics." *Review of Symbolic Logic* 5:513–37.

Williams, J. Robert G. 2014. "Decision-making under indeterminacy." *Philosophers' Imprint* 14:1–34.

Williams, J. Robert G. forthcoming. "Probability and non-classical logic." In A. Hájek and C. Hitchcock (eds.), *The Oxford Handbook of Philosophy of Probability*. Oxford: Oxford University Press.

Williamson, Timothy. 2000. *Knowledge and its Limits*. Oxford: Oxford University Press.

Index